LITTLE,
BROWN

LARGE
PRINT

Also by Annie Jacobsen

Area 51: An Uncensored History of America's
Top Secret Military Base

OPERATION PAPERCLIP

THE SECRET INTELLIGENCE PROGRAM THAT BROUGHT NAZI SCIENTISTS TO AMERICA

ANNIE JACOBSEN

LITTLE, BROWN AND COMPANY

LARGE PRINT EDITION

Little, Brown and Company
Hachette Book Group
237 Park Avenue, New York, NY 10017
littlebrown.com

First Edition: February 2014

Little, Brown and Company is a division of Hachette Book Group, Inc. The Little, Brown name and logo are trademarks of Hachette Book Group, Inc.

The publisher is not responsible for websites (or their content) that are not owned by the publisher.

The Hachette Speakers Bureau provides a wide range of authors for speaking events. To find out more, go to hachettespeakersbureau.com or call (866) 376-6591.

Library of Congress Cataloging-in-Publication Data

Jacobsen, Annie.
    Operation Paperclip : the secret intelligence program that brought Nazi scientists to America / Annie Jacobsen. — First edition.
        pages cm.
    Includes bibliographical references and index.
    ISBN 978-0-316-22104-7 (hardcover) / 978-0-316-23982-0
(large print)   1. World War, 1939–1945 — Technology.   2. Brain drain —
Germany — History — 20th century.   3. Scientists — Germany —
Recruiting — History — 20th century.   4. Scientists — United
States — Recruiting — History — 20th century.   5. Physicians —
Germany — Recruiting — History — 20th century.   6. Nazis —
History — 20th century.   7. War criminals — Germany — History —
20th century.   8. Intelligence service — United States — History —
20th century.   9. Military research — History — 20th century.
10. German Americans — History — 20th century.   I. Title.   II. Title:
secret intelligence program that brought Nazi scientists to America.
    D810.S2J43 2013
    940.54'867308850943—dc23   2013028255

10 9 8 7 6 5 4 3 2 1

RRD-C

Printed in the United States of America

For Kevin

CONTENTS

Prologue   ix

Part I

Chapter 1 The War and the Weapons   3
Chapter 2 Destruction   31
Chapter 3 The Hunters and the Hunted   53
Chapter 4 Liberation   70
Chapter 5 The Captured and Their
               Interrogators   98

Part II

Chapter 6 Harnessing the Chariot of
               Destruction   131
Chapter 7 Hitler's Doctors   162
Chapter 8 Black, White, and Gray   200
Chapter 9 Hitler's Chemists   213
Chapter 10 Hired or Hanged   249

Part III

Chapter 11 The Ticking Clock   285
Chapter 12 Total War of Apocalyptic
            Proportions   327
Chapter 13 Science at Any Price   369
Chapter 14 Strange Judgment   397

Part IV

Chapter 15 Chemical Menace   415
Chapter 16 Headless Monster   446
Chapter 17 Hall of Mirrors   481
Chapter 18 Downfall   521
Chapter 19 Truth Serum   545

Part V

Chapter 20 In the Dark Shadows   561
Chapter 21 Limelight   589
Chapter 22 Legacy   620
Chapter 23 What Lasts?   637

Acknowledgments   661
Principal Characters   669
Notes   685
Author Interviews and Bibliography   785

PROLOGUE

This is a book about Nazi scientists and American government secrets. It is about how dark truths can be hidden from the public by U.S. officials in the name of national security, and it is about the unpredictable, often fortuitous, circumstances through which truth gets revealed.

Operation Paperclip was a postwar U.S. intelligence program that brought German scientists to America under secret military contracts. The program had a benign public face and a classified body of secrets and lies. "I'm mad on technology," Adolf Hitler told his inner circle at a dinner party in 1942, and in the aftermath of the German surrender more than sixteen hundred of Hitler's technologists would become America's own. What follows puts a spotlight on twenty-one of these men.

Under Operation Paperclip, which began in May of 1945, the scientists who helped the Third Reich wage war continued their weapons-related work for the U.S.

government, developing rockets, chemical and biological weapons, aviation and space medicine (for enhancing military pilot and astronaut performance), and many other armaments at a feverish and paranoid pace that came to define the Cold War. The age of weapons of mass destruction had begun, and with it came the treacherous concept of brinkmanship—the art of pursuing dangerous policy to the limits of safety before stopping. Hiring dedicated Nazis was without precedent, entirely unprincipled, and inherently dangerous not just because, as Undersecretary of War Robert Patterson stated when debating if he should approve Paperclip, "These men are enemies," but because it was counter to democratic ideals. The men profiled in this book were not nominal Nazis. Eight of the twenty-one—Otto Ambros, Theodor Benzinger, Kurt Blome, Walter Dornberger, Siegfried Knemeyer, Walter Schreiber, Walter Schieber, and Wernher von Braun—each at some point worked side by side with Adolf Hitler, Heinrich Himmler, or Hermann Göring during the war. Fifteen of the twenty-one were dedicated members of the Nazi Party; ten of them also joined the ultra-violent, ultra-nationalistic Nazi Party paramilitary squads, the SA (Sturmabteilung, or Storm Troopers) and the SS (Schutzstaffel, or Protection Squadron); two wore the Golden Party Badge, indicating favor bestowed by the Führer; one was given an award of one million reichsmarks for scientific achievement.

Six of the twenty-one stood trial at Nuremberg, a seventh was released without trial under mysterious circumstances, and an eighth stood trial in Dachau for regional war crimes. One was convicted of mass murder and slavery, served some time in prison, was granted clemency, and then was hired by the U.S. Department of Energy. They came to America at the behest of the Joint Chiefs of Staff. Some officials believed that by endorsing the Paperclip program they were accepting the lesser of two evils — that if America didn't recruit these scientists, the Soviet Communists surely would. Other generals and colonels respected and admired these men and said so.

To comprehend the impact of Operation Paperclip on American national security during the early days of the Cold War, and the legacy of war-fighting technology it has left behind, it is important first to understand that the program was governed out of an office in the elite "E" ring of the Pentagon. The Joint Intelligence Objectives Agency (JIOA) was created solely and specifically to recruit and hire Nazi scientists and put them on weapons projects and in scientific intelligence programs within the army, the navy, the air force, the CIA (starting in 1947), and other organizations. In some cases, when individual scientists had been too close to Hitler, the JIOA hired them to work at U.S. military facilities in occupied Germany. The JIOA was a subcommittee of the Joint Intelligence Committee (JIC), which provided national security information for the Joint Chiefs of Staff. The JIC remains the least known and least studied U.S. intelligence agency of the twentieth century. To understand the mind-set of the Joint Intelligence Committee, consider this: Within one year of the atomic bombings of Hiroshima and Nagasaki, the JIC warned the Joint Chiefs of Staff that the United States needed to prepare for "total war" with the Soviets—to include atomic, chemical, and biological warfare—and they even set an estimated start date of 1952. This book focuses on that uneasy period, from 1945 to 1952, in which the JIOA's recruitment of Nazi scientists was forever on the rise, culminating in Accelerated Paperclip, which allowed individuals previously deemed undesirable to be brought to the United States—including Major General Dr. Walter Schreiber, the surgeon general of the Third Reich.

Operation Paperclip left behind a legacy of ballistic missiles, sarin gas cluster bombs, underground bunkers, space capsules, and weaponized bubonic plague. It also left behind a trail of once-secret documents that I accessed to report this book, including postwar interrogation reports, army intelligence security dossiers, Nazi Party paperwork, Allied intelligence armaments reports, declassified JIOA memos, Nuremberg trial testimony, oral histories, a general's desk diaries, and a Nuremberg war crimes investigator's journal. Coupled with exclusive interviews and correspondence with children and grandchildren of these Nazi scientists, five of whom shared with me the personal papers and unpublished writings of their family members, what follows is the unsettling story of Operation Paperclip.

All of the men profiled in this book are now dead. Enterprising achievers as they were, just as the majority of them won top military and science awards when they served the Third Reich, so it went that many of them won top U.S. military and civilian awards serving the United States. One had a U.S. government building named after him, and, as of 2013, two continue to have prestigious national science prizes given annually in their names. One invented the ear thermometer. Others helped man get to the moon.

How did this happen, and what does this mean now? Does accomplishment cancel out past crimes? These are among the central questions in this dark and complicated tale. It is a story populated with Machiavellian connivers and men who dedicate their lives to designing weapons for the coming war. It is also a story about victory, and what victory can often entail. It is rife with Nazis, many of whom were guilty of accessory to murder but were never charged, and lived out their lives in prosperity in the United States. In the instances where a kind of justice is delivered, it rings of half-measure.

Or perhaps there is a hero in the record of fact, which continues to be filled in.

PART I

"Only the dead have seen the end of war."
—Unknown

CHAPTER ONE

The War and the Weapons

It was November 26, 1944, and Strasbourg, France, was still under attack. The cobblestone streets of this medieval city were in chaos. Three days before, the Second French Armored Division had chased the Germans out of town and officially liberated the city from the Nazis, but now the Allies were having a difficult time holding the enemy back. German mortar rounds bombarded the streets. Air battles raged overhead, and in the center of town, inside a fancy apartment on Quai Klébar, armed U.S. soldiers guarded the Dutch-American particle physicist Samuel Goudsmit as he sat in an armchair scouring files. The apartment belonged to a German virus expert named Dr. Eugen Haagen, believed to be a key developer in the covert Nazi biological weapons program. Haagen had apparently fled his apartment in a hurry just a few days prior, leaving behind a framed photograph of Hitler on the mantel and a cache of important documents in the cabinets.

Goudsmit and two colleagues, bacteriological warfare experts Bill Cromartie and Fred Wardenberg, had been reading over Dr. Haagen's documents for hours. Based on what was in front of them, they planned to be here all night. Most of Strasbourg was without electricity, so Goudsmit and his colleagues were reading by candlelight.

Samuel Goudsmit led a unit engaged in a different kind of battle than the one being fought by the combat soldiers and airmen outside. Goudsmit and his team were on the hunt for Nazi science—German weaponry more advanced than what the Allies possessed. Goudsmit was scientific director of this Top Secret mission, code-named Operation Alsos, an esoteric and dangerous endeavor that was an offshoot of the Manhattan Project. Goudsmit and his colleagues were far more accustomed to working inside a laboratory than on a battlefield, and yet here they were, in the thick of the fight. It was up to these men of science to determine just how close the Third Reich was to waging atomic, biological, or chemical warfare against Allied troops. This was called A-B-C warfare by Alsos. An untold number of lives depended on the success of the operation.

Samuel Goudsmit had qualities that made him the mission's ideal science director. Born in Holland, he spoke Dutch and German fluently. At age twenty-three he had become famous among fellow physicists for identifying the concept of electron spin. Two years later he earned his PhD at the University of Leiden and moved to America to teach. During the war, Goudsmit worked on weapons development through a government-sponsored lab at MIT. This gave him unique insight into the clandestine world of atomic, biological, and chemical warfare and had put him in this chair, reading quickly in the flickering candlelight. Just days before, Goudsmit's team had captured four of Hitler's top nuclear scientists and had learned from them that the Nazis' atomic bomb project had been a failure. This was an unexpected intelligence coup for Alsos—and a huge relief. The focus now turned to the Reich's biological weapons program, rumored to be well advanced.

Goudsmit and his team of Alsos agents knew that the University of Strasbourg had been doubling as a biological warfare research base for the Third Reich. Once a bastion of French academic prowess, this four-hundred-year-old university had been taken over by the Reich Research Council, Hermann Göring's science organization, in 1941. Since then, the university had been transformed into a model outpost of Nazi science. Most of the university's professors had been replaced with men who were members of the Nazi Party and of Heinrich Himmler's SS.

On this November night, Goudsmit made the decision to have his team set up camp in Professor Haagen's apartment and read all the documents in a straight shot. Alsos security team members set their guns aside, organized a meal of K-rations on the dining room table, and settled in to a long night of cards. Goudsmit and the biological weapons experts Cromartie and Wardenberg sat back in Professor Haagen's easy chairs and worked on getting through all the files. Night fell and it began to snow, adding confusion to the scene outside. Hours passed.

Then Goudsmit and Wardenberg "let out a yell at the same moment," remembered Goudsmit, "for we had both found papers that suddenly raised the curtain of secrecy for us." There in Professor Haagen's apartment, "in apparently harmless communication, lay hidden a wealth of secret information available to anyone who understood it." Goudsmit was not deciphering code. The papers were not stamped Top Secret. "They were just the usual gossip between colleagues...ordinary memos," Goudsmit recalled. But they were memos that were never meant to be found by American scientists. The plan was for the Third Reich to rule for a thousand years.

"Of the 100 prisoners you sent me," Haagen wrote to a colleague at the university, an anatomist named Dr. August Hirt, "18 died in transport. Only 12 are in a condition suitable for my experiments. I therefore request that you send me another 100 prisoners, between 20 and 40 years of age, who are healthy and in a physical condition comparable to soldiers. Heil Hitler, Prof. Dr. E. Haagen." The document was dated November 15, 1943.

For Samuel Goudsmit the moment was a stunning reveal. Here, casually tucked away in a group of Haagen's personal papers, he had discovered one of the most diabolical secrets of the Third Reich. Nazi doctors were conducting medical experiments on healthy humans. This was new information to the scientific community. But there was equally troubling information in the subtext of the letter as far as biological weapons were concerned. Haagen was a virus expert who specialized in creating vaccines. The fact that he was involved in human medical experiments made a kind of twisted sense to Goudsmit in a way that few others could interpret. In order to successfully unleash a biological weapon against an enemy force, the attacking army had to have already created its own vaccine against the deadly pathogen it intended to spread. This vaccine would act as the shield for its own soldiers and civilians; the biological weapon would act as the sword. The document Goudsmit was looking at was a little more than a year old. How much vaccine progress had the Nazis made since?

As Goudsmit stared at the documents in front of him, he was faced with a troubling reality. Once, Eugen Haagen had been a temperate man—a physician dedicated to helping people. In 1932 Dr. Haagen had been awarded a prestigious fellowship by the Rockefeller Foundation, in New York City, where he had helped to develop the world's first yellow fever vaccine. In 1937 he had been a contender for the Nobel Prize. Haagen had been one of Germany's leading men of medicine. Now here he was testing deadly vaccines on once healthy prisoners from concentration camps supplied to him by Himmler's SS. If a leading doctor like Haagen had been able to conduct these kinds of research experiments with impunity, what else might be going on?

Goudsmit and his colleagues scoured Dr. Haagen's papers, paying particular attention to the names of the doctors with whom Haagen corresponded about his prisoner shipments, his vaccine tests, and his future laboratory plans. Goudsmit started putting together a list of Nazi scientists who were now top priorities for Alsos to locate, capture, and interview. Dr. Eugen Haagen would never become a Paperclip scientist. After the war he would flee to the Soviet zone of occupation in Germany and work for the Russians. But among the names discovered in his apartment were two physicians important to Operation Paperclip. They were Dr. Kurt Blome, deputy surgeon general of the Third Reich, and Surgeon General Walter Schreiber. Dr. Blome was in charge of the Reich's biological weapons programs; Dr. Schreiber was in charge of its vaccines. The sword and the shield.

Before Hitler rose to power, Blome and Schreiber had been internationally renowned physicians. Had Nazi science also made monsters of these men?

Almost two weeks after the Alsos mission's discovery at Strasbourg, three hundred miles to the north, in Germany, a party was under way. There, deep in the dark pine forests of Coesfeld, a magnificent moated eight-hundred-year-old stone castle called Varlar was being readied for a celebration. The castle was a medieval showpiece of the Münster region, resplendent with turrets, balustrades, and lookout towers. On this night, December 9, 1944, the banquet hall had been decorated in full Nazi Party regalia. Trellises of ivy graced the podium. Flags featuring Germany's national eagle-and-swastika emblem hung from walls, a motif repeated in each china place setting where the guests of the Third Reich celebrated and dined.

Outside, on Castle Varlar's grounds, the snow-covered fields were also being readied. For centuries the castle had been a monastery, its broad lawns used as sacred spaces for Benedictine monks to stroll about and consider God. Now, in the frigid December cold, army technicians made last-minute adjustments to the metal platforms of portable rocket-launch pads. On each sat a missile called the V-2.

The giant V-2 rocket was the most advanced flying weapon ever created. It was 46 feet long, carried a warhead filled with up to 2,000 pounds of explosives in its nose cone, and could travel a distance of 190 miles at speeds up to five times the speed of sound. Its earlier version, the V-1 flying bomb, had been raining terror down on cities across northern Europe since the first one hit London, on June 13, 1944. The V-2 rocket was faster and more fearsome. No Allied fighter aircraft could shoot down the V-2 from the sky, both because of the altitude at which it traveled and the speed of its descent. The specter of it crashing down into population centers, annihilating whoever or whatever happened to be there, was terrifying. "The reverberations from each [V-2] rocket explosion spread up to 20 miles," the *Christian Science Monitor* reported. The V-weapons bred fear. Since the start of the war, Hitler had boasted about fearsome "hitherto unknown, unique weapons" that would render his enemies defenseless. Over time, and with the aid of Propaganda Minister Joseph Goebbels, references to these mysterious weapons had been consolidated in a singular, terrifying catchphrase: Nazi wonder weapons, or *Wunderwaffe*. Now, throughout the summer and fall of 1944, the V-weapons made the threat a reality. That the Nazis had unfurled a wonder weapon of such power and potential this late in the war made many across Europe terrified about what else Hitler might have. Plans to evacuate one million civilians from London's city center were put in place as British intelligence officers predicted that a next generation of V-weapons might carry deadly chemical or biological weapons in the nose cone. England issued 4.3 million gas masks to its city dwellers and told people to pray.

Major General Walter Dornberger was the man in charge of the rocket programs for the German army's weapons department. Dornberger was small, bald on top, and when he appeared in photographs alongside Himmler he often wore a long, shin-length leather coat to match the Reichsführer-SS. He was a career soldier — this was his second world war. He was also a talented engineer. Dornberger held four patents in rocket development and a degree in engineering from the Institute of Technology in Berlin. He was one of four honored guests at the Castle Varlar party. Later, he recalled the scene. "Around the castle in the dark forest were the launching positions of V-2 troops in [our] operation against Antwerp." It had been Dornberger's idea to set up mobile launch pads, as opposed to firing V-2s from fortified bases in the Reich-controlled part of France — a wise idea, considering

Allied forces had been pushing across the continent toward Germany since the Normandy landings in June.

Antwerp was Belgium's bustling, northernmost port city, located just 137 miles away from the V-2 launch pads at Castle Varlar. For a thousand years it had been a strategic city in Western Europe, conquered and liberated more than a dozen times. In this war Belgium had suffered terrible losses under four long years of brutal Nazi rule. Three months prior, on September 4, 1944, the Allies liberated Antwerp. There was joy in the streets when the British Eleventh Armored Division rolled into town. Since then, American and British forces had been relying heavily on the Port of Antwerp to bring in men and matériel to support fighting on the western front and also to prepare for the surge into Germany. Now, in the second week of December 1944, Hitler intended to reclaim Antwerp. The Führer and his inner circle were preparing to launch their last, still secret counteroffensive in the Ardennes Forest, and for this the German army needed Antwerp shut down. The job fell to the V-2. The party at Castle Varlar was to be a night of warfare and celebration, with one 42,000-pound liquid-fueled rocket being fired off at the enemy after the next, while the guests honored four of the men who had been instrumental in building the wonder weapon for the Reich.

The man at the scientific center of the V-2 rocket program was a thirty-two-year-old aristocrat and wunderkind-physicist named Wernher von Braun. Von Braun was at Castle Varlar to receive, alongside Dornberger, one of Hitler's highest and most coveted noncombat decorations, the Knight's Cross of the War Service Cross. Also receiving the honor were Walther Riedel, the top scientist in the rocket design bureau, and Heinz Kunze, a representative from the Reich's armaments ministry. These four medals were to be presented by Albert Speer, Hitler's minister of armaments and war production.

Armaments are the aggregate of a nation's military strength, and as minister of weapons, Speer was in charge of scientific armaments programs for the Third Reich. He joined the Nazi Party in 1931, at the age of twenty-six, and rose to power in the party as Hitler's architect. In that role he created buildings that symbolized the Reich and represented its ideas and quickly became a favorite, joining Hitler's inner circle. In February 1942 Hitler made Speer his minister of armaments and war production after the former minister, Fritz Todt, died in a plane crash. By the following month Speer had persuaded Hitler to make all other elements of the German economy second to armaments production, which Hitler did by decree. "Total productivity in armaments increased by 59.6 per cent," Speer claimed after the war. At the age of thirty-seven, Albert Speer was now responsible for all science and technology programs necessary for waging war.

Of the hundreds of weapons projects he was involved in, it was the V-2 that he favored most.

Like von Braun, Speer was from a wealthy, well-respected German family, not quite a baron but someone who wished he was. Speer liked to exchange ideas with youthful, ambitious rocket scientists like Wernher von Braun. He admired "young men able to work unhampered by bureaucratic obstacles and pursue ideas which at times sounded thoroughly utopian."

As for General Dornberger, the Castle Varlar celebration was a crowning moment of his career. The pomp and power thrilled him, he later recalled. "It was a scene," Dornberger said after the war—the excitement of the evening, "[t]he blackness of the night...." At one point during the meal, in between courses, the lights inside the castle were turned off and the grand banquet hall was plunged into darkness. After a moment of anticipatory silence, a tall curtain at the end of the long hall swung open, allowing guests to gaze out across the dark, wintry lawns. "The room suddenly lit [up] with the flickering light of the rocket's exhaust and [was] shaken by the reverberations of its engines," remembered Dornberger. Outside, perched atop a mobile rocket-launch pad, the spectacle began. An inferno of burning rocket fuel blasted out the bottom of the V-2, powering the massive rocket into flight, headed toward Belgium. For

Dornberger, rocket launches instilled "unbelievable" feelings of pride. Once, during an earlier launch, the general wept with joy.

On this night the excitement focus alternated—from a rocket launch to award decoration, then back to a rocket launch again. After each launch, Speer decorated one of the medal recipients. The crowd clapped and cheered and sipped champagne until the banquet hall was again filled with darkness and the next rocket fired off the castle lawn.

This particular party would end, but the celebrations continued elsewhere. The team returned to Peenemünde, the isolated island facility on the Baltic Coast where the V-weapons had been conceived and originally produced, and on the night of December 16, 1944, a party in the Peenemünde's officer's club again honored the men. Von Braun and Dornberger, wearing crisp tuxedos, each with a Knight's Cross from Hitler dangling around his neck, read telegrams of congratulations to Nazi officials as the group toasted their success with flutes of champagne. In the eyes of the Reich, Hitler's rocketeers had good reason to celebrate. In Antwerp at 3:20 p.m., a V-2 rocket had smashed into the Rex Cinema, where almost 1,200 people were watching a Gary Cooper film. It was the highest death toll from a single rocket attack during the war—567 casualties.

*　*　*

The Allies were obsessed with the Nazis' V-weapons. If they had been ready earlier, the course of war would have been different, explained General Dwight D. Eisenhower, Supreme Allied Commander in Europe. "It seemed likely that, if the German had succeeded in perfecting and using these new weapons six months earlier than he did, our invasion of Europe would have proved exceedingly difficult, perhaps impossible," Eisenhower said. Instead, circumstance worked in the Allies' favor, and by the fall of 1944, Allied forces had a firm foothold on the European continent. But back in Washington, D.C., inside the Pentagon, a secret, U.S.-only rocket-related scientific intelligence mission was in the works. Colonel Gervais William Trichel was the first chief of the newly created Rocket Branch inside U.S. Army Ordnance. Now Trichel was putting together a group of army scientists to send to Europe as part of Special Mission V-2. The United States was twenty years behind Germany in rocket development, but Trichel saw an opportunity to close that gap and save the U.S. Army millions of dollars in research and development costs. Trichel's team would capture these rockets and everything related to them for shipment back to the United States. The mission would begin as soon as the U.S. Army arrived in the town of Nordhausen, Germany.

The British had the lead on intelligence regarding

16

V-weapons. Their photo interpreters had determined exactly where the rockets were being assembled, at a factory in central Germany in the naturally fortified Harz Mountains. Trying to bomb this factory from the air was useless, because the facility had been built underground in an old gypsum mine. While the Americans made plans inside the Pentagon, and while von Braun and his colleagues drank champagne at Peenemünde, the men actually assembling the Reich's V-2 rockets endured an entirely different existence. Nazi science had brought back the institution of slavery all across the Reich, and concentration camp prisoners were being worked to death in the service of war. The workers building rockets included thousands of grotesquely malnourished prisoners who toiled away inside a sprawling underground tunnel complex known by its euphemism, Mittelwerk, the Middle Works. This place was also called Nordhausen, after the town, and Dora, the code name for its concentration camp.

To average Germans the Harz was a land of fairy tales, of dark forests and stormy mountains. To those who read Goethe, here was the place where the witches and the devil collided at Brocken Mountain. Even in America, in Disney's popular film *Fantasia,* these mountains had meaning. They were where forces of evil gathered to do their handiwork. But at the end of the Second World War, the Reich's secret, subterranean penal colony at Nordhausen was fact, not fiction. The Mittelwerk was a place where ordinary citizens—of France, Holland, Belgium, Italy, Czechoslovakia, Hungary, Yugoslavia, Russia, Poland, and Germany—had been transformed into the Third Reich's slaves.

The underground factory at Nordhausen had been in operation since late August 1943, after a Royal Air Force attack on the Peenemünde facility up north forced armaments production to move elsewhere. The day after that attack, Heinrich Himmler, Reichsführer-SS, paid a visit to Hitler and proposed they move rocket production underground. Hitler agreed, and the SS was put in charge of supplying slaves and overseeing facilities construction. The individual in charge of expanding Nordhausen from a mine to a tunnel complex was Brigadier General Hans Kammler, a civil engineer and architect who, earlier in his career, built the gas chambers at Auschwitz-Birkenau.

The first group of 107 slave laborers arrived at the Mittelwerk in late August 1943. They came from the Buchenwald concentration camp, located fifty miles to the southeast. The wrought-iron sign over the Buchenwald gate read *Jedem das Seine*, "Everyone gets what he deserves." Digging tunnels was hard labor, but the SS feared prisoners might revolt if they had mining tools, so the men dug with their bare hands. The old mine had been used by the German army as a fuel storage facility. There were two long tunnels running parallel into the mountain that needed to be widened now for railcars. There were also smaller cross-tunnels every few meters that needed to be lengthened to create more workspace. In September 1943 machinery and personnel arrived from Peenemünde. Notable among the staff, and important to Operation Paperclip, was the man in charge of production, a high school graduate named Arthur Rudolph.

Rudolph's specialty was rocket engine assembly. He had worked under von Braun in this capacity since 1934. Rudolph was a Nazi ideologue; he joined the party before there was any national pressure to do so, in 1931. What he lacked in academic pedigree he made up for as a slave driver. As the Mittelwerk operations director, Rudolph worked with the SS construction staff to build the underground factory. Then he oversaw production on the assembly lines for V-weapons scientific director Wernher von Braun.

The prisoners worked twelve-hour shifts, seven days a week, putting together V-weapons. By the end of the first two months there were eight thousand men living and working in this cramped underground space. There was no fresh air in the tunnels, no ventilation system, no water, and very little light. "Blasting went on day and night and the dust after every blast was so thick that it was impossible to see five steps ahead," read one report. Laborers slept inside the tunnels on wood bunk beds. There were no washing facilities and no sanitation. Latrines were barrels cut in half. The workers suffered and died from starvation, dysentery, pleurisy, pneumonia, tuberculosis, and phlegmasia from beatings. The men were walking skeletons, skin stretched over bones. Some perished from ammonia burns to the lungs. Others died by being crushed from the weight of the rocket parts they were forced to carry. The dead were replaceable. Humans and machine parts went into the tunnels. Rockets and corpses came out. Workers who were slow on the production lines were beaten to death. Insubordinates were garroted or hanged. After the war, war crimes investigators determined that approximately half of the sixty thousand men eventually brought to Nordhausen were worked to death.

The Mittelwerk wasn't the first slave labor camp created and run by the Third Reich. The SS recognized the value of slave labor in the mid-1930s. Humans could be selected from the ever-growing prisoner populations at concentration camps and put to work in quarries and factories. By 1939 the SS had masterminded a vast network of state-sponsored slavery across Nazi-occupied Europe through an innocuous-sounding division called the SS Business Administration Main Office. This office was overseen by Heinrich Himmler but required partnerships. These included many companies from the private sector, including IG Farben, Volkswagen, Heinkel, and Steyr-Daimler-Puch. The most significant partner was Albert Speer's Ministry of Armaments and War Production. When Speer took over as armaments minister in February 1942 his first challenge, he said, was to figure out how to galvanize war production and make it more efficient. Speer's solution was to get rid of bureaucracy and use more slave laborers. He himself had been connected to the slave labor programs with the SS for years, including when he was an architect. Speer's buildings required vast amounts of stone, which was quarried by concentration camp laborers from Mauthausen and Flossenbürg.

The SS Business Administration Main Office specialized in engineering dangerous and fast construction projects, as was the case with the V-2 facility at Nordhausen. "Pay no attention to the human victims," Brigadier General Hans Kammler told his staff overseeing construction in the tunnels. "The work must proceed and be finished in the shortest possible time." In the first six months of tunnel work, 2,882 laborers died. Albert Speer praised Kammler for what he considered to be a great achievement in engineering, setting things up so efficiently and so fast. "[Your work] far exceeds anything ever done in Europe and is unsurpassed even by American standards," wrote Speer.

There were other reasons why the use of slave labor

was so important to wonder weapons production, namely, the secrecy it ensured. The V-2 was a classified weapons project; the less Allied intelligence knew about it, the better for the Reich. When Albert Speer and Heinrich Himmler met with Hitler in August of 1943 to brief him on the benefits of using slave labor, Himmler reminded the Führer that if the Reich's entire workforce were to be concentration camp prisoners, "all contact with the outside world would be eliminated. Prisoners don't even receive mail."

In the spring of 1944, V-2 production had accelerated to the point where the SS provided Mittelwerk managers with their own concentration camp, Dora, which in turn grew to include thirty subcamps. The man in charge of "personnel" at the Mittelwerk, its general manager, was a forty-six-year-old engineer named Georg Rickhey, an ardent Nazi and party member since 1931. On Rickhey's résumé, later used by the Americans to employ him, Rickhey described himself as "Mittelwerk General Manager, production of all 'V' and rocket weapons, construction of underground mass-production facilities, director of entire concern." As general manager of the sprawling, subterranean enterprise, Rickhey was in charge of "renting" slaves from the SS. As a former Demag Armor Works executive, he had already overseen the creation of more than 1.5 million square feet of underground tunnels around Berlin, all dug by slaves. With this experience Rickhey had become a veteran negotiator between private industry and the SS Business Administration Main Office in the procurement of slaves. "The SS began, in effect, a rent-a-slave service to firms and government enterprises at a typical rate of four marks a day for unskilled workers and six marks for skilled ones," writes V-weapon historian Michael J. Neufeld. The slaves were disposable. When they died they were replaced. At Nordhausen the SS gave Rickhey a discount, charging the Mittelwerk between two and three reichsmarks per man, per day.

On May 6, 1944, days after becoming general manager of the Mittelwerk, Rickhey called a meeting to discuss how best to acquire more prisoners for slave labor. Wernher von Braun, Walter Dornberger, and Arthur Rudolph were all present. It was decided that the SS should enslave another eighteen hundred skilled French workers to fill the shoes of those who had already been worked to death. The record indicates that von Braun, Dornberger, and Rudolph showed no objection to Rickhey's plan.

In August, the same problem was again at issue. This time Wernher von Braun initiated the action himself. On August 15, 1944, von Braun wrote a letter to a Mittelwerk engineer, Albin Sawatzki, describing a new laboratory he wanted to set up inside the tunnels. Von Braun told Sawatzki that to expedite the process, he had taken it upon himself to procure the slave laborers from the Buchenwald concentration camp.

"During my last visit to the Mittelwerk, you proposed to me that we use the good technical education of detainees available to you [from] Buchenwald," wrote von Braun. "I immediately looked into your proposal by going to Buchenwald [myself], together with Dr. Simon [a colleague], to seek out more qualified detainees. I have arranged their transfer to the Mittelwerk with Standartenführer [Colonel] Pister," the commandant of Buchenwald.

In December of 1944, with slave laborers dying by the thousands in the Mittelwerk tunnels and V-2 rockets crashing into civilian population centers, causing mayhem and terror across Europe, it would have been hard to imagine that some of those directly responsible would ever be regarded as individuals of great value to the United States. And yet in less than a year Arthur Rudolph, Georg Rickhey, Wernher von Braun, Major General Walter Dornberger, and other rocket engineers would secretly be heading to America to work. In the last days of World War II few would ever have believed such a thing.

But the war's last days were coming. Just three weeks after the celebration at Castle Varlar, Albert Speer found himself with a lot less to celebrate. Visiting the Belgian border town of Houffalize, accompanied by an SS armored force commander named Josef

"Sepp" Dietrich, Speer had what he would describe in his 1969 memoir as a realization. Gazing upon the bodies of hundreds of dead German soldiers killed in a recent Allied bomber attack, Speer decided the war was over for the Reich. The German war machine could no longer compete against the force and will of the Allied offensive. "Howling and exploding bombs, clouds illuminated in red and yellow hues, droning motors, and no defense anywhere—I was stunned by this scene of military impotence which Hitler's military miscalculations had given such a grotesque setting," Speer wrote. Standing there in Houffalize, Speer—Hitler's minister of armaments and war production—decided to flee from the danger zone.

At 4:00 a.m. on the morning of December 31, under cover of darkness, Speer and an aide climbed into a private car and hurried east, headed for the comforts of a sprawling mountaintop castle outside Frankfurt called Schloss Kransberg, or Castle Kransberg. Built on a steep, rocky cliff in the Taunus Mountains, the castle was one of Hermann Göring's Luftwaffe (German air force) headquarters. Just as many of Hitler's scientists would soon become American scientists, so would many of the Reich's headquarters and command posts become key facilities used for Operation Paperclip. Castle Kransberg also had a storied past in the history of warfare. The structure dated from the eleventh century, but its original foundation had been built on top of the ruins of a ring-wall fortification constructed in the time of the Roman Empire. Battles had been waged in this region, on and off, for over two thousand years.

Castle Kransberg was grand and splendid, built piecemeal over the centuries to include watchtowers, half-timbered meeting halls, and stone walls. It had 150-odd rooms, including a wing that had been redesigned and renovated by Albert Speer in 1939, when Speer was still Hitler's architect. At Hitler's behest Speer added several state-of-the art defense features to Kransberg Castle, including a twelve-hundred-square-foot underground bunker complex, complete with poison gas air locks designed to protect inhabitants from a chemical warfare attack. Now here was Speer, having fled from the front lines to hide out in this citadel. The next time he would live here it would be as a prisoner of the Americans.

Hitler had his own headquarters just a few miles away. Adlerhorst, or the Eagle's Nest, had also been designed by Speer. It was a series of small cement bunkers at the edge of a long stretch of valley near the spa town of Bad Nauheim. Few knew it was there. From Adlerhorst, Hitler had been directing the Ardennes campaign — the Battle of the Bulge.

Arriving at Kransberg Castle late at night after fleeing Houffalize, Speer and his aide were shown to their quarters, where they freshened up before driving over to Adlerhorst to celebrate the coming year—1945—with Hitler. When Speer arrived at the Eagle's Nest at 2:30 a.m. on January 1, Hitler, who never drank, appeared drunk. "He was in the grip of a permanent euphoria," remembered Speer. Hitler made a toast and promised that the present low point in the war would soon be overcome. "His [Hitler's] magnetic gifts were still operative," Speer later recalled. In the end Germany would be victorious, Hitler said. This was enough for Speer to change his mind about losing the war.

Two weeks later, on January 15, with the war allegedly still winnable, Adolf Hitler boarded his armored train and began the nineteen-hour trip to Berlin, where he would spend the rest of his life living underground in the Führerbunker (Führerhauptquartier, or FHQ) in Berlin. The bunker was a fortress of engineering prowess, built beneath the New Reich Chancellery. Its roof, buried under several tons of earth, was sixteen feet thick. Its walls were six feet wide. Living inside the Führerbunker, with its low ceilings and cryptlike corridors, was "like being stranded in a cement submarine," said one of Hitler's SS honor guards, Captain Beermann. Beermann described a "bat-like routine of part-time prisoners kept in a cave. Miserable rats in a musty cement tomb in Berlin." Not everyone shared the sentiment. Months later, when Mittelwerk general manager Georg Rickhey was angling to get a job from the Americans, he would boast to army officers that he'd overseen construction of the grand Führerbunker in Berlin.

Now, in mid-January 1945, with Hitler moving back to Berlin, it was decided that Speer should head east, to Silesia, in Poland. There, he was to survey what was going on. Important chemical weapons factories had been built in Poland, armaments ventures jointly pursued with IG Farben, a chemical industry conglomerate. The location of these facilities was significant; Poland was, for the most part, out of reach of Allied bombing campaigns. But a new threat was bearing down. The Soviets had just launched their great offensive in Poland, a final military campaign that would take the Red Army all the way to Berlin. Germany was being invaded from both sides—east and west—squeezed as in a vise.

The same day that Hitler left the Eagle's Nest, Speer was driven to Poland. There, he witnessed first-hand what little was left of the Reich's war machine. On January 21, he went to the village of Oppeln to check in with Field Marshal Ferdinand Schörner, newly appointed commander of the army group, and learned that very little of the Wehrmacht's fighting forces remained intact. Nearly every soldier and every war machine had been captured or destroyed. Burnt-out tank hulls littered the snow-covered roads. Thousands of dead German soldiers lay in ditches along the roadsides, but many more dead soldiers swung eerily from trees. Those who dared desert the German army had been killed by Field Marshal Schörner, a ferocious and fanatical Nazi Party loyalist who had earned his moniker, Schörner the Bloody. The dead German soldiers had placards hanging around their necks. "I am a deserter," they read. "I have declined to defend German women and children and therefore I have been hanged."

During Speer's meeting with Schörner, he was told that no one had any idea exactly how far the Red Army was from overtaking the very spot where they were standing, only that the onslaught was inevitable. Speer checked into an otherwise empty hotel and tried to sleep.

For decades, this night remained vivid in Albert Speer's mind. "In my room hung an etching by Käthe Kollwitz: La Carmagnole," remembered Speer as an old man. "It showed a yowling mob dancing with hate-contorted faces around a guillotine. Off to one side a weeping woman cowered on the ground. . . . The weird figures of the etching haunted my fitful sleep," wrote Speer. There, in his Oppeln hotel room, Albert Speer was overcome by a thought that had to have been preoccupying many Nazis' minds. After Germany, what will become of me? The guillotine? Will I be torn apart by a yowling mob?

Jedem das Seine. Could it be true? Does everyone get what he deserves?

The following week, on January 30, 1945, Speer wrote a memorandum to Hitler outlining the huge losses in Silesia. "The war is lost," is how Speer's report began.

Widespread destruction of evidence would now begin.

CHAPTER TWO

Destruction

Ninety miles southeast of Speer's Oppeln hotel room, chaos was unfolding at the Auschwitz-Birkenau death camp. Speer's chairman of the ultra-secret Committee-C for chemical weapons, a chemist named Dr. Otto Ambros, had documents to destroy. It was January 17, 1945, and every German in a position of power at Auschwitz, from the army officers to the IG Farben officials, was trying to flee. Not Ambros. He would not leave the labor-extermination camp for another six days.

Otto Ambros was a fastidious man. His calculations were exact, his words carefully chosen, his fingernails always manicured. He wore his hair neatly oiled and parted. In addition to being Hitler's favorite chemist, Ambros was the manager of IG Farben's synthetic rubber and fuel factory at Auschwitz. With the

Red Army bearing down, the Farben board members had ordered the destruction and removal of all classified paperwork; Ambros and colleague Walther Dürrfeld were at Auschwitz to do the job. In addition to being the plant manager of this hellish place, Ambros was the youngest member on Farben's board of directors.

All over the camp, SS guards were destroying evidence. Crematoria II and III were being dismantled, and a plan to dynamite Crematoria V was in effect. Some of the SS officers were already fleeing on horseback, while others were preparing to evacuate prisoners for the death march. Whips cracked. Dogs barked. Tanks painted white for camouflage outside the camp rolled through the muddy streets. Rumors swirled: The Red Army was only a few miles away. A female chemist in Farben's Buna-Werke Polymerization Department asked the prisoner and future world-renowned writer Primo Levi, a chemist by training, to fix her bicycle tire. After the war, Levi recalled how strange it was to hear a Farben employee use the word "please" with a Jewish inmate like himself.

Auschwitz was the Reich's largest extermination center. As a concentration camp it consisted of three separate but symbiotic camps: Auschwitz I, the main camp; Auschwitz II, the Birkenau gas chambers and crematoria; and Auschwitz III, a labor-concentration camp run by the chemical giant IG Farben. Since

April 7, 1942, IG Farben had been building the Reich's largest chemical plant at Auschwitz, using a workforce of slave laborers selected from the Auschwitz train car platforms. Farben called their facility "IG Auschwitz."

IG Auschwitz was the first corporate concentration camp in the Third Reich. The barracks in which the slave laborers lived and died was officially called Monowitz. "Those of us who lived there called it Buna," recalls Gerhard Maschkowski, a survivor of the camp. Maschkowski was a nineteen-year-old Jewish boy spared the gas chamber because he was of use to IG Auschwitz as an electrician. That he was still alive in the second week of January 1945 was something of a miracle. He had arrived at Auschwitz on April 20, 1943, which meant that he had been there for a year and nine months. At Buna, the average lifespan of a slave laborer was three months. Many of Gerhard Maschkowski's friends had long since been worked to death or had been murdered for minor infractions, like hiding a piece of food.

Gerhard Maschkowski remembers January 18, 1945, with clarity, because it was his last day at Auschwitz. It was still dark outside when the SS burst into the barracks. "They shouted, 'Get up! March!' They had large guns, thick jackets, and dogs," Maschkowski recalls. He put on his shoes and hurried outside. There, nine thousand emaciated, starving inmates from Buna were lining up in neat rows. Maschkowski heard cannon fire in the distance and the crack of firearms close by. There was chaos all around. SS guards were burning evidence. Bonfires of papers sent ashes up into the dark sky. Snow fell fast, then faster. There was a blizzard on the way. The guards, dressed in warm coats and boots, waved submachine guns. "Dogs on leather leashes barked and snarled," Gerhard Maschkowski recalls. Wearing thin pajamas, the prisoners at Buna-Monowitz began a death march toward the German interior. Within forty-eight hours, 60 percent of them would be dead.

Primo Levi was not part of the death march. A week before, he had contracted scarlet fever and had been sent to the infectious diseases ward. "He had a high temperature and a strawberry tongue," remembered Aldo Moscati, the Italian doctor-prisoner who tended to him. "With a [104-degree] fever I was extremely feeble and could not even walk," Primo Levi explained after the war. Lying supine in the infectious diseases ward, he listened to the sounds of the emptying camp.

On January 21, a memorandum from Berlin ordered all Farben employees to leave. The last train for Germans leaving Auschwitz, transporting mostly IG Farben's female staff from the camp, left that same afternoon. Yet Otto Ambros stayed behind. Ambros's official title was plant manager of Buna-Werk IV and managing director of the fuel production facility at IG

Auschwitz. He had been involved in the facility since January 1941, back when Farben's original plans were being drawn. Ambros had chosen the site location and sketched out the original blueprints for the plant. He was also the man who invented synthetic rubber for the Reich. Tanks, trucks, and airplanes all require rubber for tires and treads, and during wartime this feat was considered so important to the Reich's ability to wage war that Ambros had been awarded 1 million reichsmarks by the Führer.

Finally, on January 23, 1945, Ambros left the concentration camp. Only random prisoners remained. Inmates like Primo Levi who were too weak to march and had not been killed by the SS lay in the infectious diseases ward. When Levi's fever finally broke and he ventured outside, he found groups of prisoners roaming around the camp looking for food. Levi found a silo filled with frozen potatoes. He made a fire and cooked his first food in days. On January 27, he was dragging a dead friend's corpse to a large grave dug in a distant field when he spotted four men on horseback approaching the camp from far away. They were wearing white camouflage clothing, but as they got closer he could see that at the center of the soldiers' caps there was a bright red star. The Red Army had arrived. Auschwitz was liberated.

Otto Ambros was already on the way to Falkenhagen, Germany, to destroy evidence in another Farben factory there. Speer headed back to Berlin. Neither man dared travel north inside Poland, where a second armaments factory the two men were jointly involved in was also in jeopardy of being captured by the Soviets. At this facility, called Dyhernfurth after the small riverside village in which it was located, IG Farben produced chemical weapons—deadly nerve agents—on an industrial scale. On January 24, 1945, the day after Ambros fled Auschwitz, Farben had given the word to evacuate Dyhernfurth and destroy whatever evidence remained there. All munitions were loaded onto railcars and trucks and sent to depots in the west.

The destruction of evidence was now becoming standard operating procedure at laboratories, research facilities, and armaments factories across the Reich. And while Nazi Germany faced imminent collapse, its scientists, engineers, and businessmen had their futures to think about.

All across Nazi-occupied Poland, German forces were retreating en masse as the Red Army continued to shred the eastern front. On February 5, 1945, one hundred and seventy miles northwest of Auschwitz, the Soviets captured the village of Dyhernfurth. Soviet soldiers took over the town's castle, built during the seventeenth and eighteenth centuries, and drank up its wine cellar. The castle, with its fairy-tale-like conical spires, quickly transformed into a scene of wild debauchery, with inebriated Russian soldiers singing rowdy victory songs. The situation got so out of control that Russian commanders suspended fighting until order could be restored. This made for a perfect opportunity for a team of Nazi commandos hiding in the forest to launch a daring and unprecedented raid.

Less than half a mile away, one of the Reich's most prized wonder weapons facilities lay hidden underground. Camouflaged in a forest of pine trees, inside a large complex of underground bombproof bunkers, a workforce of 560 white-collar Germans and 3,000 slave laborers had been mass-producing, since 1942, liquid tabun — a deadly nerve agent the very existence of which was unknown to the outside world.

Tabun was one of Hitler's most jealously guarded secrets, a true wonder weapon of the most diabolical kind. Similar to a pesticide, the organophosphate tabun was one of the most deadly substances in the world. A tiny drop to the skin could kill an individual in minutes or sometimes seconds. Exposure meant the glands and muscles would hyperstimulate and the respiratory system would fail. Paralysis would set in and breathing would cease. At Dyhernfurth, where accidents had happened, a human's death by tabun gas resembled the frenetic last moments of an ant sprayed with insecticide.

Like the synthetic rubber and fuel factory at

Auschwitz, the nerve agent production facility at Dyhernfurth was owned and operated by IG Farben, and here the Speer ministry worked with Farben to fill aerial bombs with tabun that could eventually be deployed from Luftwaffe planes. No one in the inner circle knew for sure when, or if, Hitler would finally concede to many of his ministers' wishes and allow for a chemical weapons attack against the Allies. But as evidenced at Dyhernfurth, the opportunity was real. Enough poison gas had been produced here to decimate the population of London or Paris on any given day.

Despite the overwhelming onslaught of Red Army troops to this region in February of 1945, Hitler remained determined to keep the secrets of tabun out of enemy hands. On the morning of February 5, 1945, Major General Max Sachsenheimer of the Reich's Seventeenth Field Replacement Battalion and his troops were hiding in the forest along the banks of the Oder River. Sachsenheimer's commando force was made up of several hundred soldiers, but he also had a unique secondary contingent of men, namely, eighty-two scientists and technicians. Some were army scientists but most were IG Farben employees and chemical weapons experts. Sachsenheimer's mission was to protect the scientists while they scrubbed Farben's facility of any and all traces of tabun.

The Dyhernfurth complex was a sprawling, state-of-the-art production plant. Speer's Armaments and

War Production Ministry had paid Farben nearly 200 million reichsmarks to build and operate it. The facility had been secretly and skillfully designed and managed by Otto Ambros. As he had done with IG Auschwitz, Ambros had overseen every element of this chemical weapons factory dating from the winter of 1941, when the thick forest here was first cleared of pine trees by 120 concentration camp slaves.

As Nazi Germany blended industry, war making, and genocide, few corporations were as central a player as IG Farben. The chemical concern was the largest corporation in Europe and the fourth largest corporation in the world. IG Farben owned the patent on Zyklon B. And perhaps no single person at Farben was as central a figure in this equation as Otto Ambros had been. For his work as chairman of Committee-C, the chemical weapons committee inside the Speer Ministry, Ambros was given the prestigious title of military economy leader (Wehrwirtschaftsführer). He was awarded the War Merit Cross, 1st and 2nd Class, and the Knight's Cross of the War Merit Cross, which was similar to the award bestowed on Dornberger and von Braun.

There was a second scientist who played an important role in chemical weapons—a man who, like Otto Ambros, would be targeted for Operation Paperclip. This was SS-Brigadeführer Dr. Walter Schieber, a chemist by training, Speer's deputy and director of the

Armaments Supply Office. Schieber was a hard-core Nazi ideologue and a member of Reichsführer-SS Himmler's personal staff. Unusually corpulent for an SS officer, his official car and airplane required retrofitting to accommodate his 275-pound frame. With Hitler's physician, Karl Brandt, Schieber was in charge of gas mask production, a requisite for troop defense if chemical warfare was to be waged. As with biological weapons, the sword needs a shield, and by January of 1945 Schieber had overseen the production of 46.1 million gas masks. Their reliability had been tested at Dyhernfurth on concentration camp prisoners. The stories that would emerge during the Nuremberg trials about such tests were ghoulish, including locking prisoners in glass rooms and spraying them with nerve agent. War crimes investigators would later debate whether or not these actions were pilot programs for the gas chambers. The full story of how, and to what extent, Dr. Walter Schieber worked for the U.S. military after the war—and also for the CIA—has never been fully explained until this book.

Only weeks before the Red Army took Dyhernfurth, overran its castle, and drank all its wine, thousands of concentration camp slave laborers had been toiling away at Farben's secret chemical weapons plant performing the deadliest of jobs. Wearing double-layer rubber suits and bubble-shaped helmets, prisoners filled artillery shells and bomb casings with nerve agent, marking each munition in a secret code indicating tabun nerve agent: three green rings of paint. The prisoners' suits worked similar to deep-sea diving suits. Attached to the back of each helmet was a tube delivering breathable air. But the air tube was short and gave workers very little room to move. If a man accidentally detached from the air source, he would be exposed to the lethal vapors through the breathing tube and die.

But on the morning of February 5, 1945, the facility was empty. Not a chemist or a slave laborer remained. The munitions had been moved, documentary evidence destroyed, and all the IG Farben employees had fled. The prisoners had been evacuated by their SS guards three weeks before. Wearing prisoner pajamas and ill-fitting wooden work shoes, the Dyhernfurth laborers were marched west toward the German interior. Witnesses in nearby villages described a column of three thousand walking corpses. Temperatures in the area reached -18 degrees Fahrenheit. By the time the prisoners reached the Gross-Rosen concentration camp, fifty miles to the southwest, two-thirds of them had died of exposure.

Now, working on Hitler's orders, a technical team was preparing to return to Farben's secret facility for a final scrub of tabun residue. In the freezing predawn air, the two chemists, eighty technicians, and a group of German commandos donned double-layered rubber suits, pulled gas masks over their heads, and stole down the banks of the Oder River. They moved quietly across a partially bombed-out railroad bridge, walked slowly down the railroad tracks, and headed to the chemical weapons plant. The protective suits were cumbersome, and it took the technical team sixty-five minutes to travel less than a half a mile. Reaching the plant, the group made its way into the production facility that housed massive silver-lined kettles in which the nerve agent was made. Each kettle sat inside an operating chamber enclosed in double-glass walls encircling a complex ventilation system of double-walled pipes. While one group got to work decontaminating the chambers with ammonia and steam, another group scoured the surfaces of the munitions-filling factory where so many slave laborers had met death. The commandos kept guard.

While the technicians scrubbed, two and a half miles downriver Major General Sachsenheimer's remaining platoon of soldiers sprang into action. They launched artillery shells at sleeping Russian forces in a diversionary feint. The Red Army reorganized and retaliated, and by lunchtime eighteen Soviet tanks were engaged in fierce fighting with Sachsenheimer's troops. The battle, a footnote in the annals of the war, lasted just long enough for IG Farben's technical team to get in, get out, and disappear.

Not for several days did the Red Army finally

stumble upon Farben's chemical weapons facility at Dyhernfurth. By then it was void of people and tabun but otherwise entirely intact. As the Russians examined the facility, scrubbed immaculately clean, it became clear to the commanders that whatever this facility had produced must have been considered of great value to the Reich. The laboratory layout bore signs of chemical weapons production, and the Soviet army called in their own chemical weapons experts from the Sixteenth and Eighteenth Chemical Brigades. The factory was dismantled, crated, and shipped back to the Soviet Union for future use. What had been produced here in the forest remained a mystery to its new owners—the Russians—for a little over a year. By 1946 the entire chemical weapons factory at Dyhernfurth would be reassembled in a little town outside Stalingrad called Beketovka, and the plant given the Russian code name, Chemical Works No. 91. The Soviets themselves then began producing tabun nerve agent on an industrial scale. By 1948, the Soviet *Military Chemical Text Book* would list tabun as part of the Red Army's stockpile. But 1948 was so far away. So much would happen with America and Hitler's chemists between 1945 and 1948, most of it predicated on the emerging Soviet threat.

Back in Berlin, Hitler read the first page of Speer's report and ordered it filed away. Then Hitler became enraged. His furor was likely exacerbated by a conference taking place on the Crimea Peninsula, at Yalta, beginning February 4, 1945. It was to last for eight days. There, Roosevelt, Churchill, and Stalin were confirming their commitment to demand Germany's unconditional surrender. There would be no bargaining, the three heads of state declared. No deals made with the Nazis. The end of the war would mean the end of the Third Reich. War criminals would be tried, justice meted out.

The idea of what defined justice varied dramatically from power to power. British Prime Minister Winston Churchill wanted Nazis leaders to be treated as "outlaws." He argued that they should be lined up and shot rather than put on trial. The premier of the Soviet Union, Joseph Stalin, rather unexpectedly argued for "no executions without trial." President Franklin D. Roosevelt wanted a war crimes trial. What all parties agreed on was that, after the surrender, Germany would be broken up into three zones of occupation. (Soon, France's involvement would make it four.) When Hitler learned of the Allies' plan to divvy up the spoils of Germany, he became furious. Then he called on Albert Speer.

"If the war is lost," Hitler famously told Speer, "the nation will also perish. This fate is inevitable." If Germany loses the war, Hitler said, the people deserved existential punishment for being weak. "What will remain after this struggle will be in any case only the inferior ones, since the good ones have fallen," Hitler said. Following this logic, Hitler decreed that Speer's ministry need not provide German citizens with even the most basic necessities, including shelter and food. He told Speer that if Speer were ever to give him another memorandum saying the war was unwinnable, and that negotiations with the enemy should be considered, Hitler would consider this an act of treason punishable by death.

Hitler then issued a nationwide "scorched-earth" policy. Speer was to help organize the complete destruction of all German infrastructure, military and civilian—from its transportation and communication systems to its bridges and dams. Officially entitled Demolitions on Reich Territory, this order became known as the Nero Decree, or Nerobefehl, invoking the Roman emperor Nero, who allegedly engineered the Great Fire of 64 AD and then watched Rome burn.

In central Germany, in the naturally fortified Harz Mountains, production of V-weapons continued at a frenzied pace despite every indication that Germany would soon lose the war. By late February 1945, conditions inside the Nordhausen tunnels had reached a cataclysm. It was bitterly cold, thousands were starving to death, and there was barely any food; watery broth was all the prisoners had to live on. The

Nordhausen-Dora concentration camp was being overrun with new prisoners arriving on the death marches from Auschwitz and Gross-Rosen in the east. Some came on foot and others came in cattle cars. Many arrived dead. Dora's crematorium was overwhelmed. In this climate, Wernher von Braun and General Dornberger pressed on with plans to create a greater number of rockets each day.

Since moving his offices from the Peenemünde facility to the Harz Mountains, von Braun had been given a promotion. Now he was head of what was called the Mittelbau-Dora Planning Office, a division within Himmler's SS. Von Braun lived just a few miles from the Nordhausen complex, in a villa that had been confiscated from a Jewish factory owner years before. Each day he drove to his office in Bleicherode, eleven miles from the tunnel complex, where he drafted designs for new and better test stands and launch ramps for the V-2. Despite the reality that the war was lost, the area was abuzz with new armaments factories being built, "dug by workers from some of the forty-odd subcamps now tied to Mittelbau-Dora," Michael J. Neufeld explains. Von Braun's vision was to escalate rocket production from two or three rockets a day to two hundred rockets a day. To prepare for this ambitious expansion, he commandeered factories, schools, and mines throughout the region. But rocket assembly was dependent on workers, and the slave laborers in Nordhausen were now dying at an ever-increasing rate. Von Braun had to have known this; he visited the underground tunnels in an official capacity ten times during the winter of 1945.

For the emaciated slave laborers who had managed to stay alive, trying to assemble missiles in filthy, unfinished tunnels without food, water, or sanitation in the bitter cold of winter had become more and more difficult, and it showed in their work. In skies across Europe these hastily constructed rockets began breaking apart in flight. And in the pine forests of northern Germany, including those around the Castle Varlar, quickly assembled rockets were exploding on their mobile launch pads. The managers at Mittelwerk suspected sabotage. To send a message, public hangings were held.

"Prisoners were hanged up to 57 in one day," read one war crimes report. "They were hanged in the tunnels with the help of an electrically controlled crane, a dozen at a time, their hands bound behind their back, a piece of wood was put in their mouth...to prevent shouting." The hangings were carried out directly above the V-2 production lines. Laborers were forced to watch their fellow prisoners suffer an agonizingly slow death. In solidarity, a group of Russian and Ukrainian prisoners staged a revolt. The suspects were rounded up. Mittelwerk managers and SS guards decided to make an example of them. After these men were hanged, their bodies were left dangling for a day. Only after Arthur Rudolph, the Mittelwerk operations director, received a memo from one of his German engineers asking when they were going to get their crane back were the bodies taken down.

In addition to von Braun's recent promotion to head of the Mittelbau-Dora Planning Office, he was also promoted to SS-Sturmbannführer, or SS major. One of the benefits that came with this position was a chauffeur-driven car to shuttle von Braun back and forth between Nordhausen and Berlin. It was in the backseat of this car on the night of March 12, 1945, that von Braun was nearly killed. As his car was speeding down the autobahn headed for Berlin, his driver nodded off at the wheel. The car veered off the road and hurtled down a forty-foot embankment until it crashed on its side near a railroad track. The driver was knocked unconscious. Von Braun broke his arm. Both men lay bleeding in the cold, dark night when two of von Braun's colleagues from Nordhausen, facilities designer Bernhard Tessmann and architect Hannes Lührsen, happened to drive by and spot the smashed-up car. They called for a military ambulance, which came to the scene and transported von Braun and his driver to a hospital.

While recuperating, von Braun received a visit from his personal aide, Dieter Huzel, and Bernhard Tessmann, one of the two men who had saved von

Braun's life. Tessmann and Huzel told von Braun that the arrival of the U.S. Army was imminent. Any day now, they said, V-2 operations would cease. Word was going around that every man not considered a valuable scientist was going to be assigned to an infantry unit, handed a weapon, and ordered to fight the Americans on the front lines.

Von Braun was now ready to concede that Germany would lose the war. What he was unwilling to do was relinquish his career. He needed a bargaining chip to use against the Americans after he was captured. Von Braun told Tessmann and Huzel where he kept the most valuable classified V-2 documents. Because von Braun was bedridden, he needed his two subordinates to crate up these documents and hide them in a remote, secure place where the Allies would never find them on their own. Von Braun told Tessmann and Huzel that if they could do this they would be included in von Braun's future negotiations with the Allies. General Dornberger would also be part of the team. Von Braun told Tessmann and Huzel that he would personally bring the general up to speed.

Mittelwerk laborers continued to produce missiles until the end of March. The last V-2s were fired on March 27 and the last V-1s were fired the following day. On April 1, Dornberger received an order from SS-General Kammler demanding that Dornberger evacuate his staff from the Mittelwerk at once.

Kammler had selected five hundred key scientists and engineers who were to board his private train car, parked in Bleicherode and nicknamed the Vengeance Express, and then travel four hundred miles south into the Bavarian Alps to hide out. Huzel and Tessmann were on Kammler's list, but after colluding with General Dornberger they were able to stay behind to complete the document stash. Von Braun, still requiring medical attention and encumbered by a heavy cast, was taken to the Alps in a private car. Dornberger and his staff drove themselves, fleeing the Harz in a small convoy.

When night fell on Dörnten, a small mining community at the northern edge of the Harz, the village was in a blackout. The local gauleiters (Nazi Party district leaders) had ordered villagers to shutter their windows, turn off the lights, and stay home. It was April 4, 1945, and American forces were reportedly camped out just thirty miles to the west. The village's cobblestone streets were empty, save a lone truck driving slowly with its lights off, navigating by the moon. In the front seat sat Tessmann and Huzel. In the back of the truck were seven German soldiers wearing blindfolds. Also in the truck were dozens of crates filled with classified V-2 information.

The truck passed through town and headed up a winding rural road leading into the mouth of an abandoned mine. Huzel and Tessmann parked the truck and shook hands with the caretaker, Herr Nebelung, a loyal Nazi, who sold the two engineers space in a large antechamber in the back of the Dörnten mine. The seven soldiers were told it was okay to take off their blindfolds now and get to work.

The group unloaded the crates, placed them onto flatcars, and oversaw them as they were driven down a long tunnel by an electric-battery-powered locomotive. At the end of the tunnel, behind an iron door, was a small, dry room. The crates of V-2 documents were packed inside, the door shut and locked. Outside, a soldier lit a stick of dynamite in front of the doorway to create a huge pile of rubble and keep the important Nazi documents hidden from the outside world.

Huzel and Tessmann had sworn not to tell anyone what they'd done, but just as they were leaving for Bavaria to rendezvous with Dornberger and von Braun, the two men decided to make an exception. Karl Otto Fleischer had served as the army's business manager for the Mittelwerk slave labor enterprise. Fleischer's name was not on General Kammler's list, and he'd gone home to his house in Nordhausen to blend in. Karl Fleischer was a loyal Nazi who reported directly to General Dornberger during V-2 production, and Huzel and Tessmann believed Fleischer would keep their secret safe. Fleischer was local. He could keep an eye on things and monitor if anyone started poking around the Dörnten mine. Huzel and Tessmann told Karl Fleischer about the secret location of the V-2 document stash, and with the U.S. Army just a few days outside Nordhausen, Huzel and Tessmann sped away.

CHAPTER THREE

The Hunters and the Hunted

In France, Samuel Goudsmit and his team of Operation Alsos scientists had been waiting patiently since November to launch their next scientific intelligence mission. It had been four months since the team hit intelligence gold, inside the Strasbourg apartment of the Reich's virologist Dr. Eugen Haagen. Now, in the last week of March 1945, Goudsmit and his military commander, Colonel Boris Pash, were finally getting ready to seize their next targets: IG Farben factories over the border in Germany, believed to be the locus of the Nazis' chemical weapons program.

Since November, Haagen's apartment in Strasbourg had been serving as Alsos headquarters. There had been long delays. In December, Hitler's counteroffensive in the Ardennes Forest meant that Alsos scientists could not conduct frontline missions as planned. But as serendipity would have it, the entire first floor below Haagen's Strasbourg apartment belonged to IG Farben, the primary chemical weapons supplier of the Third Reich, and Alsos seized a considerable cache of documents Farben kept there. Alsos knew Farben was involved in weapons-related vaccine research and suspected they were involved in medical experiments on prisoners. The two Farben factories they had in their sights were located only about eighty miles away, in the German cities of Ludwigshafen and Mannheim. For this, Operation Alsos had put together their largest task force to date: ten civilian scientists, six military scientists, and eighteen security operators. But venturing into German-held territory on their own was considered too dangerous for the American scientists. They had been waiting for Allied forces to cross over the Rhine River into Germany, at which point they, too, would be allowed in. Now, in the third week of March 1945, this long-awaited crossing appeared imminent.

On March 23, 1945, British Field Marshal Bernard Montgomery began a colossal troop offensive across the Rhine River, code-named Operation Plunder. Adding to the namesake's subtext of seizure and pillage were Montgomery's famous words: "Over the Rhine, then, let us go. And good hunting to you all on the other side." Sacking a vanquished country at the end of a war was as old as war itself, but the Hague Conventions of 1899 and 1907 prohibited looting beyond token items claimed as "trophies of war." What, exactly, did Montgomery mean? On a tactical level, the Rhine River crossing meant that the Allies would smash open more than five hundred miles along the western front. From an intelligence collection standpoint, it truly meant plunder.

As soldiers pushed forward into Germany, accompanying them were more than three thousand scientific and technical experts with the Combined Intelligence Objectives Subcommittee, or CIOS, the joint British-American program that had been established in London the summer before. CIOS teams reported to Supreme Headquarters Allied Expeditionary Forces, or SHAEF, located in Versailles and staffed by experts: scientists, engineers, doctors, and technicians accompanied by linguists and scholars to translate and interpret the documents that were seized. Representing the United States from CIOS were men from the War Department General Staff, the navy, the Army Air Forces, the State Department, the Foreign Economic Administration, the Office of Strategic Services (OSS), and the Office of Scientific Research and Development. The British sent experts from the Foreign Office, the Admiralty, the Air Ministry, and the Ministries of Supply, Aircraft Production, Economic Warfare, Fuel and Power. All CIOS staff worked from a list of targets, similar to those used by Alsos, which became known as Black Lists. Frontline requests for specific CIOS teams were relayed to SHAEF from the combat zone. An appropriate CIOS team would be dispatched into the field.

Assisting CIOS teams were security forces called T-Forces, identifiable by helmets with a bright red *T* painted on the front. These small squadrons of elite soldiers operated at the army group level but worked independently of traditional combat units. Their job was to recognize potentially valuable scientific targets and then secure them until personnel from CIOS arrived.

The goal of Combined Intelligence Objectives Subcommittee was to investigate all things related to German science. Target types ran the gamut: radar, missiles, aircraft, medicine, bombs and fuses, chemical and biological weapons labs. And while CIOS remained an official joint venture, there were other groups in the mix, with competing interests at hand.

Running parallel to CIOS operations were dozens of secret intelligence-gathering operations, mostly American. The Pentagon's Special Mission V-2 was but one example. By late March 1945, Colonel Trichel, chief of U.S. Army Ordnance, Rocket Branch, had dispatched his team to Europe. Likewise, U.S. Naval Technical Intelligence had officers in Paris preparing for its own highly classified hunt for any intelligence regarding the Henschel Hs 293, a guided missile developed by the Nazis and designed to sink or damage enemy ships. The U.S. Army Air Forces (AAF) were still heavily engaged in strategic bombing campaigns, but a small group from Wright Field, near Dayton, Ohio, was laying plans to locate and capture Luftwaffe equipment and engineers. Spearheading Top Secret missions for British intelligence was a group of commandos called 30 Assault Unit, led by Ian Fleming, the personal assistant to the director of British naval intelligence and future author of the James Bond novels. Sometimes, the members of these parallel missions worked in consort with CIOS officers in the field. Certainly, they took full advantage of all the information CIOS made available, including its Black Lists. But each mission almost always put its individual objectives and goals first. The result, some officers joked, was CHAOS for CIOS.

What began as a gentlemanly collaboration among Allies quickly transformed into one of the greatest competitions for information about weapons-related research in the history of war. Once the Rhine River was crossed, the hunt for Nazi science became a free-for-all.

On its search for chemical weapons, Alsos scientists crossed the Rhine on the heels of the Third Army.

They traveled in a small convoy of army jeeps. "The area was not in good shape," Colonel Pash recalled after the war. "Buildings of the town shook from the shock waves generated by the gun explosions," and "stalled or broken-down vehicles added to the confusion created by the wrecked armor and trucks of the retreating Nazis." Entering Ludwigshafen, one of the Alsos jeeps became separated from its convoy and ran smack into the line of fire of a German antitank gun. Hardly expecting a single jeep to appear on the road without a tank accompanying it, the Germans were apparently caught off-guard. "A single salvo and the few machine-gun bursts went wide," Pash recalled after the war.

The Alsos scientists had heard that T-Forces units were close behind them and were making arrangements for a large team of technical experts to arrive under the CIOS banner. The Alsos scientists were determined to get to the IG Farben factory first. But what they ended up finding in Ludwigshafen was disappointing to them. Not only had the factory been heavily damaged by Allied bombing raids, but filing cabinets were empty. Paperwork had been destroyed or removed. There were no chemical weapons found.

The following morning, March 24, 1945, while at breakfast, the Alsos team met up with the T-Forces and the CIOS team. Both had come to inspect the Farben factory. "It was interesting to watch the effect of bombshells casually dropped by our scientists," Pash recalled, "sounding something like...'Oh sure, You'll be able to go to the Farben plant tomorrow, after the T-Forces secures it. You'll find it interesting. We know, because we were there yesterday.'" The rivalry between the teams was evident.

The two men leading the CIOS chemical weapons team were an American officer named Lieutenant Colonel Philip R. Tarr and a British officer named Major Edmund Tilley. These men would soon become critical players in the Operation Paperclip story. Each was considered an expert in chemical warfare, but Major Tilley had a leg up on his American counter-part—at least in the field—in that he spoke fluent German. In addition to his role as a CIOS leader, Colonel Tarr was chief officer in the U.S. Army's Intelligence Division of the U.S. Chemical Warfare Service, Europe. Unknown to Major Tilley at this time was that Colonel Tarr's U.S. Army objective would soon supersede his loyalty to their work together as a CIOS team.

With American forces now on German soil, the fear that Hitler might unleash a devastating chemical weapons attack in a last-ditch attempt to make good on his wonder weapons promise was real, but Tarr and Tilley had very few leads as to where the chemical weapons might be stashed. German counterintelligence agents had done a brilliant job concealing the Reich's nerve agent programs from foreign intelligence agencies during the war. Tabun had been given various code names, including Trilon 83, Substance 83, and Gelan 1. Even its raw materials were coded. Ethanol was "A4," and sodium was "A-17," making identification all but impossible. In 1942, a U.S. intelligence report entitled "New German Poison Gas" concluded that the possibility that Germany had new chemical weapons was "no longer seriously regarded." Only in May of 1943, after capturing a German chemist in North Africa, did British agents learn of a colorless nerve agent of "astounding properties" being developed by IG Farben chemists in Berlin. The British interrogation officers found the German chemist's information to be credible and wrote up a ten-page secret report for the Chemical Defense Experimental Establishment, which operated out of Porton Down. But the captured scientist knew only the substance's code name, Trilon 83. Without further information, no action was taken by the British. Now, in March 1945, Tarr and Tilley were on the hunt for this mysterious Trilon 83 and anything else like it.

Leading the CIOS team into Germany, Tarr and Tilley inspected IG Farben factories at Ludwigshafen, Mannheim, and Elberfeld. At each location the officers noted with suspicion how remarkably little each town's Farben scientists claimed to know. As far as intelligence collection was concerned, the scenario at each seized Farben factory was always strikingly similar: Where there should have been huge troves of company records, there were empty cabinets instead. Farben scientists who were taken into custody and questioned always said the same thing: IG Farben made chemical products for domestic use—detergent, paint, lacquer, and soap. And none of the scientists interviewed by CIOS officers claimed to have any idea where the bosses had gone.

In a CIOS memorandum, Tilley and Tarr expressed mounting frustration. One scientist after the next "lied vigorously about his activities," the men wrote in their intelligence report, entitled "Interrogation of German Scientific Personnel." Good, actionable intelligence remained out of reach. Then, as circumstance would have it, Alsos agents caught a huge break farther north in the city of Bonn.

Alsos had been trailing the Third Armored Division since it first rolled into Ludwigshafen on March 23. Several days later the soldiers liberated the city of Cologne and then set their sights on Bonn, some fifteen miles to the north. Scouting soldiers reported seeing men who might be professors burning caseloads of documents in Bonn University courtyards. What they couldn't see was that inside university bathrooms, professors were also desperately flushing documents down the toilet, hoping to destroy evidence that might implicate them in war crimes. When the Allies finally secured the university, a Polish lab technician approached a British soldier to say that he had salvaged a large pile of documents that did not properly flush down a toilet bowl, as had apparently been intended.

These papers looked important, the lab technician said. And indeed they were. The man had turned over to British intelligence a classified list of the Reich's top scientists. The officer handed the list over to Samuel Goudsmit of Operation Alsos. This group of documents would lead to what would become known as the Osenberg List.

Dr. Werner Osenberg, a mechanical engineer, was a dedicated Nazi and member of the SS; he was also a high-ranking member of the Gestapo, the secret police. In June 1943 Osenberg was assigned by Göring to run the so-called Planning Office of the Reich Research Council, which was dedicated to warfare. Per a Führer decree of June 9, 1942, the Research Council's charter read: "Leading men of science above all, are to make research fruitful for warfare by working together in their special fields."

From the council's Planning Office, Osenberg's job was to coordinate a Who's Who list of German scientists, engineers, doctors, and technicians. With bureaucratic precision Osenberg set to work, tracking down and cataloguing every scientist in Germany. Osenberg's mission was to put these men into service for the Reich. In short order he had compiled a list of fifteen thousand men and fourteen hundred research facilities. All across Germany scientists, engineers, and technicians were recalled from the front lines, an act Hitler called the Osenberg Action. This led to the release of five thousand scientists from the German Armed Forces. After being screened for skill level, the men of science were set up in appropriated universities and institutions across Reich territory where they could work on weapons-related research programs in service of the war effort.

The mystery of how the Osenberg List ended up in a Bonn University toilet was never solved, but for Alsos it was an intelligence gold mine. Not only was this list a record of who had been working on what scientific project for the Reich, but it contained addresses, including one for Werner Osenberg himself. Goudsmit dispatched a team to a little town near Hannover. There, Alsos agents captured Osenberg and his complete outfit.

The papers found in the toilet were valuable, but it was an index of cards in Osenberg's office that was priceless. "The primary index consists of a four-drawer cabinet containing approximately two-thousand large printed cards (10" x 7") adapted for multiple entry on both sides of the card," reported Alsos. Secondary indexes included three additional sets of "approximately one thousand cards [each], 6" x 4" . . . containing the same information but classified from a different standpoint and facilitating searches along different lines." This was an overwhelming trove of information, too much for any one organization to handle. Alsos shared the Osenberg List with CIOS; there were thousands of leads that needed following up on. Osenberg's card catalogue would allow the various teams to begin piecing together how science programs worked under the Reich and who had been in charge.

The Alsos officers packed up Osenberg's office and took him to Paris, where he was put to work organizing information. Goudsmit was appalled by the hubris Osenberg displayed after he was installed in an office in a guarded facility in Versailles. "Here, Osenberg had set up business as usual; he merely had his secretary change the address on his letterhead to…"at Present in Paris," Goudsmit explained after the war. He also became exasperated when Osenberg tried repeatedly to convince Goudsmit of his sworn loyalty to the Allies. "I became impatient," Goudsmit explained, and told Osenberg, "[O]ne cannot trust you.…You were in charge of the scientific section of the Gestapo, which you never revealed to us and you burned all the relating papers." Osenberg was enraged by the accusation. "No, I did not burn those papers," he told Goudsmit. "I buried them and, moreover, I was not the chief of the scientific section of the Gestapo, I was merely the second in command." After that it was easy for

Goudsmit to find out from Osenberg "where those papers were buried and where the missing Berlin papers were stored."

But no one from the Allied forces was going into Berlin just yet. In these last days of March 1945, Berlin continued to be the heart of the Reich, and it was being fiercely defended. It would remain German-held territory for another month, until the last day of April 1945.

It was a city in ruins, and Berliners' morale was sinking with each passing day. With nearly 85 percent of the city destroyed, Berlin had been reduced to rubble piles. The majority of buildings still standing had broken windows. Everyone was cold. The underground shelters were vastly overcrowded. Heating fuel was scarce. "During early April," explains historian Anthony Beevor, "as Berlin awaited the final onslaught [of the Red Army,] the atmosphere in the city became a mixture of febrile exhaustion, terrible foreboding and despair."

Nazi radio broadcasts reminded Berliners that they were required to fight to the end. According to Goebbels, "A single motto remains for us: 'Conquer or die.'" Hitler blamed everything on the Jews and the Slavs. "The deadly Jewish-Bolshevik enemy with his masses is beginning his final attack," he told troops on the eastern front, in a last appeal on April 15. The Russians were determined to exterminate all German people, Hitler promised, "the old men and children will be murdered, women and girls will be degraded as barracks whores." Everybody else would be marched off to Siberia.

In the heart of Berlin, in the Wilhelmstrasse district, where the Reich's ministries were located, Colonel Siegfried Knemeyer tried to maintain order over the vast group of personnel he was responsible for at the Air Ministry. It was a miracle that the building was still standing. Constructed of reinforced concrete, the Air Ministry was a prized symbol of Nazi Party intimidation architecture, seven stories tall, with twenty-eight hundred rooms, four thousand windows, and corridors that totaled four miles in length. At the height of the air war, the place bustled with four thousand Nazi bureaucrats and their secretaries. This was no longer the case. The Luftwaffe was in ruins.

Since 1943, Siegfried Knemeyer had served as chief of all air force technical developments for the Luftwaffe. His title was technical adviser to Reichsmarschall Göring, who adored Knemeyer, calling him "my boy." From aircraft engines to instruments, if a new component was being developed for the Luftwaffe, Göring wanted to know what Knemeyer had to say about it before he gave the project the go-ahead. But now, in the second week of April 1945, most orders that arrived at the Air Ministry were impossible for Knemeyer to fulfill. In the second half of 1944, the German Air Force had lost more than twenty thousand airplanes. The Speer ministry had since managed to produce approximately three thousand new aircraft, but they were of little use now. Across the Reich, German aircraft sat stranded on tarmacs. The Allies had bombed Luftwaffe runways, and there was barely any aviation fuel left. The plan for IG Farben to make synthetic fuel at its Buna factory had ended when the Red Army liberated Auschwitz on January 27, 1945. Germany's fuel sources in Hungary and Romania were tapped out. The German jet was superior to conventional Allied fighter planes in the air, but that didn't mean much anymore, with most Luftwaffe aircraft stuck on the ground.

Soon Knemeyer would flee Berlin, but not before he completed a final task assigned to him by Speer. According to Nuremberg trial testimony, Speer instructed Knemeyer to hide Luftwaffe technical information in the forest outside Berlin. Stashing official documents was a treasonable offense, but according to Knemeyer's personal papers, Speer and Knemeyer had agreed that Germany's seminal scientific progress in aeronautics could not, under any circumstances, fall into Russian hands. There was also a second unofficial job that Speer tapped Knemeyer for, one that was not discussed at Nuremberg but which Speer admitted to decades later. Speer asked Knemeyer to help plot his escape.

Speer's plan to flee Germany had been in the works

for some time; it was the details he needed help ironing out. Ever since Speer had seen the film *S.O.S. Iceberg,* starring Leni Riefenstahl and Ernst Udet, he knew he wanted to escape to Greenland should Germany lose the war. In Greenland Speer could set up camp and write his memoirs, he later explained. Of course he'd need a pilot, which is where Knemeyer fit in.

Knemeyer was an aeronautical engineer, but he was also one of the Luftwaffe's most revered pilots, ranked among the top ten aviators in all of Germany. His specialty, back when he flew missions in the earlier stages of the war, had been espionage. From 1938 to 1942, Knemeyer flew a number of the most dangerous Abwehr (military intelligence) missions on record, including ones over England and Norway. And it was Knemeyer who made the first high-altitude sortie over North Africa, flying at forty-four thousand feet. But Knemeyer was also a pragmatist. He knew, apparently more so than Speer, that attempting to fly out of Germany and into Greenland transporting one of the most wanted war criminals of the Third Reich during the final days of the war would be a near-to-impossible feat. There was brutal weather in Greenland and fierce terrain. The mission would require a very specific aircraft that could handle the harsh conditions and difficult landing, namely, the Bv 222, designed by Blohm & Voss. Only thirteen had ever been built. There was only one man who had access to that kind of airplane, and that was Knemeyer's friend Werner Baumbach, a twenty-eight-year-old dive-bomber pilot whom Hitler had made general of the bombers.

Knemeyer knew that bringing Baumbach on board was imperative for a successful Greenland escape. During the war, Baumbach had flown missions between Norway and a German weather station located in Greenland. Speer agreed and Baumbach was brought into the plan. In secret, Baumbach and Knemeyer began gathering "food, medicine, rifles, skis, tents, fishing equipment and hand grenades" at Speer's behest. At the Travemünde airfield, north of Berlin, Baumbach earmarked a Bv 222 for use. The only thing that remained was Speer's command for the group to flee. Time was running out. Berlin was nearing its downfall.

CHAPTER FOUR

Liberation

A ll across Germany, the liberations were beginning. In one location after the next, prisoners in
concentration camps and slave labor factories
were being freed by Allied soldiers who stormed across
Germany in tanks and jeeps and on foot. The action
had begun in western Germany and continued steadily
as the Allies marched east, headed toward Munich
and Berlin. Alongside these liberations, soldiers also
discovered Reich laboratories and research facilities,
one after the next. After each discovery a team of
CIOS scientists was called in to investigate. During
the second week of April in 1945, four key facilities
were seized — at Nordhausen, Geraberg, Völkenrode,
and Raubkammer — each of which would lead to the
capture of key scientists who would in turn become
part of Operation Paperclip.

On the morning of April 11, 1945, a unit of American soldiers with the 104th Infantry Division, also known as the Timberwolves, entered the slave tunnels at Nordhausen. Among the liberating soldiers was an infantry sharpshooter, a private first class named John Risen Jones Jr. In his bag he carried a camera, a gift given to him by his family before he shipped off to war. Expensive and sleek-looking, Jones's Leica III was one of the first portable 35 mm cameras ever made.

It had been seven months since John Risen Jones landed in France, back in September 1944. He had spent 195 days on the continent thus far, many of them engaged in fierce combat, all the while pushing though snow, sleet, and mud—much of it on foot. Jones had walked across France, Belgium, and Holland, and now here he was, in the deep mountains of central Germany, the Harz. He had lost friends in battle and taken many photographs of the war. When his unit arrived in this little mountain town, he imagined the day would pass like the one before. Just one step closer to the end of this brutal war.

No amount of fighting prepared John Risen Jones for what he saw through the lens of his Leica when his unit entered Nordhausen. The photographs he took documented the tragedy that had befallen thousands of V-2 rocket laborers condemned to die as slaves in the tunnels here. Hundreds of corpses were stretched out across the tunnel floors. Equally disturbing was the condition of hundreds more still alive: emaciated humans covered with bruises and sores, too weak to even stand. "It was a fabric of moans and whimpers of delirium and outright madness," recalled fellow soldier Staff Sergeant Donald Schulz. John Risen Jones would not speak of it for fifty-one years.

Following along behind the soldiers was a team of seven war crimes investigators. Among them was a young Dutch officer working for the U.S. Army, William J. Aalmans. Like John Risen Jones, Aalmans was deeply affected by what he saw and smelled. "Stench, the tuberculosis and the starved inmates," he told journalist Tom Bower after the war. "Four people were dying every hour. It was unbelievable." Aalmans and his team began taking witness statements from prisoners, who sipped watered-down milk for strength. The job facing the war crimes investigators was overwhelming, and their schedule was intense. After five days in Nordhausen they were ordered to move on. Most of the official paperwork regarding rocket production had been hidden or destroyed, but Aalmans and his team found a single sheet of paper inadvertently left behind, tacked to the wall. It was the Mittelwerk telephone list; a directory of who was in charge. At the very top were two names: Georg Rickhey, director of production, and Arthur Rudolph, deputy production manager. Aalmans found the document interesting enough to staple it to the report.

Although it would take years to come to light, this single sheet of paper would eventually lead to the downfall of Rudolph and Rickhey and threaten to expose the dark secrets of Operation Paperclip.

Seventy-five miles south of Nordhausen, in the Thuringian Forest at the edge of the Harz, Allied soldiers liberated the town of Geraberg. Here, they came upon a curious-looking research facility concealed in a thick grove of trees. Clearly the place had recently been abandoned. It comprised a laboratory, an isolation block, animal houses, and living quarters for fourteen men. Part of the facility was still under construction. Word was sent to SHAEF headquarters in Versailles that a team of bacteriologists was needed in Geraberg. Alsos scientists were dispatched to investigate. One of the first biological warfare experts to arrive was Bill Cromartie, who had been hunting for evidence of Hitler's biological weapons program since the mission began back in Strasbourg, France. Back in November, Cromartie had been one of the men scouring files with Samuel Goudsmit, inside the apartment of Dr. Eugen Haagen, when Alsos agents first learned that the Reich was testing deadly vaccines on prisoners in concentration camps. Arriving at Geraberg, Cromartie determined that the laboratory here was a significant lead.

"The building and sites were on either side of a

small valley and constructed under tall trees," read Cromartie's classified report. "On one side there was a building [that] was to have been the experimental laboratory," he surmised, suggesting that this facility was designed to produce experimental vaccines to protect German soldiers against a biological weapons offensive.

A local villager provided Cromartie and a colleague, J. M. Barnes, with two key pieces of information in the biological weapons puzzle. The villager explained that an SS man named Dr. Karl Gross had been overseeing work at this facility. Gross kept dozens of trunks and boxes locked in the upper floors of a local schoolhouse, and, while he had recently disappeared, he had left the trunks behind. The villager took the American scientists to the schoolhouse to investigate.

An inventory was taken of Dr. Gross's possessions, mostly laboratory equipment. "There were crates of test tubes and small flasks and large numbers of test tube racks. There were two incubators and an autoclave. There were two boxes of gas mask filters and some rubber hoods and gowns," read the report. Everything was military-grade protective gear, marked as having "been obtained from the Hygiene Institute of the Waffen SS." There was also a large collection of books, "several boxes of periodicals all dealing with infectious diseases." The ones that really caught the scientists' attention were "Russian contributions on plague."

Next, the villager took the Alsos agents to the nearby boardinghouse where Dr. Gross had been renting a room. The place was cleared out and void of personal possessions. "His landlady said she believed he had burnt a lot of papers the night before he left," the Alsos scientists noted in their report. But Dr. Gross was only the intermediary, the landlady said. There was another man who came to the facility and appeared to be in charge. He was an older man, about fifty, five foot nine, with a mustache and black hair. On his upper lip he had a pronounced dueling scar. He had to have been of high rank, because everyone on the staff deferred to him. When cross-referenced by Alsos against the Osenberg List, the situation became even clearer: Dr. Karl Gross worked under Dr. Kurt Blome, the individual in charge of biological weapons research and the deputy surgeon general of the Third Reich.

All indicators pointed to the idea that Geraberg was a Reich facility for biological weapons research. Alsos agents photographed the site: the animal house, the vaccine station, the experimental laboratory, and the isolation hospital. They typed up a report and filed it away for future use. Now, near the top of the biological weapons Black List was the name Dr. Kurt Blome.

Sixty miles north of Nordhausen, a battalion of American soldiers with the First U.S. Infantry moved cautiously through the forest on the western edge of a small city called Braunschweig. It was April 13, 1945, when they came upon a compound of about seventy buildings. Great care had gone into camouflaging this place, the soldiers noted. Thousands of trees had been planted closely so that the area would appear from the air to be dense forest. The buildings in the compound had been designed to look like simple farmhouses. Traditional gardens had been planted and tended to. Stork nests covered the rooftops.

Inside the buildings, soldiers discovered state-of-the-art aircraft laboratories, including entire warehouses filled with airplane parts and rocket fuel. There were wind and weapons tunnels that were radically more advanced than anything the Army Air Forces had at Wright Field. The oldest division in the United States Army had unexpectedly happened upon the most scientifically advanced aeronautics laboratory in the world. It was called the Hermann Göring Aeronautical Research Center at Völkenrode. The Allies had never heard of it before. It didn't appear on any CIOS Black List. It was an incredible find.

At first, it seemed as if the place had been abandoned. But after an hour of looking around, the soldiers came upon the institute's scientific director, a man named Adolf Busemann. Busemann told the soldiers that this facility was called Völkenrode for short, and that it had been up and running for ten years. A team of Army Air Forces technical intelligence experts, working as part of a mission called Operation Lusty and stationed in Saint-Germain, France, was dispatched to investigate. By now the U.S. Strategic Air Forces in Europe had destroyed the Luftwaffe, and the bombing campaign had essentially stopped. Its commander, General Carl A. "Tooey" Spaatz, had just received a fourth star for his success commanding the largest fleet of combat aircraft ever assembled for war. Now Spaatz had a new mission for his field commanders. "Operation Lusty," Spaatz wrote in a memo, was "in effect," and everyone "not engaged in critical operational duties" was to seek out "technical and scientific intelligence [that would be] of material assistance in the prosecution of the war against Japan." The man Spaatz chose to lead the hunt for Luftwaffe scientists and engineers was Colonel Donald L. Putt.

When Putt arrived at Völkenrode on April 22, 1945, he was thrilled by what he saw. All he could think about was how quickly he could get all this equipment back to the United States. Putt was a legendary test pilot who had been at Wright Field since 1933, assigned to various branches, including the Flying Branch. He had walked away from a deadly air crash that killed his colleagues and left him with second-degree burn scars on his face and neck. Putt was a hard-charging, Type A personality—a tiger among men. "He displayed the ability to withstand great emotional shock, to absorb it, and take in stride," explained a colleague from Wright Field. Putt was also intellectually gifted, with a degree in electrical engineering from the Carnegie Institute of Technology and a master's in aeronautical engineering from the California Institute of Technology. As an older man, Donald Putt recalled the prewar mind-set regarding pilots who were also engineers. "Then, the philosophy was, 'Don't put an engineering pilot in the cockpit, because he tries to figure out why things happen.'" But when the air wars in Europe and Japan escalated, the Army Air Forces found itself in need of fast, out-of-the-box thinking from American pilot-engineers like Putt. The Army Air Forces put the old philosophy to the side and Putt's expertise to use. In 1944, Putt's career milestone arrived when he was put in charge of modifying a B-29 bomber so that it could deliver an extraordinarily heavy, Top Secret payload on Japan. This payload was eventually revealed to be the atomic bombs dropped on Hiroshima and Nagasaki. After Putt finished his B-29 bomber retrofitting jobs, in January 1945, he was sent overseas as director of technical services for the Air Service Command. Now here he was in the last week of April 1945 at Völkenrode.

With his engineer's expertise, Putt was able to clearly judge the revolutionary nature of the technology he was looking at. Most astonishing to Putt were

Völkenrode's seven wind tunnels that had allowed the Luftwaffe to study how a swept-back wing would behave at the speed at which an aircraft broke the sound barrier. This transition place, between Mach 0.8 and Mach 1.2, was still unknown to American fliers in 1945. When Putt learned from Völkenrode's director, Adolf Busemann, that the sound barrier had already been breached by German scientists in these wind tunnels, he was amazed. Putt knew immediately that the facility had "the most superb instruments and test equipment" in the world.

Putt was taking orders from the U.S. Army, European Theater of Operations, Directorate of Intelligence, Exploitation Division, which meant that he had access to stripped-down B-17s and B-24s if he needed them, a means to transport much of this equipment to the United States, which Putt very much wanted to do. He wrote to his boss, Major General Hugh Knerr, deputy commander of U.S. Strategic Air Force in Europe, outlining his proposition and suggesting a second idea: Why not also fly scientists like Adolf Busemann out of Germany, along with the captured Luftwaffe equipment? "If we are not too proud to make use of this German-born information, much benefit can be derived from it and we can advance where Germany left off," Putt wrote. The German scientists "would be of immense value in our jet engine and airplane development program." Putt and Knerr both knew that the War Department General Staff was filled with individuals who were wary of Germans in general and totally opposed to making deals of any kind with the very Nazi scientists who had helped to prolong the war. But if anyone could get the War Department to bend, Knerr and Putt believed they could.

Major General Knerr sent a memo to the War Department in Washington, D.C., explaining that using Luftwaffe technology to fight the war in Japan was imperative. He added that the scientists' Nazi Party membership needed to be overlooked. "Pride and face saving have no place in national insurance," wrote Knerr.

The War Department General Staff was not so easily convinced—at least not now. Colonel Putt was informed that the equipment could come out of Völkenrode immediately but that getting German scientists to Wright Field would take some more time. Putt oversaw the massive airlift of German aircraft and rocket parts from Völkenrode to the United States; five thousand scientific documents were also shipped. Meanwhile, he and his staff rounded up as many Luftwaffe personnel as they could, tracking down leads and making deals with scientists and engineers in their homes. Putt informed the Germans that he could not offer them U.S. Army contracts just yet but that he would most likely be able to do so soon. In the meantime, he arranged for dozens of Luftwaffe scientists and engineers to be quartered in the Hotel Wittelsbacher Hof, in the spa town of Bad Kissingen, and made sure that the men had plenty to eat, drink, and smoke. Wait here, the scientists were told. The U.S. Army contracts are on the way.

Colonel Putt and Major General Knerr would then put their heads together and figure out a way to convince the War Department that their point of view was best for the United States.

Around this time, the single largest cache of chemical weapons discovered to date was found seventy-five miles west of Hannover. On April 16, 1945, British soldiers from Montgomery's Twenty-first Army Group pulled up to the entrance of an abandoned German army proving ground called the Robbers' Lair, or Raubkammer. The place appeared to be abandoned, but Waffen-SS snipers still were known to be hiding in the woods. The soldiers exercised caution as they drove their armored personnel cars through a pair of entrance pillars adorned with Reich eagles and swastikas.

At first glance the facility looked like a standard military proving ground—a place where bombs were exploded and blast measurements recorded. Raubkammer was located in a rural forested area called Münster-Nord, and it extended more than seventy-six square miles. Large craters in open fields suggested that Luftwaffe airplanes had practiced dropping bombs here. There was fancy housing for hundreds of officers. There were large administrative buildings and an officers' mess hall. Then the soldiers came upon the zoo.

It was a large zoo, capable of housing a vast array of animals of all sizes. There were cages for mice, cats, and dogs, as well as large pens and stables for farm animals like horses, cows, and pigs. There were also monkey cages. But it was the discovery of a massive, round wooden cylinder—most likely an aerosol chamber—that triggered alarm. The structure was sixty-five feet tall and a hundred feet wide, and it was ringed with a network of scaffolding, pipes, and ventilator fans. Between the zoo and the large chamber, the soldiers were now relatively certain that Raubkammer was no ordinary military proving ground. The Robbers' Lair bore the hallmarks of a field-testing facility that likely involved poison gas. An urgent memo was sent to SHAEF asking for a team of chemical warfare experts to be dispatched to Raubkammer. Two teams descended, one from the British Chemical Defense Experimental Establishment at Porton Down and another from CIOS, including Major Tilley and Colonel Tarr.

At the same time, a second unit of British troops working just a few miles to the southwest of the Robbers' Lair came upon two bunker clusters totaling almost two hundred structures. The area had been artfully concealed from overhead view by dense forest cover. The first cluster consisted of several dozen small wooden buildings intermittently spaced between similarly sized concrete blockhouses. The soldiers inventoried the contents with caution. Inside one set of bunkers they found thousands of bombs, stacked in neat piles. Each bomb was marked with a single yellow ring painted around the sides of the munition. This was the standard marking to denote mustard gas, the chemical weapon used by both sides in World War I. The British soldiers took inventory and counted one hundred thousand mustard gas shells.

The second munitions depot was marked as belonging to the Luftwaffe. Here, 175 bunkers were filled with bombs that were unidentifiable to the Allies. Each bomb had been marked with three green rings painted around its sides. Montgomery's soldiers sent an urgent memo to SHAEF asking for a team of chemical weapons experts to come investigate the munitions in the forest.

The scientists from CIOS and Porton Down were nervous about what might be inside the mysteriously marked bombs. They decided that it was best to try to locate German scientists in the area who might be familiar with the contents of the bombs before they opened the casings themselves — so they began knocking on the doors of the nicer houses in the vicinity of Robbers' Lair. As they had suspected they would,

CIOS officers located a number of individuals who confessed to being German army scientists and having worked at Raubkammer. While each scientist claimed to have no idea what kind of weapons testing had been going on at the military facility, CIOS officers were able to persuade several of the German scientists to assist them in extracting the liquid substance from the center of the bombs.

By this time, chemists with the U.S. Army's Forty-fifth Chemical Laboratory Company had arrived, bringing with them a mobile laboratory unit and cages filled with rabbits. The original thought was that the substance marked by three green rings was some kind of new Nazi blister agent—similar to, but perhaps more powerful than, mustard gas. The chemists were wrong. Extractions were made, and when tested on the rabbits in the mobile laboratory, whatever this liquid substance was killed a warm-blooded rabbit five times faster than anything that British or American scientists had ever seen, or even heard about, before. Even more alarming, the liquid substance did not have to be inhaled to kill. A single drop on the rabbit's skin killed the animal in just a few minutes. The millions of gas masks England had distributed to city dwellers during the war would have offered no defense against a chemical weapon as potent as whatever this killing agent was.

CIOS field agents wrote up a Top Secret report for

their superiors at SHAEF. A menacing new breed of chemical weapons had been discovered. Aerial bombs "found to contain a markedly potent and hitherto unknown organophosphorus nerve agent" had been developed by the Nazis during the war and stashed in two hundred bunkers in the forest nearby. No chemical this lethal to man had ever been developed before. CIOS agents did not know it yet, but this nerve agent was tabun. The three green rings had been painted on the Luftwaffe bombs at Farben's Dyhernfurth facility in Poland.

Allied chemical weapons experts were suddenly in possession of one of the most dangerous wonder weapons—and one of the best-kept secrets—of the Third Reich. That these weapons were never used was astonishing.

Who were the scientists who had discovered this nerve agent, and where were they now?

Underground in the center of Berlin, in a makeshift hospital set up in a subway tunnel beneath the Reich Chancellery, Major General Dr. Walter Schreiber performed emergency surgeries on wounded Wehrmacht soldiers. He gave blood transfusions, performed amputations—whatever it was that needed to be done. As surgeon general of the Third Reich, Schreiber was not a hospital doctor accustomed to dealing with triage. But, as he would later testify at Nuremberg, all of his physician-colleagues had fled Berlin; this was not necessarily the truth. Schreiber was a short, squat man, five foot six, with blond hair, blue eyes, and a nose that ended in a fleshy point. A man of great willpower and stamina, he prided himself as setting his mind to a task and getting it done.

Today was the Führer's fifty-sixth birthday, April 20, 1945. And because it was Hitler's birthday, this Berlin morning began with Propaganda Minister Joseph Goebbels's Happy Birthday broadcast, calling on all Germans to trust Hitler and to follow him faithfully to the bitter end. While Major General Dr. Schreiber performed surgeries in his makeshift underground hospital, a group of his Nazi Party colleagues had gathered for a party almost directly above where he was located, in the half-destroyed Reich Chancellery building. Around noon, Hitler's inner circle made its way into a cavernous room with polished marble walls and floor-to-ceiling doors: Speer, Göring, Himmler, SA-Obergruppenführer and police and Waffen-SS general Ernst Kaltenbrunner, Nazi Party foreign minister Joachim von Ribbentrop, Grand Admiral Karl Dönitz of the navy, Generalfeldmarschall Wilhelm Keitel, military commander Alfred Jodl, and SS-Brigadeführer Hans Krebs. The men gathered around an enormous table covered with bottles of champagne and a spread of food. Hitler said a few words and promised that the Russians would soon suffer their most crushing defeat yet.

That morning, the Red Army had in fact begun its final assault on Berlin. Before dawn German soldiers had retreated from the Seelow Heights, fifty-five miles from the center of Berlin, leaving no front line. The Soviet operation to capture Berlin was colossal, involving 2,500,000 Red Army soldiers, 41,600 guns and mortars, 7,500 aircraft, and more than 6,000 tanks.

Rumor, panic, and chaos enveloped the city at an unstoppable pace. The majority of Berliners were now living underground, in cellars and air raid shelters, appearing aboveground only to scavenge for food. Every road out of Berlin leading west was overwhelmed with refugees. Casualties were skyrocketing. To the south, a detachment of Hitler Youth fighting near the Buckow Forest became trapped in a forest fire; most were burned alive. Major General Dr. Walter Schreiber's makeshift hospital could not keep up with the wounded. During the course of the next twelve days the Red Army would fire 1.8 million shells on the city.

Hitler's birthday was also the day Knemeyer and Baumbach would flee Berlin for good. In the morning, Baumbach had received a cryptic message from Göring, who instructed him to go meet with SS-Brigadeführer Walter Friedrich Schellenberg, the notorious chief of military intelligence and Himmler's number-two man. The end was near and everyone seemed to know it, so what did Schellenberg want from Baumbach now? Schellenberg told Baumbach that a warrant had been issued for his arrest and that he was scheduled to be taken into custody during the Führer's birthday party. Baumbach should leave the city immediately, Schellenberg said. Anyone close to Hitler who was arrested this late in the war—usually suspected of treason—faced a quick execution. This was happening all across Berlin. Had Hitler found out about the Greenland escape plan?

"Schellenberg, whom I had known for years," Baumbach later explained, "was a clever man." Baumbach interpreted the tip-off to mean only one thing: Schellenberg needed Baumbach alive to help facilitate an escape. "It was known among a core group of SS officers that Himmler had been trying to use concentration camp inmates as bargaining chips for clemency, through an intermediary, the Swedish Red Cross," Baumbach explained. Of course Himmler would never be granted clemency; Baumbach figured that Schellenberg and Himmler wanted him alive so he could help them escape somewhere overseas.

Baumbach located Knemeyer and the two pilots agreed to flee Berlin immediately. They got in Baumbach's BMW and headed north to the Travemünde airfield, two hundred miles north of Berlin, on the Baltic Sea. The long-range aircraft they planned to use for their escape with Speer sat stocked and fueled on the tarmac. "We were supplied with everything we needed for six months," Baumbach explained after the war. Knemeyer and Baumbach found many Luftwaffe officers packing up their belongings, stripping themselves of military identification, and preparing to disappear among civilians. An aide delivered an urgent message to Baumbach. This time it was from Himmler himself. The Reichsführer-SS wanted to see Baumbach immediately. Baumbach was to come to Mecklenburg, halfway back to Berlin, where Himmler was staying. Baumbach asked Knemeyer to accompany him.

The road leading to Mecklenburg was swamped with refugees. This part of Germany was one of the only regions still in German control. SS guards herded concentration camp prisoners along the roads like cattle in a last-ditch effort to keep them out of the liberators' hands. The roads were almost impassable, and it took five hours for Baumbach and Knemeyer to drive a hundred miles. When they finally arrived at a large country home referred to as the manor of Dobbin, Himmler's SS-guards escorted them inside.

"The Reichsführer will receive you now," a guard said. Knemeyer was told to wait outside Himmler's office, while Baumbach was led down a long, narrow corridor, up a winding staircase, and into Himmler's study. Behind the desk, Himmler sat alone. He wore a gray field uniform covered with SS insignia, the death's head (Totenkopf). The sleeves on Himmler's uniform were too long, and Baumbach noted a cheap ring on the pinkie finger of his left hand. Himmler sized up the general of the bombers from behind his signature pince-nez and got to the point.

"I've sent for you to clear up some Luftwaffe problems," Himmler said, as recalled by Baumbach after the war. "The war has entered the final stage and there are some very important decisions I shall have to take." Baumbach listened. "In the very near future I expect to be negotiating with our enemies probably through some neutral country," Himmler said. "I've heard that all aircraft suitable for this purpose are under your command." Baumbach looked out the windows and across the carefully pruned gardens outside, considering his response. Yes, Baumbach told Himmler, he had aircraft at his disposal, ready at any time. Himmler assumed an even friendlier tone, Baumbach recalled, and asked where he could get hold of Baumbach in the coming days. At the Travemünde airfield, Baumbach said. An aide interrupted to announce the arrival of Field Marshal Keitel. Baumbach was dismissed.

Baumbach made his way back to the sitting room, where Knemeyer waited. By now Knemeyer had figured out whose fancy manor Himmler was living in. The home once belonged to Sir Henry Deterding, the

English lord known as the Napoleon of Oil. Next to Knemeyer on a side table were two portraits in silver frames. One showed Göring wearing a medieval hunting costume and holding a large knife. It read, "[T]o my dear Deterding in gratitude for your noble gift of Rominten Reichs Hunting Lodge," a detail Knemeyer shared with his son decades after the war. The second photograph was a portrait of Hitler. "Sir Henry Deterding," it read, "in the name of the German people for the noble donation of a million reichsmarks. Adolf Hitler."

Knemeyer and Baumbach headed outside. The SS officer posted to guard the hallway gave the two men a stiff salute. He told them that the Reichsführer-SS had arranged a tray of coffee and sandwiches for them to enjoy before they headed back to Travemünde.

The reason that the Greenland escape plan was still on hold was because Speer decided to visit Hitler one last time at the Führerbunker, compelled by an "overwhelming desire to see him once more." Driving alone in his private car from Hamburg back into Berlin, Speer was fifty-five miles outside the city when the road became impassable, clogged with what Speer later recalled to be "a ten-thousand vehicle traffic jam." No one was driving into Berlin anymore; everyone was getting out. All lanes in both directions were being used for travel west. "Jalopies and limousines, trucks and delivery vans, motorcycles and even Berlin fire trucks" blocked the road. Unable to advance, Speer turned off the road and drove to a divisional staff headquarters, in Kyritz, where he learned Soviet forces had encircled Berlin.

He also learned that there was only one landing strip inside Berlin that remained under German control, Gatow airport, on the bank of the Havel River. Speer decided that he would now fly into Berlin. But the nearest aircraft with fuel were parked on the tarmac at the Luftwaffe's Rechlin test site, near Mecklenburg. Jet fuel was now as rare as hen's teeth, and the aircraft was undoubtedly needed for other things. Speer insisted that the commandant at Rechlin locate a pilot capable of flying him into Berlin. The commandant at Rechlin explained that from Gatow, Speer would never be able to get to the Führerbunker if he traveled by car or by foot; the Russians controlled the way there. In order to get to the Führerbunker under the New Reich Chancellery, Speer would need a second, smaller aircraft to fly him from Gatow to the Brandenburg Gate. He would need a short takeoff and landing aircraft, or STOL, like the Fieseler Fi 156 Storch (Stork).

"Escorted by a squadron of fighter planes, we flew southward at an altitude of somewhat over 3,000 feet, a few miles above the battle zone," remembered Speer. "Visibility was perfect....All that could be seen were brief, inconspicuous flashes from artillery or exploding shells." The airfield at Gatow was deserted when they landed, with the exception of one of Hitler's generals who was fleeing Berlin. Speer and his pilot climbed into a waiting Stork and flew the short distance over Berlin, landing amid rubble piles directly in front of the Brandenburg Gate. Speer commandeered an army vehicle and had himself driven to the Chancellery, or what was left of it.

American bombers had reduced the building to ruins. Speer climbed over a pile of rubble that had once been a ceiling and walked into what used to be a sitting room. There, Hitler's adjunct, Julius Schaub, stood drinking brandy with friends. Speer called out. Schaub appeared stunned by the sight of Speer. The companions dispersed. Schaub hurried off to inform Hitler that Speer had come to see him. Speer waited next to a rubble pile. Finally, he heard the words he had come to hear: "The Führer is ready to see you now."

Speer walked down into the bunker, where he was met by Martin Bormann, "Hitler's Mephistopheles," holding court. Bormann wanted to know if Speer had come to try to get Hitler to fly with him out of Berlin. Propaganda Minister Joseph Goebbels and his wife, Magda, were also in the bunker, plotting the murder-suicide of their six children and themselves. Hitler's girlfriend, Eva Braun, invited Speer into her quarters to eat cake and drink Moët & Chandon until Hitler was ready to see him. At 3:00 a.m. Speer was told he could come in. "I was both moved and confused," Speer later recalled. "For his part he [Hitler] showed no emotion when we confronted one another. His words were as cold as his hand."

"So you're leaving?" Hitler asked Speer. Then he said, "Good. Auf Wiedersehen." Good-bye.

Speer felt scorned. "No regards to my family, no wishes, no thanks, no farewell." For a moment, Speer lost his composure and mumbled something about coming back. But Hitler dismissed his minister of war and weapons, and Speer left.

Six days after Speer's final meeting with Hitler, the U.S. Army liberated Dachau, a concentration camp located twelve miles outside Munich. It was 7:30 in the morning on April 29 when fifty tanks from the Seventh Army, Third Battalion of the 157th Infantry Regiment, pulled up to what at first seemed like an ordinary military post, located adjacent to an SS training camp. The weather was cold and there was a dusting of snow. The post was surrounded by high brick walls, an electrified barbed-wire fence, and a deep ditch. Seven fortified guard towers loomed overhead. The large iron front gates were closed and locked. A few American soldiers scaled the fence, cut the locks, and opened the gates. The soldiers rushed inside. A brief exchange of rifle fire ensued. Turkish newspaper correspondent Nerin E. Gun, imprisoned in Dachau for his reports on the Warsaw Ghetto, bore witness as some of the SS guards in the watchtowers began shooting at prisoners. But the American soldiers, Gun said, put a quick end to that. "The SS guards promptly came down the ladders, their hands raised high in surrender." Other accounts describe brutal acts of vengeance inflicted by prisoners against their former SS guards. More gunfire ensued as a second unit, the Forty-fifth Thunderbird Division, approached Dachau from the southwest. They discovered fifty open freight cars abandoned just outside the garrison. Each train car was filled with emaciated bodies. There were several thousand corpses in all.

Dachau, the first Nazi concentration camp, had been established by Himmler on March 20, 1933. It was originally a place where Communists and other political enemies of National Socialism, the ideology of the Nazi Party, were sent. The name came from the fact that prisoners could be "concentrated" in a group and held under protective custody following Nazi law. Quickly, this changed. Himmler made concentration camps "legally independent administrative units outside the penal code and the ordinary law." Dachau had served as a training center for SS concentration camp guards and became a model for how hundreds of other concentration camps were to be set up and run. It was also a model for Nazi medical research programs involving doctors who would later become part of Operation Paperclip.

A young U.S. Army lieutenant and physician, Dr. Marcus J. Smith, arrived at Dachau early on the morning of April 30, 1945, and in his journal Smith noted how cold and gloomy the thousand-year-old city was. Before noon it started to hail. Dr. Smith was the sole medical officer attached to a ten-man displaced persons team sent to the concentration camp the day after it was liberated. He and his fellow soldiers had instructions to do what they could to help the 32,000 starved, diseased, and dying camp survivors as they waited for Red Cross workers to arrive. The newly liberated suffered from dysentery, tuberculosis, typhus, pneumonia, scabies, and other infectious diseases in early, late, and terminal stages, Smith wrote. "Even my callous, death-hardened county-hospital exterior begins to crack.... One of my men weeps."

During breaks, Dr. Smith walked around Dachau's gas chamber to try to make sense of what had gone on there. "I cannot believe this is possible in this enlightened age," he wrote. "In the rear of the crematorium is [a] sign, depicting a man riding a monstrous pig. 'Wash your hands,' says the caption. 'It is your duty to remain clean.'"

In his spare time, Dr. Smith wandered through the camp. "On one of these walks I enter a one-story building that contains laboratory counters and storage shelves," Smith wrote, "almost everything in it has been smashed: I step over broken benches and drawers, twisted instruments and shattered glassware. In the debris, I am surprised to find a few specimen jars and bottles intact, filled with preserved human and insect tissues." Smith asked questions around the concentration camp to try to learn more. Prisoners told him that the laboratory had served Nazi doctors as an experimental medical ward, and that everyone was afraid of it because it was a place "where selected prisoners [were] used as experimental subjects without their consent."

Although it was not yet known by American or British intelligence at the time, what Dr. Marcus Smith had come upon at Dachau was the place where a group of Luftwaffe doctors had been conducting medical research experiments on humans. This work took place in a freestanding barracks, isolated from the others, and was called Experimental Cell Block Five. Many of the Reich's elite medical doctors passed through the laboratory here. The work that was performed in Experimental Cell Block Five was science without conscience: bad science for bad ends. That at least six Nazi medical doctors involved in this research at Dachau would be among the first scientists given contracts by the U.S. Army would become one of the darkest secrets of Operation Paperclip.

CHAPTER FIVE

The Captured and Their Interrogators

The capturing of Nazi scientists would now become a watershed. One by one, across the Reich, Hitler's scientists were taken into custody and interrogated. The day after Dachau was liberated, 375 miles to the north, Soviet commanders planned their final assault on the iconic Reichstag building, in Berlin. Sometime around 3:30 in the afternoon, inside the Führerbunker, Hitler fired a bullet into his head. The Russians were just five hundred meters from the Führerbunker's emergency exit door. Around the corner, under the Reich Chancellery, Red Army soldiers took over the underground subway tunnels, including the one Major General Dr. Walter Schreiber had been using as a hospital. Soviet film footage—alleged by Schreiber to have been filmed days later as a reenactment — shows Schreiber coming out of a cellar with his hands over his head.

Up north, Siegfried Knemeyer was captured by the British. Baumbach, Knemeyer, and Speer never escaped to Greenland after all. Shortly after Hitler killed himself, Baumbach was ordered by Grand Admiral Dönitz to go to the small town of Eutin, forty miles north of the city of Hamburg. Hitler had named Dönitz his successor; Dönitz set up his new government in the naval barracks at Eutin because it was one of the few places not yet controlled by the Allied forces.

Siegfried Knemeyer hadn't been invited to join the new inner circle. Baumbach let him keep the BMW and Knemeyer fled west. On a country road outside Hamburg, Knemeyer spotted a vehicle filled with British soldiers on approach. He knew that the BMW he was driving would be recognized as belonging to a senior military officer, so he pulled off the highway, ditched the car, and fled on foot. British soldiers found him hiding under a bridge and arrested him. Knemeyer was taken to a newly liberated concentration camp outside Hamburg, where hundreds of other German officers and Nazi Party officials were held. He was a prisoner of war now and was accordingly stripped of his valuables and military insignia. Years later, Knemeyer would share with his son that he managed to hide his one remaining meaningful possession in his shoe: a 1,000 Swiss franc note given to him by Albert Speer.

Von Braun and Dornberger were not captured. So confident were they as to their future use by the U.S. Army that they turned themselves in. Since departing from Nordhausen several weeks before, von Braun, Dornberger, and hundreds of other men from the rocket program had been hiding out in a remote ski village in the Bavarian Alps. Their resort, Haus Ingeburg, was located at an elevation of 3,850 feet along a windy mountain road then called the Adolf Hitler Pass (known before and after the war as the Oberjoch Pass). Thanks to the resources of the SS, the scientists had plenty of fine food and drink. There was a sun terrace and, as von Braun reflected after the war, little for any of them to do but eat, drink, sunbathe, and admire the snow-capped Allgäu Alps. "There I was living royally in a ski hotel on a mountain plateau," von Braun later recalled, "the French below us to the west, and the Americans to the south. But no one, of course, suspected we were there."

On the night of May 1, 1945, the scientists were listening to the national radio as it played Bruckner's Symphony No. 7 when, at 10:26, the music was interrupted by a long military drumroll. "Our Führer, Adolf Hitler, fighting to the last breath against Bolshevism, fell for Germany this afternoon in his operational headquarters in the Reich Chancellery," the radio announcer declared. The fight was a fabrication. But Hitler's death spurred Wernher von Braun to action. Von Braun approached General Dornberger, suggesting that they move quickly to make a deal with the Americans. "I agree with you, Wernher," Dornberger was overheard saying late that night. "It's our obligation to put our baby into the right hands."

At Haus Ingeburg the rocket scientists had been using a network of German and Austrian intelligence sources to keep track of U.S. Army developments in the area. Von Braun and Dornberger knew that a unit of U.S. soldiers had set up a base at the bottom of the mountain on the Austrian side. The two men agreed it was best to send von Braun's younger brother, Magnus, down the mountain to try to make a deal with the Americans. Magnus was trustworthy. He understood what could be said about the V-2 and what could never be said. Magnus von Braun had been in charge of overseeing slave labor production of the gyroscopes that each rocket required, and he understood why the subject of slave labor was to be avoided at all cost. He also was the best English speaker in the group.

On the morning of May 2, Magnus von Braun climbed onto a bicycle and began pedaling down the steep mountain pass through the bright alpine sunshine. Shortly before lunchtime he came upon an

American soldier manning a post along the road. It was Private First Class Fred Schneikert, the son of a Wisconsin farmer, now a soldier with the Forty-fourth Infantry Division of the U.S. Army. When Private Schneikert spotted a lone German on a bicycle, he ordered the man to drop the bike and raise both hands. Magnus von Braun complied. In broken English, he tried explaining to the American soldier that his brother wanted to make a deal with regard to the V-2 rocket. "It sounded like he wanted to 'sell' his brother to the Americans," Private Schneikert recalled.

Private Schneikert escorted Magnus von Braun farther down the mountain so he could speak with a superior at the Forty-fourth Division's U.S. Counter Intelligence Corps (CIC) headquarters, located in Reutte, just over the border in Austria. There, CIC contacted SHAEF headquarters in Versailles, which contacted a CIOS team. The CIOS Black List for rocket research included one thousand names of scientists and engineers slated for interrogation. Wernher von Braun was at the top of that list.

It was May 2, 1945, and although Hitler was dead, the German Reich had not yet surrendered. Allies feared members of a fanatical resistance group, the Werewolves, were lurking in Bavaria, planning a final attack. Thinking that the von Braun brother could be part of a trap, the Counter Intelligence Corps told Magnus to go tell his brother Wernher to come down and surrender himself. Magnus headed back up the mountain with the news.

At the Haus Ingeburg ski resort, Wernher von Braun and General Dornberger had selected a small group to join their deal-making team. They were Magnus von Braun; General Dornberger's chief of staff, Herbert Axster; the engine specialist Hans Lindenberg; and the two engineers who had hidden the V-2 documents inside the Dörnten mine, Dieter Huzel and Bernhard Tessmann. The men stuffed their personal belongings into three gray passenger vans and headed down the Adolf Hitler Pass. Heavy snow gave way to driving rain.

When the group of seven arrived in Reutte later that night, they found First Lieutenant Charles Stewart doing paperwork by candlelight. Their welcome was, by many accounts, warm. "I did not expect to be kicked in the teeth," von Braun told an American reporter years later. "The V-2 was something we had and you didn't have. Naturally, you wanted to know all about it." The rocket scientists were served fresh eggs, coffee, and bread with real butter. Fancy, but not quite as good as what was being provided at Haus Ingeburg. The scientists were given private rooms to sleep in, with pillows and clean sheets. In the morning, the press had arrived. The "capture" of the scientists and engineers behind the deadly V-2 was a big story for the international press. The group posed for photographs, and in the pictures they are all smiles. Von Braun boasted about having invented the V-2. He was "its founder and guiding spirit," von Braun insisted. Everyone else was secondary to him.

Some members of the Forty-fourth Division Counter Intelligence Corps found von Braun's hubris appalling. "He posed for endless pictures with individual GIs, in which he beamed, shook hands, pointed inquiringly at medals and otherwise conducted himself as a celebrity rather than a prisoner," noted one member, "treat[ing] our soldiers with the affable condescension of a visiting congressman." Second Lieutenant Walter Jessel was the American intelligence officer originally in charge of interrogating von Braun. His first and most lasting impression was the lack of remorse. "There is recognition of Germany's defeat, but none whatsoever of Germany's guilt and responsibility." So confident were von Braun and Dornberger about their value to the U.S. Army, they demanded to see General Eisenhower, whom they called "Ike."

Another observer noted, "If we hadn't caught the biggest scientist in the Third Reich, we had certainly caught the biggest liar."

Hitler's chemists — sought after as they were — were nowhere to be found. It was early May, and the Seventh Army was in control of the beautiful old city of Heidelberg, located on the Neckar River. Twenty-five agents attached to the U.S. military government's Cartels Division, including clerks with OSS and the Foreign Economic Division, had descended on the town looking for board members from IG Farben. In addition to being wanted for war crimes, the IG Farben board of directors was being investigated for international money laundering schemes. A number of high-ranking Farben executives were known to have houses in Heidelberg, but to date no one had been able to find Hermann Schmitz, the company's powerful and secretive CEO. Schmitz was also a director of the Deutsche Reichsbank, the German central bank, and director of the Bank for International Settlements in Geneva. He was believed to be the wealthiest banker in all of Germany. The reason no one had been able to locate Hermann Schmitz was not because he was hiding out or had fled but because officers were going around Heidelberg looking for "Schmitz Castle." Despite the vast wealth he had accumulated during the war, Hermann Schmitz was actually a miser. He lived in a modest, if not ugly, little house. "No one would associate the legend of Schmitz with the house he lived in," Nuremberg prosecutor Josiah DuBois recalled after the war.

Working on a tip, and as part of a door-to-door search for suspected war criminals, a group of enlisted soldiers knocked on the door of a "stucco pillbox of a house" overlooking the city where a short man with a red face and a thick neck answered the door. Behind him, on a placard nailed to the wall, it was written that God was the head of this house. Schmitz had dark eyes and a goatee and was accompanied by his wife, described by soldiers as "a dumpy Frau in a crisp gingham dress." Frau Schmitz offered the soldiers coffee, but Schmitz intervened and told her no. Schmitz said he had no interest in answering the questions of the enlisted men whom he considered beneath him. If an officer came to speak with him, Schmitz said he might have something to say.

The men conducted a cursory search of the house. Schmitz's office was plainly furnished and contained nothing expensive or of any obvious value. Searching though his desk, however, the soldiers learned that Schmitz had friends in high places. They found a collection of birthday telegrams sent from Hitler and Göring, both of whom addressed Schmitz as "Justizrat," Doctor of Laws.

"Doctor of Laws Schmitz," the soldiers asked, mocking him. "How much money do you have in this house, and where is it?"

Schmitz declined to say, and the soldiers were only able to locate a small stash of about 15,000 reichsmarks, or about half the annual salary of a field marshal. So they left, letting Schmitz know that they would return the following day. On the second day, Schmitz let the soldiers back in. This time the soldiers found an air raid shelter behind the house, where Schmitz had hidden a trunk filled with IG Farben documents. There was still not enough evidence to justify arresting Schmitz. It would be a few more days until an incredible discovery was made.

When CIOS team leader Major Tilley learned that Hermann Schmitz had been located, he rushed to Heidelberg. Tilley and Tarr had been leading the CIOS chemical weapons mission across Germany. Ever since they had discovered the tabun nerve agent cache hidden in the forest outside the Robbers' Lair, they had been looking for Farben executives. Now they had the man at the top.

If anyone could skillfully interrogate Hermann Schmitz, Major Tilley could. Not only did he speak fluent German, but he was deeply conversant on the subject of chemical warfare. In Heidelberg, Tilley went directly to Schmitz's house. He suggested that the two men discuss a few things in Herr Schmitz's private study. Schmitz said that would be fine.

Tilley asked the Farben CEO a series of banal questions, all the while tapping on the walls of Schmitz's study. Slowly, Tilley made his way around the room this way, listening for any inconsistencies in the way the walls were built. Schmitz grew increasingly uncomfortable. Finally he began to cry. Tilley had found what he was looking for: a secret safe buried in Schmitz's office wall.

Hermann Schmitz was one of the wealthiest bankers in Germany and one of the most important players in the economics of the Third Reich. What secret was contained in his safe? Major Tilley instructed Schmitz to open it. Inside, lying flat, was a photo album. "The photographs were in a wooden inlaid cover dedicated to Hermann Schmitz on his twenty-fifth jubilee, possibly as a Farben director," Tilley explained in a CIOS intelligence report. Tilley lifted out the photo album from its hiding place, flipped open the cover, and began reviewing the pictures. On page 1 of the scrapbook, the word "Auschwitz" was written. Tilley's eyes scanned over a picture of a street in a Polish village. Next to the photograph was a cartoonish drawing "depicting individuals who had once been part of the Jewish population who lived there, portrayed in a manner that was not flattering to them," Tilley explained. The caption underneath the cartoon read: "The Old Auschwitz. As it Was. Auschwitz in 1940."

At this point, Tilley wrote in his CIOS report, he was surprised at how "highly emotional" Schmitz became. What Tilley did not yet know was that he was looking at Schmitz's secret photo album that chronicled the building history of Farben's labor concentration camp, IG Auschwitz, from the very start. In May of 1945, almost no one, including Major Tilley, had any idea what really had happened at Auschwitz—that at least 1.1 million people had been exterminated there. The facts about the camp had not yet come to light. On January 27, 1945, Soviet troops liberated Auschwitz, and Red Army photographers took film footage and photographs of the atrocities they discovered there. But that information had not yet been shared openly with the rest of the world. A short report about the extermination camp had appeared in *Stalinskoe Znamya,* the Red Army's newspaper, on January 28, 1945. Stalin was waiting to release the bulk of information until after Germany surrendered. What was clear to Major Tilley was that this photo album was important to Schmitz, and that he wanted it to remain hidden. Why, exactly, Major Tilley had no idea.

As CIOS team leader, Major Tilley was on a hunt for Farben chemists who had developed nerve gas. Hermann Schmitz, while important to Farben in the bigger picture, was not a chemist. He claimed to have no idea where the Farben chemists had gone. The scrapbook was taken into evidence, and Tilley moved on in his search. Meanwhile, in southern Germany, in an Austrian border town called Gendorf, the man Major Tilley was really looking for—Dr. Otto Ambros—had just been located by U.S. Army soldiers. The soldiers had no idea who Ambros really was.

When American soldiers rolled into the town of Gendorf, about sixty miles southeast of Munich, they noticed one man in particular because he stood out like a sore thumb. This first encounter with Ambros, later recounted at the Nuremberg trials, stuck out in soldiers' minds because Ambros had been dressed in a fancy suit to greet the victors. The man hardly looked like he'd been through a war. The soldiers asked the man his rank and serial number.

"My name is Otto Ambros," he said, smiling. He added that he was not a military man but "a plain chemist."

Was he German, the soldiers asked?

"Yes, I am German," Ambros replied, and made a joke. He said that he had so many French friends he could almost be considered a Frenchman. In fact, his true home was in Ludwigshafen, on the border with France. He told the soldiers that the reason he was here in southern Bavaria was because he was the director of a large business concern called IG Farben. The company had a detergent factory here in Gendorf, Ambros explained. As a Farben board member he'd been asked to oversee production. German society might be experiencing a collapse, he told the soldiers, but everyone needed to stay clean.

The soldiers asked to be taken to the detergent factory. Inside, they inspected huge vats of soap and other cleaning products. Work at the factory appeared to have been uninterrupted by the war. Ambros took the soldiers to his office, where someone had taped a rainbow of color spectrum cards to the wall. In addition to cleaning products, the facility made lacquers, Ambros explained. The soldiers looked around, thanked Ambros for the tour, and asked him not to leave town.

I have "no reason to flee," said Ambros. The soldiers noted how much he smiled.

Over the next few days, more soldiers arrived in Gendorf. These filthy, grime-covered American GIs were delighted when the so-called plain chemist offered them free bars of soap. Some of the soldiers hadn't washed in more than a month. The chemist's generosity did not stop there. Otto Ambros gave the soldiers powerful cleaning solvents so that they could wash their mud-covered armored tanks.

Soldiers interviewed Otto Ambros a second time. This time, Ambros voluntarily offered up character witnesses. Working at the Farben factory in Gendorf were skinny men with shaved heads. Ambros said they were war refugees and that they could vouch for his kindness as a boss. They were from Poland, just across the border to the east. Ambros told the soldiers that he had personally brought these poor workers here to Gendorf. He'd handpicked the men and trained them how to work hard. This way, when the refugees went home, they would have skills that could help them earn a living, Ambros explained. The skinny refugees were quiet and said nothing to dispute the plain chemist's claims.

Some of them even helped the American soldiers wash their tanks.

Otto Ambros was a talkative man. He regaled the Americans with stories about the joys of chemistry. For example, did the soldiers realize what a miracle it was that man could make one hundred wonders from a single chemical compound like ethylene oxide? Or how amazing rubber was? Ambros told the soldiers that he had been to Ceylon, where the rubber plant grows. Rubber had so much in common with man, Ambros said. He was a rubber expert, so he knew this to be fact. Rubber was civilized. Neat and perfect if kept clean. Ambros told the soldiers that a rubber factory and a man must always be very clean. A single flake of dust or dirt in a vat of liquid rubber could mean a blowout on the autobahn one day. IG Farben had synthetic rubber factories and, like natural rubber, the laboratories and factories must always be kept perfectly clean. Ambros talked a lot, but he did not mention anything about the rubber factory he had built and managed at Auschwitz. The soldiers thanked him for his generosity with the soap and the cleaning agents. Before they left, they reminded Ambros again that it was important he not leave town. He was technically under house arrest.

When American officials of higher rank finally arrived in Gendorf a few days later, they had more specific questions for Ambros. Why was part of the

Farben detergent factory built underground? It would take months for CIOS investigators to learn that the factory here in Gendorf produced chemical weapons during the war—and that, after Ambros had fled Auschwitz in late January 1945, he and his deputy, Jürgen von Klenck, had come to Gendorf to destroy evidence, hide documents, and disguise the factory so that it appeared to produce only detergents and soap.

In Munich, on May 17, 1945, U.S. soldiers at a checkpoint were conducting a routine identification request when a well-dressed man—134 pounds, five foot nine, with dark black hair, hazel eyes, and a pronounced dueling scar on the left side of his face between his nose and his upper lip—presented a German passport bearing the name Professor Doctor Friedrich Ludwig Kurt Blome.

Dr. Blome's name triggered an alert: "Immediate arrest. 1st Priority." Samuel Goudsmit and the entire team of biological warfare experts with Operation Alsos had been on the hunt for Dr. Blome. Agent Arnold Vyth, with the army's Counter Intelligence Corps, made the arrest. Agent Vyth completed the necessary paperwork while the prisoner was processed. Dr. Blome was sent to the Twelfth Army Group Interrogation Center for questioning. Several days later, a document arrived via teletype from the Office of Strategic Services (OSS), America's wartime espionage agency and the precursor to the CIA. They, too, had been searching for Dr. Blome.

The War Crimes Office had considerable information about Dr. Kurt Blome. He was deputy surgeon general of the Third Reich and vice president of the Reich's Physicians' League, Reichsärztekammer. He was believed to have reported directly to Göring and maybe even to Himmler, or to both. Blome had been named head of Reich cancer research in 1942. Alsos and OSS presumed that this was a cover name for biological weapons work. Blome was a dedicated and proud Nazi. His book *Arzt im Kampf* (Doctor in Battle) compared a doctor's struggle with the struggle of the Third Reich. Soldiers, officers, and doctors weren't all that different, each constantly in battle against invading forces and disease.

Investigators were trying to piece together the labyrinthine medical hierarchy of the Third Reich so as to understand who was in charge of what organization. Particularly interesting to the interrogators was the fact that Dr. Kurt Blome had been part of a top-tier group of Nazi doctors who focused on "hygiene." This word connoted disease control but was also believed to have been used by the Reich as a euphemism for ethnic cleansing and extermination of Jews. Alsos was in possession of correspondence between Blome and Himmler that discussed giving certain groups of sick individuals—in this case tubercular Poles—"special treatment" (*Sonderbehandlung*). What exactly did special treatment mean? At the time of Blome's capture and interrogation, Allied intelligence agencies believed that there was only one physician higher than Blome in the hierarchy of the Reich Hygiene Committee, and that was the notorious Reich Health Leader (Reichsgesundheitsführer), Leonardo Conti.

Dr. Blome spoke fluent English with his first army interrogator. He described himself as a "good Nazi" — obedient — and promised that he was willing to cooperate with the Allies. At first his interrogators were thrilled by the prospect of learning more about Reich medicine from such a big fish as Dr. Blome.

Why was he cooperating? Blome was asked.

"[I] can not approve of the way new advances in medical science have been used for atrocities," declared Dr. Blome.

What kind of atrocities? Blome's investigator wanted to know.

Blome stated that in his capacity as deputy surgeon general of the Reich he had "observe[d] new scientific studies and experiments which led to later atrocities e.g. mass sterilization, gassing of Jews." It was an astonishing admission. Until Dr. Blome gave up this information so freely, no physician in the inner circle had admitted to having known about such wide-scale atrocities as mass murder and sterilization programs. That Blome was willing to talk was extremely promising news. Blome was "cooperative and intelligent," noted his interrogator. Most important, he was "willing to supply information."

But the U.S. investigators' excitement did not last long. By his next interrogation, Dr. Kurt Blome had shut down entirely. He told his interrogating officer, Major E. W. B. Gill, that he had only ever been an administrator for the Reich; that he did nothing "hands-on." Major Gill pressed Blome for information about his direct superior, Dr. Leonardo Conti. Blome said he knew nothing about Conti's job.

"When I pointed out that the deputy must presumably know something about his chief's job," Major Gill wrote in his report, "he said the organization was extremely complicated and really he would like to draw me a diagram on it." Gill lost his temper. "I told Blome I didn't want his dammed diagrams, but an answer to a simple question. How did he take Conti's place if he [Conti] were absent or ill if he knew nothing of the job?"

Blome repeated his position. That it was all too complicated to explain to a man like Major Gill. Outraged by the sidestepping, Gill kept at it. But by the end of Blome's Alsos interrogation, Major Gill had been unable to get even a scrap of new information from Blome. He claimed never to have heard of the majority of the names of fellow doctors that Major Gill asked him about. Instead, Blome insisted that he knew nothing about the medical chain of command inside the Third Reich or the SS, despite the fact that he had personally met with Himmler five times since 1943. Gill asked how Blome, a "cancer expert," had been put in charge of the Reich's bioweapons program, a subject he claimed to know very little about. Blome said he had no answer for that.

"On my suggestion that a most important branch of war research would not be assigned to a complete ignoramus he, after endless explanations of the complexity of the German world, finally said it must have been because, as an undergraduate, he wrote on BW [biological weapons] as his thesis for a doctorate." Major Gill felt for certain that Dr. Blome was lying. But there was nothing he could do except present Blome with information and evidence that Alsos had compiled about him since they had seized Dr. Eugen Haagen's apartment six months before.

Gill told Blome that in a series of interrogations with sixteen Reich doctors also involved in bioweapons-related research, Alsos officers had learned about many horrific medical crimes. Gill explained that Alsos had documents that tied Blome to the crimes. For example, Alsos had found letters inside the apartment of Dr. Eugen Haagen that linked Dr. Blome to Dr. Haagen and also to an SS colleague named Dr. August Hirt. These letters made clear that someone was providing Reich doctors with human guinea pigs.

Who exactly was in charge of this program, Gill asked Blome? Gill needed a name.

Blome denied having any idea what Gill was referring to. Major Gill told Blome he had a letter that implicated Blome. In another letter, Gill said, Dr. Blome had instructed Dr. Hirt to conduct research on "the effect of mustard gas on living organisms." The phrase "living organisms" was a code name for people, wasn't it? Gill asked Blome. Dr. Blome kept stonewalling. "On the whole subject of SS research, his attitude was always that it was so secret that not even [the] Reich chief medical advisor knew anything about it," Major Gill wrote in his report.

Gill was convinced that Dr. Kurt Blome was lying. He felt certain that Doctors Haagen, Hirt, Blome, and the SS were connected to medical research on prisoners at concentration camps.

"This interrogation was extremely unproductive," a frustrated Major Gill summarized in his report. "Although I do not wish to be definitive my first impression is that Blome is a liar and a medical charlatan."

Down south in the Bavarian Alps, while the V-2 rocket scientists angled for a deal with the U.S. Army, Georg Rickhey, former general manager of the Mittelwerk, tried to blend in. Rickhey had taken a job ninety miles from Nordhausen, running operations in a salt mine.

For several weeks, no one was looking for him. Then Colonel Peter Beasley, of the U.S. Strategic Bombing Survey (USSBS), arrived in the area on a mission from the War Department. Beasley's job was to locate the engineers who had built the fortified underground weapons facilities in the Harz. These bombproof bunkers were extraordinary engineering feats, and the USSBS was impressed with how so many of them had withstood relentless Allied air bombing campaigns. The rocket facility at Nordhausen was of particular importance to USSBS officers, and Colonel Beasley set up shop in an abandoned barracks just north of the former Mittelwerk factory, in a town called Ilfeld, to investigate. As circumstance would have it, the barracks he chose to occupy was the building in which the former office of Georg Rickhey was located. From documents and equipment left behind, Colonel Beasley learned that Rickhey possessed extremely valuable information about how the tunnel factory had been built. Beasley asked around, but none of the locals claimed to know where Rickhey had gone.

"I made daily visits to the jails in the small towns to see if I could locate anyone who might interest me," Beasley wrote in a report. Eventually he found a man who gave him a tip. Georg Rickhey was running operations at a salt mine in the Black Forest, the man said. Colonel Beasley sent two officers into the field to track Rickhey down.

Meanwhile, Beasley and his team followed another lead. "In Blankenburg," Beasley wrote, "we found a school building with some miscellaneous papers bearing the Speer Ministry insignia." From these documents Beasley learned that Georg Rickhey was the liaison between the Mittelwerk and the Ministry of Armaments. When Beasley's two officers returned with Georg Rickhey in custody, Beasley placed Rickhey under arrest and began to interrogate him. He was "a nervous little man who smoked incessantly and always brought the conversation back to scientific or technical matters," Beasley recalled after the war, but in the end he "was a most profitable catch."

"I've got a job for you," Beasley told Rickhey. "I want you to begin right now writing out a full description of yourself and all the activities of the V-2 factory, and what your people were working on." Rickhey complied. When the task was complete, Beasley told the former general manager of the Mittelwerk, "[W]e accept you as an official of the German Government; we have patience and time and lots of people—you have lost the war and so as far as I am concerned you are a man who knows a lot about rockets. As an American officer, I want my country to have full possession of all your knowledge. To my superiors, I shall recommend that you be taken to the United States."

Rickhey embraced this news with open arms. He told Beasley that he was a scientist and only wanted to work in pleasant surroundings, like the United States. He agreed to tell Beasley where some important records had been hidden. Rickhey took Colonel Beasley to a cave several miles away. There, forty-two boxes of worksheets, engineering tables, and blueprints relating to Nordhausen and the V-2 had been stashed. This was certainly not Wernher von Braun's documents stash, but for the USSBS, it was more than they possessed up to this point. Now that he was in possession of a huge trove of documents, Colonel Beasley realized that he needed to have them translated by someone with technical expertise. He had promised Rickhey a recommendation for a job in the United States, but first he needed Rickhey to come with him to London to translate and analyze these documents for him.

Albert Speer, one of the most wanted Nazi war criminals in the world, was finally captured on the morning of May 23, 1945. He was standing in one of the bathrooms of a friend's castle, Schloss Glücksburg, near Flensburg, in north Germany. Hitler's successor, Grand Admiral Dönitz, had by now moved his new government from Eutin to Flensburg, which was located just a few miles from the Danish border. Speer, a Dönitz cabinet member, had been making the daily six-mile drive from Schloss Glücksburg to the new government's headquarters. The way Speer told the story of his capture, he had been shaving when he heard the sounds of heavy footsteps and loud orders being delivered in English. Sensing that the end of his freedom had arrived, Speer opened the bathroom door a little, his face half-covered in shaving cream, and saw the British soldiers standing there.

"Are you Albert Speer, sir?" a British sergeant asked.

"Yes, I am Speer," he answered in English.

"Sir, you are my prisoner," the sergeant said.

Speer got dressed and packed a bag. Outside on the castle lawn, a unit of British soldiers with antitank guns had surrounded Schloss Glücksburg. Speer was arrested and taken away.

By the time the British arrested Speer, American officials had known for nearly two weeks where he had been hiding out. Speer's previous eleven days in the castle had been spent in discussions with American officials with the U.S. Strategic Bombing Survey (USSBS). The head of that organization, Paul Nitze, had managed to be the first person in an international manhunt to track down Albert Speer. Nitze considered Speer his most desired intelligence target, and on May 12, 1945, he boarded his DC-3 from where he was stationed in London and headed to Castle Glücksburg "before you could say 'knife,'" Nitze recalled after the war. Because Flensburg was in the British zone of control, it was the British who needed to arrest Albert Speer. Until then, Nitze, an American, felt at liberty to get as much information from Speer as he could. "We were looking for absolutely vital information and knowledge and he was literally the only person in Germany who was in a position to provide it," Nitze recalled years later.

Specifically, Nitze's organization wanted to know which Allied bombing campaigns had proved the most devastating against Germany during the war. America was still at war with the Japanese, and the USSBS believed Speer could provide them with information that might help America defeat them. Nitze was joined at the castle by two of his colleagues, George Ball and John Kenneth Galbraith. For the next eleven days the three men questioned Speer. From inside an elegant sitting room wallpapered in red and gold brocade, the men discussed which Allied bombing campaigns had done the most damage to Nazi Germany and which had had the least effect. Of particular interest to Nitze, Ball, and Galbraith was how the Reich's armaments industry had been able to hold out for so long. Speer explained that at his initiative the majority of the Reich's weapons facilities had been moved underground. These weapons complexes had proved to be impervious to even the heaviest bombing campaigns. They were engineering triumphs, their construction spearheaded largely by Franz Dorsch and Speer's deputy Walter Schieber, Speer said. Speer's secretary, Annemarie Kempf, took notes. The only interruption was when the cook for the castle summoned everyone for lunch.

Speer did not mention that his deputy, SS-Brigade-führer Schieber, a chemist, also worked with Speer in chemical weapons production; that would be opening up a can of war crimes–related worms. The Americans were not interested in pressing Speer about his involvement in war crimes, and Speer was certainly not offering up any incriminating evidence against himself. Mostly he boasted about his ministry's feats. George Ball recalled that only once, maybe twice, during the USSBS questioning was Speer asked about the concentration camps. "I asked him what he knew about the extermination of the Jews. He said he couldn't comment because he hadn't known about it, but he added that it was a mistake not to have found out," Ball told Speer's biographer, Gitta Sereny, after the war.

John Kenneth Galbraith was the only one of the three men from the U.S. Strategic Bombing Survey who had toured a liberated concentration camp before interviewing Speer. Galbraith had seen the atrocities at Dachau and Buchenwald. He explained, "One was just beginning to hear rumors about Auschwitz." Did Galbraith believe that Speer did not know about the extermination of the Jews? "No, I don't believe he didn't know," Galbraith told Sereny. "Certainly he knew about all the slave laborers. I remember him saying to us, 'You should hang Saukel' " — Speer's deputy in charge of slave labor — "and then a few weeks later, Saukel said to us, 'You should hang Speer.' Nice people, weren't they?"

After eleven days of discussions with the Americans, the British located and arrested Speer. They drove him the six miles to Flensburg, where the remaining members of Hitler's government were also arrested. Under an escort of more than thirty armored vehicles, the prisoners were driven to waiting aircraft. There, in a field of grass, the men of Hitler's inner circle were loaded onto two airplanes and flown to a Top Secret interrogation center code-named Ashcan.

That same afternoon, one hundred miles south of Flensburg, at the Thirty-first Civilian Interrogation Camp near Lüneburg, a former Wehrmacht sergeant was making a lot of noise. The officer in charge of Camp 31, Captain Thomas Selvester, found the man's behavior odd. Wehrmacht soldiers who were prisoners rarely did anything to draw attention to themselves. Captain Selvester sent for the agitated man, whom he described as a short, "ill-looking" person in civilian clothing with a black eye patch over his left eye. Face-to-face with Captain Selvester, the small, ugly man ceremoniously pulled off the eye patch, revealing a pale, unshaven face. The man then produced a pair of horn-rimmed glasses from his pocket and put them on his face.

"Heinrich Himmler," the prisoner announced in a quiet voice.

With the glasses on, Captain Selvester recognized Heinrich Himmler at once. Before him stood a man many considered the most powerful man in the Third Reich after Hitler. Himmler was Reichsführer-SS and chief of the German police, commander of the Reserve Army of the Wehrmacht, and Reich minister of the interior. That face. The cleft chin and the sinister, smiling eyes. Ever since a drawing of Himmler had appeared on the cover of *Time* magazine, on October 11, 1943—portraying the "Police Chief of Nazi Europe" in front of a mountain of corpses—he had become synonymous with evil. Now that the little round glasses were on, Selvester was certain this person was indeed Heinrich Himmler. Still, Captain Selvester followed protocol and asked for signature verification. When Himmler had been captured days before he'd presented forged military papers that identified him as a Wehrmacht sergeant named Heinrich Hitzinger.

The signature matched, and Captain Selvester sent for the most senior interrogator in Camp 31, a Captain named Smith. Once Smith arrived, Selvester ordered Himmler searched again. This time British soldiers found two vials of poison hidden in Himmler's clothes. It was medicine to treat stomach cramps, Himmler said. Captain Smith ordered a second physical exam of the prisoner, and the camp's doctor, Captain Clement Wells, spotted a blue-tipped object—hidden in the back of Himmler's mouth. When Dr. Wells tried to remove it, Himmler jerked his head back and bit down. The vial contained poison. Within seconds, the prisoner collapsed. Now Heinrich Himmler was dead. An assistant to Dr. Wells noted in his diary, "[T]his evil thing breathed its last breath at 23:14."

The war in Europe was over. Germans called it *die Stunde Null,* zero hour. Cities lay in ruins. Allied bombing had destroyed more than 1.8 million German homes. Of the 18.2 million men who had served in the German army, navy, Luftwaffe, and the Waffen-SS, a total of 5.3 million had been killed. Sixty-one countries had been drawn into a war Germany started. Some 50 million people were dead. The Third Reich was no more.

Heinrich Himmler and Adolf Hitler were dead. Albert Speer was in custody. So were Siegfried Knemeyer and Dr. Kurt Blome. Otto Ambros was under house arrest in Gendorf, with no one in CIOS or Alsos yet having figured out who he really was. Wernher von Braun, Walter Dornberger, and Arthur Rudolph were in custody, working toward contracts with the U.S. Army. Georg Rickhey had a job in London, translating documents for the U.S. Strategic Bombing Survey.

The future of war and weapons hung in the balance. What would happen to the Nazi scientists? Who would be hired and who would be hanged? In May 1945 there was no official policy regarding what to do with any of them. "The question who is a Nazi is often a dark riddle," an officer with the Third Army, G-5, wrote in a report sent to SHAEF headquarters in May. "The question what is a Nazi is also not easy to answer."

Over the next few months, critical decisions about what to do with Hitler's former scientists and engineers would be made, almost always based on an individual military organization's needs and justified by perceived threats. Official policy would follow, one version for the public and another for the Joint Chiefs of Staff (JCS). A headless monster called Operation Paperclip would emerge.

PART II

"The scale on which science and engineering have been harnessed to the chariot of destruction in Germany is indeed amazing. There is a tremendous amount to be learnt in Germany at the present time."

—W. S. Farren, British aviation expert with the Royal Aircraft Establishment

CHAPTER SIX

Harnessing the Chariot of Destruction

What to do about Hitler's former scientists? The fighting had stopped, and the Allied forces were transitioning from a conquering army to an occupying force. Germany was to be disarmed, demilitarized, and denazified so its ability to make war would be reduced to nil, and science and technology were at the very heart of the matter. "Clearly German science must be curbed," noted Army Air Forces Lieutenant Colonel John O'Mara, in the CIOS report he authored on the rise of the Luftwaffe. "But how?" World War I had ended with a peace treaty that, among other restrictions, "sought to prevent the rise of German Air Power by forbidding powered flight. The result," explained O'Mara, "was as ludicrous as it was tragic." By the time Germany started World War II, its air force was the most powerful in the world. The mistake could not be repeated, and the U.S. procedural guidelines for an occupied Germany, contained in a directive known as JCS (Joint Chiefs of Staff) 1076, promised to nullify Germany's appetite for war. All military research was to cease. Scientists were rounded up and taken to detention centers for extensive questioning.

Across the former Reich, SHAEF had set up internment centers where more than fifteen hundred scientists were now being held separate from other German prisoners of war. The U.S. Army had approximately 500 scientists in custody in Garmisch-Partenkirchen, in the Bavarian Alps, including the von Braun and Dornberger group; there were 444 persons of interest detained in Heidenheim, north of Munich; 200 were in Zell am See, in Austria; 30 kept at Château du Grande Chesnay, in France. The U.S. Navy had 200 scientists and engineers at a holding facility in Kochel, Germany, including many wind tunnel experts. The Army Air Forces had 150 Luftwaffe engineers and technicians in Bad Kissingen, Germany, a majority of whom had been rounded up by Colonel Donald Putt. CIOS had 50 scientists, including Werner Osenberg, in Versailles. But there was no clear policy regarding what lay ahead for the scientists, engineers, and technicians in Allied custody, and General Eisenhower sought clarification on the issue. From Supreme Headquarters Allied Expeditionary Force in France, he sent a cable to the War Department General Staff in Washington, D.C., asking for specific direction about longer-term goals. "Restraint and control of future German scientific and technical investigations are clearly indicated," General Eisenhower wrote, "but this headquarters is without guidance on the matter and is in no position to formulate long-term policy." Were these men going to be detained indefinitely? Interrogated and released?

The War Department responded to Eisenhower's cable by letting him know his query was considered a "matter of urgency." Tentative responsibility was assigned to the Captured Personnel and Materiel Branch of the Military Intelligence Service, Europe. Now that group was in charge of overseeing the scientists' basic needs, including living quarters, food, and in some cases pay. But it would be another two weeks before the War Department would get back to General Eisenhower with any kind of a statement regarding policy. In the meantime, a number of events were unfolding—in America and in Germany—that would affect the decision making of the War Department General Staff.

In the absence of policy, ideas were floated at the Pentagon. Some, like Major General Kenneth B. Wolfe, of the Army Air Forces, took matters into their own hands. General Wolfe was chief of engineering and procurement for Air Technical Service Command at Wright Field, and he supported Major General Knerr and Colonel Putt in their quest for capturing Luftwaffe spoils discovered at Völkenrode. But General Wolfe envisioned an even bigger science exploitation program and felt strongly that policy needed to be set now. Wolfe flew to SHAEF headquarters in France to meet with Eisenhower's deputy, General Lucius D. Clay, to promote his idea.

General Clay told General Wolfe that he was not opposed to such a program but that now was hardly a good time to broach it. "Besieged by the countless demands and the chaotic conditions relevant to ending the war, and the burdensome complexities of planning for the peace, [Clay] considered such efforts six months premature," explains historian Clarence Lasby. General Clay told General Wolfe to come back and talk to him in six months. Instead, Wolfe set out for Nordhausen, Germany, where his colleague at the Pentagon, Colonel Gervais William Trichel, was running Special Mission V-2, the Top Secret scientific intelligence operation for the U.S. Army Ordnance, Rocket Branch.

Inside the abandoned rocket production facility in the underground tunnel complex at Nordhausen, Special Mission V-2 was just getting under way. When General Wolfe saw the vast numbers of V-weapons left behind, he became even more convinced that a U.S. program to exploit Nazi science had to happen now.

Upon his return to Washington, D.C., General Wolfe wrote to General Clay with a revised idea. Not only did the United States military need to act immediately to capture Nazi armaments, Wolfe said, but America needed to hire the "German scientists and engineers" who had created the weapons and put them to work in America. "If steps to this end are taken, the double purpose of preventing Germany's resurgence as a war power and advancing our own industrial future may be served." Clay did not respond; he had already told General Wolfe to back off for six months. Meanwhile, the work that was going on at Nordhausen under the auspices of Special Mission V-2 would greatly influence the future of all the Nazi science programs that would follow.

The man in charge of Special Mission V-2, twenty-eight-year-old Major Robert B. Staver, was no stranger to the military significance of the Nazis' rockets. While preparing for Special Mission V-2 in London the winter before, Staver was nearly killed by one. He and a British colleague had been working inside an office at 27 Grosvenor Square one afternoon in February when a loud blast knocked both of them to the floor. Staver went to the window and saw a "big round cloud of smoke where a V-2 had exploded overhead." Watching pieces of burning metal rain down from the sky, Staver did a few calculations in his head and determined that the V-2 had likely been heading "very directly" at the building in which he was working when it blew up prematurely. A few weeks later, Major Staver was asleep in a hotel room near the Marble Arch when he was thrown out of bed by an enormous blast. A V-2 had landed in nearby Hyde Park and killed sixty-two people.

The near-death experiences made him ever more committed to Special Mission V-2. For six weeks Staver worked twelve hours a day, seven days a week, studying aerial photographs of Nordhausen supplied to him by the British and otherwise learning everything he could about the V-weapons. As soon as the Allied forces liberated the tunnel complex, Major Staver would be one of the first intelligence officers inside.

Now, finally, here he was at Nordhausen. It was May 12, 1945, and though his mission was almost complete, time was running out, because the Russians were headed into this area soon. By U.S. Army calculations, they would most likely arrive in eighteen days from Berlin.

U.S. Army Ordnance believed that the V-2 rocket could help win the Pacific war, and for nearly two weeks Staver had been hard at work. He had overseen the collection of four hundred tons of rocket parts, which had been loaded onto railcars for delivery to the port at Antwerp, from where they would be shipped to the United States. But with his degree in mechanical engineering from Stanford University, Staver knew that the V-2 rocket was a lot more than the sum of its parts. Without blueprints or technical drawings, it was highly unlikely that American engineers could simply cobble the rocket components together and make the V-2 fly. The drawings and blueprints had to have been stashed somewhere near Nordhausen. If only Major Staver could find a German scientist to bribe, he might be able to find out where the crucial documents were hidden.

For two weeks now Staver had been traveling through the Harz Mountains touring underground weapons factories, searching for a clue or a lead as to who might know more about the V-2 document stash. Locals told a wide variety of stories. Some spoke of paperwork going up in flames. Others talked about truckloads of metal trunks being hidden away in abandoned buildings, in beer gardens, and in castle walls. But this was all hearsay. No one could produce a concrete lead, and it was not exactly difficult to understand why. War crimes investigators were also in Nordhausen asking locals lots of questions. And as Staver trolled for rocket scientists, American GIs continued to dig mass graves for the thousands of corpses found at Nordhausen-Dora slave labor camp. The entire town of Nordhausen still smelled of death.

While driving around on his hunt, Staver kept boxes stashed in the back of his army jeep, filled with cigarettes, alcohol, and cans of Spam. These valuable black market goods worked well in exchange for information, and finally, Staver got the lead he was looking for. A source told him that there was a V-2 rocket scientist by the name of Karl Otto Fleischer who lived nearby. Fleischer had been an engineer inside the Nordhausen tunnels as well as the Wehrmacht's business manager, and he knew a lot more than he was letting on. Fleischer reported directly to General Dornberger; he knew things. Staver drove to the scientist's residence with a proposition more powerful than a can of Spam.

Major Staver told Karl Otto Fleischer that he could cooperate or go to jail. Important V-2 documents had been hidden somewhere around Nordhausen, Staver said. If anyone knew, Fleischer did, Staver surmised. Dieter Huzel and Bernhard Tessmann had indeed told Fleischer about the document stash in the Dörnten mine before they fled for the Bavarian Alps. But Fleischer's allegiance was to his colleagues, so he lied to Staver and said he had no idea what Staver was talking about. He pointed the finger at another colleague, an engineer and von Braun deputy named Dr. Eberhard Rees. Ask Rees, Fleischer said. He was the former chief in charge of the Peenemünde assembly line.

When interviewed by Staver, Dr. Eberhard Rees played his own disinformation card, using Major Staver's influence to help spring a third colleague from jail. Walther Riedel, chief of V-2 rocket motor and structural design, had been one of the four men honored at the Castle Varlar event the previous December. Now Riedel was receiving rough treatment in a jail eighty miles away, in Saalfeld. He had been mistaken by military intelligence as having been Hitler's biological weapons chief. Agents with the Counter Intelligence Corps had knocked out several of Riedel's front teeth. His security report listed him as "an active Nazi who wore the uniform and the party badge. Ardent." Riedel joined the Nazi Party in 1937 and was a member of five Nazi organizations.

In a series of interviews with Riedel, Major Staver found him to be a strange bird. Riedel was obsessed with outer space vehicles, which he called "passenger rockets." In one interview, Riedel insisted he'd designed these passenger rockets for "short trips around the moon," and that he'd been pursuing "space mirrors which would be used for good and possibly evil." Riedel said he knew of at least forty rocket scientists besides himself who should be brought to America to complete this groundbreaking work. If the Americans didn't act, Riedel said, the Russians surely would. Staver asked Riedel if he knew where the V-2 technical drawings were hidden. Riedel said he had no idea.

Staver was working on a number of problems, all compounded by the fact that the Russians were coming. That much was real. Nordhausen had been liberated by the Americans and was originally designated to be part of the American zone. Stalin protested, saying Russia had lost seventeen million men in the war and deserved greater reparations for greater losses sustained. The Allies agreed to turn over a large swath of American-held German territory to the Soviets on June 1. This territory included all of Nordhausen and everything in it.

But Staver had more to worry about than the Russians. On May 18, 1945, an airplane arrived carrying a physicist and ordnance expert named Dr. Howard Percy "H. P." Robertson. Robertson had been a team leader for Operation Alsos, and now he served President Eisenhower as chief of the Scientific Intelligence Advisory Section under SHAEF. Dr. H. P. Robertson told Major Staver that he intended to take rocket engineers Fleischer, Riedel, and Rees to Garmisch-Partenkirchen for interrogation, where they would be held alongside General Dornberger and Wernher von Braun until the War Department General Staff decided on a policy regarding Nazi scientists.

Major Staver refused to give up Fleischer, Riedel, and Rees. They were his charges, he told Robertson. As far as exploiting Nazi science for American use, Staver and Robertson saw eye to eye. But as far as giving Nazi scientists special privileges, the two men were on opposite sides of the aisle. The idea outraged Robertson, who saw Nazi scientists as amoral opportunists who were "hostile to the Allied cause."

Dr. Robertson was a mathematical physicist who had taken a leave of absence from a professorship at Princeton University to help in the war effort. He was a jovial, gentle man who liked crossword puzzles, Ivy League football matches, and scotch. Robertson spoke German fluently and was respected by Germany's academic elite not just for his scientific accomplishments but because he had studied, in 1925, in Göttingen and Munich. Before the war, Dr. Robertson counted many leading German scientists as his friends. World War II changed his perspective, notably regarding any German scientist who stayed and worked for Hitler.

While at Princeton, Dr. Robertson had become friendly with Albert Einstein. The two men worked on theoretical projects together and spent time discussing Hitler, National Socialism, and the war. Einstein, born in Germany, had worked there until 1933, becoming director of the Kaiser Wilhelm Institute of Physics and professor at the University of Berlin. But when Hitler came to power, Einstein immediately renounced his citizenship in defiance of the Nazi Party and immigrated to the United States. Dr. Robertson shared Einstein's core view. It had been the duty of German scientists to protest Hitler's racist policies, beginning in 1933. Anyone who had served the Reich's war machine was not going to be given a free pass by H. P. Robertson now.

Determined to keep the Nazi scientists in his custody, Staver played the Russian card. Robertson may have been anti-Nazi, but he was also deeply patriotic. With access to secret Alsos intelligence information, Robertson was well aware that Russian rocket development was a legitimate and growing threat. Both men knew that in as little as twelve days, the Russians would arrive in Nordhausen. If Staver was not able to locate the V-2 documents by then, the Russians would eventually find them. Major Staver appealed to Dr. Robertson, arguing that his keeping Fleischer, Riedel, and Rees was the army's last and best shot at locating the hidden V-2 documents. Ultimately, Dr. Robertson agreed. In a final appeal, Staver asked Robertson if there was anything Robertson could offer up that might help him in his search for the V-2 stash. Some clue or detail that Staver might be overlooking?

Indeed there was. Dr. Robertson's fluency in science and his familiarity with German scientific intelligence had thus far made him an extremely effective interrogator of the Nazi scientific and military elite. Wehrmacht generals, SS officers, and scientists were notoriously eager to speak with him. Listening to Staver, Dr. Robertson had an idea. He pulled a small writing pad out of his shirt pocket and looked over his notes. During an earlier interrogation of a rocket scientist named von Ploetz, Robertson had gotten an interesting lead. He decided to share it with Major Staver.

"Von Ploetz said that General Dornberger told General Rossman [the German army's Weapons Office department chief] that documents of V-weapon production were hidden in Kaliwerke [salt mine] at Bleicherode, walled into one of the mine shafts," read Robertson from his notes. Robertson suggested that Staver use that information to his advantage. He agreed to leave Fleischer, Riedel, and Rees with Major Staver while he headed to Garmisch-Partenkirchen to interview General Dornberger and Wernher von Braun.

At Garmisch-Partenkirchen, Robertson found the rocket scientists sunbathing in the Alps. This lovely Bavarian ski resort was the place where Adolf Hitler had hosted the Winter Olympics in 1936. Now the U.S. Army had hundreds of scientists set up in a former military barracks here. The food was plentiful, and the air was fresh and clean. "Mountain springtime," Dieter Huzel recalled in his memoir. "Trees by now had donned their fresh, new green, flowers everywhere as far as one could see from our windows and balconies. Rain was infrequent and almost every day sunbathing was possible on a lawn-covered yard."

Huzel's only complaint was that he didn't receive mail and couldn't make telephone calls.

Wernher von Braun, General Dornberger, and their group had been here since being transferred from CIC headquarters in Reutte. Isolated in the Alps, the two scientists had been frustrating their interrogators, stonewalling and withholding information. Dr. Robertson came to see if he could get any better information out of the scientists. Most of the rocket team was here, including the two men who had stashed the V-2 documents that Staver was now searching for, Dieter Huzel and Bernhard Tessmann. Neither Huzel nor Tessmann had shared with von Braun or Dornberger the fact that they'd told Karl Otto Fleischer the location of the stash in the Dörnten mine. Dornberger and von Braun were under the assumption that they held all the bargaining chips.

The intelligence officer Walter Jessel had sensed something was amiss with Dornberger and von Braun—that the two rocket scientists were playing games. "Control was exercised by Dornberger in the course of the CIOS investigations," Jessel noted in one report. Dornberger's "first instructions [to other scientists], probably under the impression of immediate transfer of the whole group to the United States, were to cooperate fully with the investigators," Jessel explained. But as the days wore on and no deal was offered by the Allies, Jessel watched Dornberger become intractable. "Sometime later, he [Dornberger] gave the word [to the others] to hold back on information and say as little as possible." The scientists were walking on the razor's edge. If they said too much, many of them could be implicated in the slave labor war crimes, as was the case with Arthur Rudolph, Mittelwerk operations director.

For Rudolph, it was best to say as little as possible. He described his time at Garmisch-Partenkirchen as enjoyable because it meant that "the horrible days of fleeing were over." Years later, he described his weeks of internment in the Alps as ones where he could finally "enjoy a few days of relief," but this relief was short-lived owing to his "restless intellect." Rudolph demanded more of himself than a suntan, he later said. "There were already rumors that the Americans would take us to the U.S.A. So, I decided I needed to learn English."

Arthur Rudolph's interrogator saw Rudolph differently than he saw himself. In military intelligence documents, Rudolph was described as "100% Nazi," a "dangerous type." There was a decision to be made: whether to use Rudolph as an intelligence source or to intern him for denazification and investigation into possible war crimes. Denazification was an Allied strategy to democratize and demilitarize postwar Germany and Austria through tribunals in local civilian courts (Spruchkammern) that were set up to determine individual defendants' standings. Each German who was tried was judged to belong in one of five categories, or classes: (1) major offenders; (2) party activists, militarists, and profiteers; (3) individuals who were less incriminated; (4) Nazi Party followers; (5) those who were exonerated. Rudolph's interrogator did not believe a committed Nazi like Arthur Rudolph would make a viable intelligence source, and he wrote, "suggest internment."

Rudolph hoped he would be hired by the Americans. He located a murder mystery in the Garmisch-Partenkirchen library, *The Green Archer,* and attempted to learn English for what he believed, correctly, would be a new job.

Back in Nordhausen, Major Staver was making headway. Working on the new tip from Dr. Robertson, Staver drove to meet with his new source, Karl Otto Fleischer, in a parking lot. This time, Staver had Walther Riedel with him. In the parking lot, Staver demonstratively pulled a notebook from his breast pocket, just as Dr. Robertson had done with him. Staver read aloud a narrative he'd composed, part truth and part fiction. "Von Braun, [Ernst] Steinhoff, and all the others who fled to the south have been interned at Garmisch," Major Staver told his two prisoners. "Our intelligence officers have talked to von Ploetz, General Dornberger, General Rossman, and

General Kammler," Staver said—also partially true. "They told us that many of your drawings and important documents were buried underground in a mine somewhere around here and that Riedel, or you[,] Fleischer, could help us find them," Staver said, which was made up.

Staver told the men that it was in their best interest to think over their next move very carefully. They could cooperate, he said, and give up the location of the V-2 documents. Or they could stonewall and risk being put in prison for withholding information. They had one night to consider the offer. Staver would meet the two men the following morning, in the same parking lot, at exactly 11:00.

When Staver arrived at the rendezvous point the next day, he was disheartened to find Riedel waiting for him but not Fleischer. Even odder, Riedel said he had a message from Fleischer to pass along: Fleischer was waiting for Major Staver in Haynrode, a nearby village, with "some very important news." Staver needed to travel to Haynrode, find a boardinghouse called the Inn of the Three Lime Trees, and ask for the concierge. Was this some kind of a trap, or just another wild goose chase?

Staver and Walther Riedel drove together to the Inn of the Three Lime Trees. There, they met up with the innkeeper, who produced a message from Fleischer. Staver and Riedel were to walk through town, pass down a long alleyway, and head to the edge of the village, where they were to go to the home of a local priest. Staver and Riedel followed the trail, finally arriving at the priest's house. There, in flawless English, the priest told Major Staver that Fleischer would see him soon. Fleischer emerged at the top of the stairs, came down, and asked Staver to follow him outside so the two men could talk privately under an apple tree. There, "in almost inaudible, somewhat apologetic tones[,] Fleischer admitted he had not been completely frank" about the whereabouts of the V-2 document stash, Staver explained. In fact, he knew where they were hidden and "believed he was the only one in Nordhausen who did." But there was a problem, Fleischer said. He described to Staver how the caretaker at the mine had dynamited a wall of rubble over the entrance so no one could find them. This man was an ardent Nazi and would never turn over the documents to an American officer like Major Staver. Fleischer said he'd take Dr. Rees with him to do the job. As unreliable as he was, Staver decided to take Fleischer at his word. He gave him passes that allowed for travel around Nordhausen as well as enough gasoline to get back and forth between Nordhausen and the mine. Fleischer and Rees succeeded in getting the mine's caretaker, Herr Nebelung, to cooperate. Local miners were paid by Fleischer, using money from the

U.S. Army, to excavate through the rubble and retrieve the documents hidden in the mine.

The stash was enormous, the crates weighing more than fourteen tons. Only now there was a new hurdle to overcome. British soldiers were set to arrive in Nordhausen on May 27 to oversee the transition to Red Army rule. This meant that Major Staver had to get the documents out fast. The original agreement between the British and the Americans was that the two Allies would share with one another everything they learned about the V-weapons. If the British found out Staver was planning to secretly ship one hundred V-2 rockets back to the United States, they would likely consider it a double-cross. Major Staver needed to get to Paris. It was the only way he could obtain access to the ten-ton trucks necessary for moving such a large cache in such a short period of time.

Staver assigned a colleague to oversee the Dörnten mine operation while he attempted to hitch a ride to Paris in a P-47 Thunderbolt. The pilot said it was impossible—that the Thunderbolt was a single-seat fighter. Staver said that his mission was urgent and offered to ride in the tiny space behind the pilot's seat. The pilot finally agreed. Avoiding terrible weather higher up, the men flew all the way to Paris at "treetop level" and arrived safely at Orly Field. Staver found a ride down the Champs-Élysées in a U.S. Army jeep.

At Ordnance Headquarters he found the exact man he was looking for, Colonel Joel Holmes, sitting at his desk. As chief of the Technical Division, Colonel Holmes had the authority to grant Major Staver the semitrailers he needed to evacuate the Dörnten mine stash before the British and the Russians arrived.

But Staver had a second plan that he had been conceiving, and, as he later explained, this moment in Paris was his prime opportunity to act. He told Colonel Holmes that there was a third element necessary to make the V-2 rocket program in America a success. Staver had been locating rocket parts and the documents necessary to assemble them correctly. But to make the rockets fly, the Americans needed the German scientists. The army needed to bring these scientists to the United States, Staver explained. Their superior knowledge could be used to help win the war in Japan.

"You write the cable and I'll sign it," Staver remembered Colonel Holmes having said. In Paris, Staver sat down and wrote a cable that would have a huge impact on the future of the Nazi scientist program. "Have in custody over 400 top research development personnel of Peenemünde. Developed V-2," Staver wrote. "The thinking of the scientific directors of this group is 25 years ahead of U.S.... Later version of this rocket should permit launching from Europe to U.S." Given the enormity of this idea in 1945, that a rocket could one day actually fly from one continent to another, Staver pushed: "Immediate action recommended to prevent loss of whole or part of this group to other interested parties. Urgently request reply as early as possible."

The cable was sent to Colonel Trichel's office at the Pentagon, and Major Staver returned to Nordhausen. The documents were loaded up and driven to Paris under armed guard. From there, they were shipped to the Foreign Documents Evaluation Center at the Aberdeen Proving Ground, in Maryland. Special Mission V-2 was declared a success. The U.S. Army Ordnance, Rocket Branch, now had one hundred rockets and fourteen tons of technical documents in its possession. But Staver did not view Special Mission V-2 as entirely complete. He still had his sights set on the rocket scientists themselves.

With the arrival of the Soviets into Nordhausen less than forty-eight hours away, Staver got the approval he'd been waiting for from SHAEF headquarters. He went to Garmisch-Partenkirchen, picked up von Braun, and returned to Nordhausen. The clock was ticking. Staver needed von Braun's help getting every last rocket scientist out of the Harz before the Soviets arrived.

"We landed running," remembered Staver's team member Dr. Richard Porter. Back at Nordhausen, the men got to work locating the scientists who were still living in the area. Staver had a stack of note cards with the names and addresses of the V-2 engineers. He instructed every available U.S. soldier he could find to round up anything with wheels—trucks, motorcycles, and donkey carts—and sent soldiers fanning out across the Harz. Scientists and their families were given an offer. They could be transported out of what would soon become the Russian zone, or they could stay.

Arthur Rudolph's wife was in Stepferhausen, eighty-five miles south of Nordhausen, when a U.S. soldier arrived. "A black GI drove into town in a truck looking for me," remembered Martha Rudolph. "He had a list of names and mine was on it. He told me that if I wanted to leave to get ready. He would be back in 30 minutes to pick me up. My friends all said, 'Go, go—the Russians are coming. Why would you want to stay here?' So I packed up what I could and left on the truck when the GI came back." The Rudolphs' daughter, Marianne, accompanied her to the train station.

The scene at the station was surreal, recalling the horrific transport of prisoners during the war but with the roles reversed, fate and outcome turned upside down. Over one thousand Germans—scientists and their families—stood on the platforms, waiting to fit themselves into boxcars and passenger cars. The train's engine had yet to be attached, and there was no announcement explaining the delay. Tension escalated, but the crowd remained calm until a mob of displaced persons flooded the station. Word had leaked out that German scientists were being evacuated out of Nordhausen in advance of the Red Army's arrival. Suddenly, many other locals wanted out of the Harz, too. The Red Army had a terrible reputation. There were stories of entire units arriving in towns drunk and seeking revenge. At the railroad station, U.S. soldiers were called to the scene. Using the threat of weapons, they prevented any displaced person who was not a scientist or an engineer from boarding the boxcars and passenger cars.

At the eleventh hour, Major Staver and Dr. Porter learned of one last potential disaster that needed to be dealt with. Right before boarding the train, General Dornberger confessed to having hidden his own stash of papers, an ace in the hole had Dornberger been double-crossed by von Braun and left out of the American deal. General Dornberger told Major Staver that he had buried five large boxes in a field in the spa town of Bad Sachsa. The boxes, which were made of wood and lined in metal, contained critical information about the V-2 rocket that would compromise the U.S. Army if it fell into Russian hands. In a last-ditch effort to find Dornberger's secret stash, Staver and Porter set out on a final mission.

The men drove sixty miles to the headquarters of the 332nd Engineering Regiment at Kassel, where they borrowed shovels, pickaxes, three men, and a mine detector. Back in a large field in Bad Sachsa, they searched the ground as if looking for a buried mine. Finally, they located Dornberger's metal-lined cases, which contained 250 pounds of drawings and documents. The stash was loaded onto a truck and driven to an army facility in the American zone.

On their way out of the Harz, Staver and Porter passed by Nordhausen to have one last look. "I wanted to blow up the whole factory at Nordhausen before we pulled out but [I] couldn't swing it legally. I was afraid at the time to do the job 'unofficially,' and have regretted it ever since," Porter recalled. He was referring to the European Advisory Commission decree, signed by General Eisenhower on June 5, 1945, in Berlin, which prohibited the destruction of military research installations in another power's zone.

The Soviets were now heirs to the Harz. Major Staver had succeeded in secretly shipping out enough parts to reassemble one hundred V-2 rockets in America. Still, thousands of tons of rocket parts remained. For all the effort and moral compromise that went into Special Project V-2, the Red Army would now have no shortage of wonder weapons parts to choose from. The underground slave labor factory at

Nordhausen was still virtually intact. Thousands of machine tools sat on the assembly lines ready to manufacture more parts.

After an eleven-day delay, the Russians finally arrived. Leading the pack were technical specialists from Soviet missile program chief Georgy Malenkov's Special Committee for rocket research. For every German scientist that had taken up the U.S. Army's offer to evacuate, between two and ten remained behind. The Soviet secret police began rounding up hundreds of former rocket scientists and engineers and put them back to work. A Soviet guidance engineer named Boris Chertok even managed to move into von Braun's old villa, the one the SS had confiscated from a Jewish businessman a few years before. Chertok oversaw the renaming of the Nordhausen tunnel complex from the Mittelwerk to the Institute Rabe, an abbreviation for Raketenantrieb Bleicherode, or Rocket Enterprise Bleicherode.

Von Braun, eighty scientists, and their families were taken to the town of Witzenhausen, forty miles from Nordhausen, in the American zone. There, they were set up in a two-story schoolhouse and paid to get to work on future rocket plans while Army Ordnance worked on a plan to bring them to the United States, to the Fort Bliss Army Base, in Texas. The Americans had been obsessed with the V-weapons during the war. Now they had the science and the scientists.

*   *   *

In Washington, D.C., officials with the War Department General Staff remained undecided on a policy regarding what to do with Nazi scientists. General Eisenhower's questions about long-term plans had not yet been answered, and Undersecretary of War Robert Patterson was asked to weigh in. Major Staver's cable from Paris regarding the four hundred rocket scientists "in custody" drew attention to the issue. In America, five Nazi scientists had already been secreted into the United States for classified weapons work. Just a few days after the German surrender, the director of naval intelligence successfully lobbied the War Department General Staff to circumvent State Department regulations so that a Nazi guided missile expert named Dr. Herbert Wagner and four of his assistants could begin working on technology meant to help end the war with Japan. The War Department gave approval, and in mid-May Dr. Wagner and his team were flown from Germany to a small airstrip outside Washington, D.C., inside a military aircraft with the windows blackened to keep anyone from seeing who was inside.

During the war, Dr. Herbert Wagner had been chief missile design engineer at Henschel Aircraft. He was the man behind the first guided missile used in combat by the Reich, the Hs 293. This remote-control bomb was the nemesis of the U.S. Navy and the British Royal Navy and had sunk several Allied ships during the war. Not only did the U.S. Navy see glide bomb technology as critically important in the fight in the Pacific, but they saw Dr. Wagner as a man with "knowledge, experience and skills unmatched anywhere in the world."

The perceived importance of having Wagner's expertise in the fight against Japan was illuminated by a dramatic event unfolding in Portsmouth, New Hampshire, just as he and his team arrived. On May 15, 1945, a Nazi submarine, identified in a *New York Times* headline as "the Japan-bound U-234," surrendered itself to the USS *Sutton* in the waters five hundred miles off Cape Race, Newfoundland. Inside the submarine, which was en route to Japan, was a cache of Nazi wonder weapons, "said to contain what few aviation secrets may be left," as well as "other war-weapon plans and pieces of equipment." One of the wonder weapons on board was Dr. Wagner's Hs 293 glider bomb, meant for use against the U.S. Navy in the Pacific. Additionally, there were drawings and plans for the V-1 flying bomb and the V-2 rocket, experimental equipment for stealth technology on submarines, an entire Me 262 fighter aircraft, and ten lead-lined canisters containing 1,200 pounds of uranium oxide—a basic material used in making an atomic bomb. The specifics of the weapons cache were not made public, but the notion that the Nazis had sold the secrets of some of their most prized wonder weapons to their Axis partner Japan was alarming. The scenario was made even more forbidding by the fact that also on board the U-234 was a top Reich scientist whose job it was to teach Japanese scientists how to copy and manufacture these Nazi wonder weapons for themselves.

The scientist in the submarine was Dr. Heinz Schlicke, director of Naval Test Fields at Kiel. To the public he was only identified as a German "technician." In fact, Dr. Schlicke was one of the most qualified Nazi scientists in the field of electronic warfare. His areas of expertise included radio-location techniques, camouflage, jamming and counterjamming, remote control, and infrared. The navy took Dr. Schlicke prisoner of war and brought him to the Army Intelligence Center at Fort Meade, in Maryland.

As for Dr. Wagner, the navy felt it needed to keep him happy so that his work would continue to bear scientific fruit. To soften the reality of his being a prisoner, his incarceration was called "voluntary detention." Wagner and his assistants required a classified but comfortable place to work, the navy noted in an intelligence report, ideally in "an ivory tower or a gilded cage where life would be pleasant, the guards courteous, the locks thick but not too obvious." The navy found what it was looking for in Hempstead House, a great stone castle on Sands Point, on the North Shore of Long Island, that was formerly the home of Daniel and Florence Guggenheim. The 160-acre estate had been donated by the Guggenheims to the navy for use as a training center. With its three stories, forty rooms, and sweeping view of the sea, the navy decided it was an ideal location. Hempstead House was given the code name the Special Devices Center, and Dr. Wagner and his assistants got to work.

There were more problems afloat in Washington, this time coming from the FBI. If Nazi scientists were going to work for the U.S. military, the Department of Justice said it needed to perform background checks. J. Edgar Hoover's FBI looked into Dr. Wagner's past, based on information collected by army intelligence in Europe, and learned that Dr. Wagner had "once belonged to the German SS," the paramilitary wing of the Nazi Party run by Himmler. This meant that Wagner was an ardent Nazi. If he had stayed in Germany, as a former SS member and per the laws of the occupying forces, he would have been arrested and subject to a denazification trial. But the FBI was made aware of how badly the navy needed Wagner, and they labeled him "an opportunist who is interested only in science." The FBI's bigger concern, read an intelligence report, was how much Dr. Wagner had been drinking lately. The FBI did not consider Wagner to be a "drunkard" but blamed his near-nightly intoxication on the recent death of his wife.

The scientist in the submarine, Dr. Heinz Schlicke,

became a prisoner at Fort Meade, where it did not take long for the U.S. Navy to learn how "eminently qualified" he was. Soon, Schlicke was giving classified lectures on technology he had developed during the war. The first was called "A General Review of Measures Planned by the German Admiralty in the Electronic Field in Order to Revive U-Boat Warfare." The navy wanted to hire Schlicke immediately, but State Department regulations got in the way. Schlicke was already in military custody in the United States as a prisoner of war. He would have to be repatriated back to Germany before he could be given a contract to work in the United States, according to the State Department. The saga of the U-234 and its passenger made one thing clear: If the War Department was going to start hiring German scientists on a regular basis, it needed to create a committee to deal with the intricacies of each specific case. Finally, on May 28, 1945, Undersecretary of War Robert Patterson weighed in on the classified subject of hiring Nazi scientists for U.S. military research.

Patterson wrote to the chief of staff to the president, Admiral William D. Leahy. "I strongly favor doing everything possible to utilize fully in the prosecution of the war against Japan all information that can be obtained from Germany or any other source," Patterson wrote. He also expressed concern. "These men are enemies and it must be assumed they are capable of sabotaging our war effort. Bringing them to this country raises delicate questions, including the strong resentment of the American public, who might misunderstand the purpose of bringing them here and the treatment accorded them." Patterson believed that the way to avoid foreseeable problems with the State Department, which handled visa approvals, was to involve the State Department in decision making now. Until a new committee was formed to deal specifically with Nazi scientists, Patterson suggested that the State-War-Navy Coordinating Committee (SWNCC) be in charge.

Patterson's letter to the president's chief of staff, which was not shared with President Truman, prompted a meeting at the Pentagon by the War Department General Staff. The group agreed on a temporary policy. Contracts would be given to a limited number of German scientists "provided they were not known or alleged war criminals." The scientists were to be placed in protective military custody in the United States, and they were to be returned to Germany as quickly as possible after their classified weapons work was complete.

A cable was sent to General Eisenhower, at SHAEF headquarters in Versailles, fulfilling his two-week-old request to be advised on longer-term policy. But what had been decided in Washington, D.C., had very little impact on the reality of what was going on in the European Theater with scientists who had spent years serving Adolf Hitler.

CHAPTER SEVEN

Hitler's Doctors

Duranting the war, physicians with the U.S. Army Air Forces heard rumors about cutting-edge research being developed by the Reich's aviation doctors. The Luftwaffe was highly secretive about this research, and its aviation doctors did not regularly publish their work in medical journals. When they did, usually in a Nazi Party–sponsored journal like *Aviation Medicine* (*Luftfahrtmedizin*), the U.S. Army Air Forces would circumvent copyright law, translate the work into English, and republish it for their own flight surgeons to study. Areas in which the Nazis were known to be breaking new ground were air-sea rescue programs, high-altitude studies, and decompression sickness studies. In other words, Nazi doctors were supposedly leading the world's research in how pilots performed in extreme cold, extreme altitude, and at extreme speeds.

At war's end, there were two American military officers who were particularly interested in capturing the secrets of Nazi-sponsored aviation research. They were Major General Malcolm Grow, surgeon general of the U.S. Strategic Air Forces in Europe, and Lieutenant Colonel Harry Armstrong, chief surgeon of the Eighth Air Force. Both men were physicians, flight surgeons, and aviation medicine pioneers.

Before the war, Grow and Armstrong had cofounded the aviation medicine laboratory at Wright Field, where together they initiated many of the major medical advances that had kept U.S. airmen alive during the air war. At Wright Field, Armstrong perfected a pilot-friendly oxygen mask and conducted groundbreaking studies in pilot physiology associated with high-altitude flight. Grow developed the original flier's flak vest—a twenty-two-pound armored jacket that could protect airmen against antiaircraft fire. Now, with the war over, Grow and Armstrong saw unprecedented opportunity in seizing everything the Nazis had been working on in aviation research so as to incorporate that knowledge into U.S. Army Air Force's understanding.

According to an interview with Armstrong decades later, a plan was hatched during a meeting between himself and General Grow at the U.S. Strategic Air Forces in Europe headquarters, in Saint-Germain, France. The two men knew that many of the Luftwaffe's medical research institutions had been located in Berlin, and the plan was for Colonel Armstrong to go there and track down as many Luftwaffe doctors as he could find with the goal of enticing them to come to work for the U.S. Army Air Forces. As surgeon general, Grow could see to it that Armstrong was placed in the U.S-occupied zone in Berlin, as chief surgeon with the Army Air Forces contingent there. This would give Armstrong access to a city divided into U.S. and Russian zones. General Grow would return to Army Air Forces headquarters in Washington, D.C., where he would lobby superiors to authorize and pay for a new research laboratory exploiting what the Nazi doctors had been working on during the war. With the plan set in motion, Armstrong set out for Berlin.

The search proved difficult at first. It appeared as if every Luftwaffe doctor had fled Berlin. Armstrong had a list of 115 individuals he hoped to find. At the top of that list was one of the Reich's most important aviation doctors, a German physiologist named Dr. Hubertus Strughold. Armstrong had a past personal connection with Dr. Strughold.

"The roots of that story go back to about 1934," Armstrong explained in a U.S. Air Force oral history interview after the war, when both men were attending the annual convention of the Aero Medical Association, in Washington, D.C. The two men had much in common and "became quite good friends." Both were pioneers in pilot physiology and had conducted groundbreaking high-altitude experiments on themselves. "We had some common bonds in the sense that he and I were almost exactly the same age, he and I were both publishing a book on aviation medicine that particular year, and he held exactly the same assignment in Germany that I held in the United States," Armstrong explained. The two physicians met a second time, in 1937, at an international medical conference at the Waldorf-Astoria hotel in New York City. This was before the outbreak of war, Nazi Germany was not yet seen as an international pariah, and Dr. Strughold represented Germany at the conference. The two physicians had even more in common in 1937. Armstrong was director of the Aero Medical Research Laboratory at Wright Field, and Strughold was director of the Aviation Medical Research Institute of the Reich Air Ministry in Berlin. Their jobs were almost identical. Now, at war's end, the two men had not seen one another in eight years, but Strughold had maintained the same high-ranking position for the duration of the war. If anyone knew the secrets of Luftwaffe medical research, Dr. Hubertus Strughold did. In Berlin, Harry Armstrong became determined to find him.

One of the first places Armstrong visited was Strughold's former office at the Aviation Medical Research Institute, located in the fancy Berlin suburb of Charlottenburg. He was looking for leads. But the once-grand German military medical academy, with its formerly manicured lawns and groomed parks, had been bombed and was abandoned. Strughold's office was empty. Armstrong continued his journey across Berlin, visiting universities that Strughold was known to be affiliated with. Every doctor or professor he interviewed gave a similar answer: They claimed to have no idea where Dr. Strughold and his large staff of Luftwaffe doctors had gone.

At the University of Berlin, Armstrong finally caught a break when he came across a respiratory specialist named Ulrich Luft, teaching a physiology class to a small group of students inside a wrecked classroom. Luft was unusual-looking, with a shock of red hair. He was tall, polite, and spoke perfect English, which he had learned from his Scottish mother. Ulrich Luft told Harry Armstrong that the Russians had taken everything from his university laboratory, including the faucets and sinks, and that he, Luft, was earning money in a local clinic treating war refugees suffering from typhoid fever. Armstrong saw opportunity in Luft's predicament and confided in him, explaining that he was trying to locate German aviation doctors in order to hire them for U.S.

Army–sponsored research. Armstrong said that, in particular, he was trying to find one man, Dr. Hubertus Strughold. Anyone who could help him would be paid. Dr. Luft told Armstrong that Strughold had been his former boss.

According to Luft, Strughold had dismissed his entire staff at the Aviation Medical Research Institute in the last month of the war. Luft told Armstrong that Strughold and several of his closest colleagues had gone to the University of Göttingen. They were still there now, working inside a research lab under British control. Armstrong thanked Luft and headed to Göttingen to find Strughold. Whatever the British were paying him, Armstrong figured he would be able to lure Strughold away because of their strong personal connection. There was also a great deal of money to be made pending the authorization of the new research laboratory. Armstrong and Grow's plan meant hiring more than fifty Luftwaffe doctors. There was a lot of work to be had, not just for Strughold but for many of his colleagues as well.

A great drama was now set in motion owing to the fact that there was a second army officer also looking for Dr. Hubertus Strughold—a medical war crimes investigator and physician named Major Leopold Alexander. Dr. Strughold's name had been placed on an army intelligence list of suspected war criminals with the Central Registry of War Criminals and

Security Suspects, or CROWCASS. Major Alexander was on a mission to locate him.

It is not known if Armstrong was aware of the allegations against Strughold and chose to ignore them or if he was in the dark as to Strughold's having been placed on the CROWCASS list. But as Armstrong forged ahead with plans to hire Dr. Strughold and to make him a partner in the U.S. Army Air Forces laboratory, he did so in direct violation of the policy that had just been set by the War Department. German scientists could be hired for U.S. military contract work "provided they were not known or alleged war criminals." The CROWCASS allegations against Dr. Strughold were serious. They included capital war crimes.

The CROWCASS list came out of the immediate aftermath of the German surrender, when public pressure to prosecute Nazis accused of war crimes had reached fever pitch. On May 7, 1945, *Life* magazine published a story on the liberation of Buchenwald, Bergen-Belsen, and other death camps, complete with graphic photographs. This was some of the first documentary evidence presented to the public. When confronted by these ghastly images, people all over the world expressed their outrage at the scale of atrocity that had been committed by the Nazis. Death camps, slave labor camps, the systematic extermination of entire groups of people—this defied the rules of war. The idea of having a war crimes trial appealed to the general public as a means of holding individual Nazis accountable for the wickedness of their crimes.

The group responsible for investigating war crimes was the United Nations War Crimes Commission (UNWCC), located in London and founded by the Allies in 1942 (it was originally called the United Nations Commission for the Investigation of War Crimes). The War Crimes Commission was not responsible for hunting down the criminals; that job was delegated to SHAEF. The War Crimes Commission had three committees: Committee I dealt with lists, Committee II coordinated enforcement issues with SHAEF, and Committee III gave advice on legal points. The commission and its committees worked in concert with CROWCASS, also located in Paris, which was responsible for gathering and maintaining information about suspected war criminals.

After the fighting stopped, SHAEF sent war crimes investigators into the field to locate German doctors with the purpose of interrogating them. One of these investigators was Major Leopold Alexander, a Boston-based psychiatrist and neurologist and a physician with the U.S. Army. Dr. Alexander had been tending to wounded war veterans at a military hospital in England when he learned about his new assignment, just two weeks after the end of the war. With this undertaking, his whole life would change, as would his understanding of what it meant to be a doctor and what it meant to be an American.

Dr. Alexander would unwittingly become one of the most important figures in the Nuremberg doctors' trial. And he would inadvertently become a central player in one of the most dramatic events in the history of Operation Paperclip. That would take another seven years. For now, at war's end, Dr. Alexander accepted his orders from SHAEF, boarded a military transport airplane in England, and headed for Germany to begin war crimes investigative work. His first stop was the Dachau concentration camp. Dr. Alexander did not yet know that it was inside Dachau, in the secret barracks called Experimental Cell Block Five, that Luftwaffe doctors had conducted some of the most barbaric and criminal medical experiments of the war.

On May 23, 1945, Dr. Alexander, 39, was seated inside an American military transport airplane flying into Munich when, approximately fifteen miles north of the airport, his plane circled low and he saw the liberated Dachau concentration camp for the first time. "Surviving inmates were waving and cheering at the plane and you could see that two American field hospitals were set up on the camp grounds," Alexander wrote in his journal late at night. American aircraft brought fresh corned beef, potato salad, and real coffee by the ton to the newly liberated prisoners, many of whom were still too weak to leave the camp. The airplanes also brought doctors and nurses with the American Red Cross and the U.S. Army Typhus Commission and also, on occasion, a medical war crimes investigator like Dr. Alexander. The Dachau concentration camp was the first stop on a long list of Reich medical facilities and institutions that Dr. Alexander was scheduled to visit, locations where medical crimes were suspected of having taken place. With him, Dr. Alexander carried SHAEF instructions that granted him "full powers to investigate everything of interest" and also gave him the authority to "remove documents, equipment, or personnel as deemed necessary."

Fate and circumstance had prepared him for the job. Like Samuel Goudsmit, the scientific director of Operation Alsos, Dr. Alexander had a unique background that qualified him to investigate German doctors and had also made things personal. A Jew, he had once been a rising star among Germany's medical elite. In 1933, Germany's race laws forbade the twenty-eight-year-old physician from practicing medicine any longer. Devastated, he left the country and wound up in America. Now, thirteen years later, he was back on German soil. His former existence here seemed like a lifetime ago.

From as far back as he could remember, Leopold Alexander longed to be a doctor, like his father, Gustav. "One of the strongest unconscious motives for becoming a physician was the strong bond of identification with my father," he once said, explaining the pull toward medicine. Gustav Alexander was an ear, nose, and throat doctor in turn-of-the-century Vienna, a distinguished scholar who published more than eighty scientific papers before Leopold was born. His mother, Gisela, was the first woman awarded a PhD in philosophy from the University of Vienna, the oldest university in the German-speaking world. From a young age Leopold led a charmed life. The Alexanders were sophisticated, wealthy professionals who lived in intellectual-bohemian splendor in a huge house with live peacocks on the lawns. Sigmund Freud was a frequent guest, as was the composer Gustav Mahler. By the time Leopold was fifteen, he was allowed to accompany his father on weekend hospital rounds. The father-son bond grew deep. On weekends they would walk through Vienna's parks or museums, always engaged in lively conversation about history, anthropology, and medicine, as Dr. Alexander later recalled.

In 1929, Leopold Alexander graduated from the University of Vienna Medical School and became a doctor, specializing in the evolution and pathology of the brain. For almost every aspiring physician in

Europe at the time, the goal was to study medicine in Germany, and in 1932 Dr. Alexander was invited to enroll at the prestigious Kaiser Wilhelm Institute for Brain Research in Berlin. There he first rubbed elbows with Germany's leading medical doctors, including Karl Kleist, the distinguished professor of brain pathology who would become his mentor. Alexander focused his studies on brain disorders and began field-work on patients with schizophrenia. Life was full of promise.

Tragedy struck in two cruel blows. In 1932, Gustav Alexander was killed by a former mental patient — murdered in cold blood on the streets of Vienna by a man who, ten years earlier, had been hospitalized and declared insane. The second tragedy occurred in January of 1933, when Adolf Hitler became chancellor of Germany. National Socialism was on the rise. For every Jew in Germany, life was about to change inexorably. Fortune blessed Dr. Alexander. On January 20, 1933, just days before Hitler took power, the ambitious physician prepared to decamp to rural China to study mental illness. "I have accepted an invitation to go for half a year to Beijing Union Medical College in Beijing (China) as an honorary lecturer in neurology and psychiatry," Dr. Alexander wrote to his professors at the Kaiser Wilhelm Institute, promising to return to Germany by October 1, 1933. It was a promise he was never to fulfill.

Within two months of Hitler's taking power, the Nazis initiated a nationwide boycott of Jewish doctors, lawyers, and business professionals. This was followed, in April 1933, by the Reich's Law for the Restoration of the Professional Civil Service. It was now illegal for non-Aryans to work as civil servants, a ban that included every university teaching position throughout Germany. In Frankfurt, where Dr. Alexander had lived, sixty-nine Jewish professors were fired. The news of Germany's radical transformation reached Dr. Alexander in China. The Alexander family lawyer, Maximilian Friedmann, wrote him a letter warning against return. "The prospects in Germany are most unfavorable," Friedmann said. An uncle, Robert Alexander, was even more candid about what was happening in Nazi Germany when he wrote to Dr. Alexander to say that the nation had "succumbed to the swastika." A Jewish colleague and friend, a neurologist named Arnold Merzbach, also penned a letter to Dr. Alexander in despair, telling him that all of their Jewish colleagues in Frankfurt had been dismissed from their university posts. "Our very existences are falling apart," Merzbach wrote. "We are all without hope."

For months, Dr. Alexander lived in denial. He clung to the fantasy that Nazi laws would not apply to him, and he vowed to return once his fellowship ended. In China, Dr. Alexander was in charge of the neurological departments of several field hospitals, where he tended to soldiers with head injuries received on the battlefield. Ignoring the Nazi mandate that now barred Jews from working as doctors or professors, he wrote a letter to Professor Kleist, his mentor in Germany, saying how much he looked forward to returning home. Kleist wrote back to say that his return to Germany was "totally impossible. . . . You as a Jew [since] you have not served as a soldier in the First World War, can not be state employed." In closing, Kleist wrote, "Have no false hopes." The letter may have saved Dr. Alexander's life.

Untethered in China, Alexander was now a nomad, a man without a home to return to. With remarkable ambition and fortitude, he pressed on. As his father had done before him, he wrote and published scientific papers; his were on mental illness, which made him a viable candidate for a fellowship in America. Fortune again favored him when, in the fall of 1933, he learned that he had been awarded a position at a state mental hospital in Worcester, Massachusetts, fifty miles outside Boston. Dr. Alexander boarded an American steamship called *President Jackson* and set sail for America by way of Japan. Out at sea, a strange event occurred. It happened in the middle of the long journey, when his ship was more than a thousand miles from land. A series of violent storms struck, sending passengers inside for days until finally the weather cleared. On the first clear day, Dr. Alexander ventured outside to play shuffleboard. Gazing out across the wide sea, he spotted an enormous single wave traveling with great speed and force, bearing down on his tiny steamship. There was no escape from what he quickly recognized as a tidal wave. Before Dr. Alexander could run back inside the ship, the *President Jackson* was lifted up by this great wave. "The ship traveled up the steep slope very slowly, further and further, until we finally reached the top," he wrote to his brother, Theo. And then, with the ship balancing precariously at the top of the wave, he described the terrifying feeling that followed. "Suddenly there was nothing behind it . . . nothing but a steep descent." The ship began to free-fall, "its nose plunged deep into the water. . . . The impact was harsh, water splashed to all sides, and things fell to this side and that in the kitchen and common rooms." The ship regained its balance, almost effortlessly, and steamed on. "The whole thing happened so unbelievably fast," Alexander wrote. "When it was over, I said to myself now I understand the meaning of the saying, 'the ocean opens up before you and swallows you whole.' "

It did not take long for Dr. Alexander to thrive in America. He was a supremely hard worker. On average he slept five hours a night. Working as a doctor at a New England mental institution was endlessly fascinating to him. He once told a reporter that what interested him most was determining what made men tick. Only a few months after his arrival in New England, he was promoted to a full-time position in the neuropsychiatric ward at Boston State Hospital. While performing hospital rounds in 1934, he met a social worker named Phyllis Harrington. They fell in love and married. By 1938 they had two children, a boy and a girl. A prolific writer, Dr. Alexander published fifty scientific papers. By the end of the decade he'd been hired to teach at Harvard Medical School. He was a U.S. citizen now. Journalists wrote newspaper articles about the "doctor from Vienna," citing his outstanding accomplishments in the field of mental illness. He had a new home; he had been accepted as one of Boston's medical elite.

In December 1941 America went to war. Dr. Alexander joined the fight and was sent to the Sixty-fifth General Hospital in Fort Bragg, North Carolina, and then to an army hospital in England. For the duration of the war, Dr. Alexander helped wounded soldiers recover from shell shock. He also collected data on flight fatigue. After the Germans surrendered, he expected to be sent home. Instead he received his unprecedented order from SHAEF. He was to go to Germany and investigate allegations of Nazi medical crimes. In doing so, he would come face-to-face with former professors, mentors, and fellow students. It was his job to figure out who might be guilty and who was not.

*   *   *

Dr. Alexander's first trip to Dachau did not produce any significant leads despite rumors that barbaric medical experiments had gone on there. On June 5, 1945, he traveled the twelve miles to Munich to visit the Luftwaffe's Institute for Aviation Medicine. This research facility was headed by a radiologist named Georg August Weltz, still working despite Germany's collapse. On paper Weltz was a man of repute. He was gentle-looking, fifty-six years old with a shock of white hair and a wrinkled, sun-tanned face. In their first interview Weltz told Alexander he had worked as a military doctor his entire life, beginning as a Balloon Corps physician in World War I.

Dr. Alexander had a SHAEF dossier on Weltz that revealed Weltz had joined the Nazi Party in 1937, after which he had moved quickly up the Reich's medical chain of command. By 1941 he reported directly to the air marshal of the Luftwaffe, Erhard Milch, who reported to Reichsmarschall Hermann Göring. By war's end, there were only a few men with more medical authority on Luftwaffe issues than Georg August Weltz — one of them being Dr. Hubertus Strughold.

In their first interview, Weltz told Dr. Alexander that it had been his job to conduct a variety of research on methods of saving Luftwaffe pilots' lives. Weltz cited what happened to Luftwaffe pilots in 1940 during the Battle of Britain. Many had been shot down over the English Channel by the British Royal Air Force and had bailed out of their crashing airplanes and initially survived. The fatalities, Weltz explained, often occurred hours later, usually from hypothermia. The bodies of many Luftwaffe pilots had been rescued from the icy waters of the channel just minutes after they had frozen to death. The Luftwaffe wanted to know if, through medical research, doctors could learn how to "unfreeze a man," to bring him back to life. Dr. Weltz told Dr. Alexander that he and his team of researchers had performed groundbreaking research in this area. Weltz declared that they had in fact made a "startling and useful discovery." The results, said Weltz, were simply "astounding."

Dr. Alexander asked, "What kind of results?"

Weltz hesitated to provide details but promised that the U.S. Army would be very interested in the knowledge he possessed. Weltz asked Dr. Alexander if a deal could be made. Weltz said that he was interested in securing a grant with the Rockefeller Foundation. Dr. Alexander explained that he had no authority with any private-sector foundation and that, before anything else, he needed Weltz to tell him about this so-called "astounding" discovery.

Weltz said he and his team had solved an age-old riddle: Can a man who has frozen to death be brought back to life? The answer, Weltz confided, was yes. He had proof. He and his team had solved this medical conundrum through a radical rewarming technique they'd invented. Alexander asked Weltz to be more specific. Weltz said success was dependent upon precise body temperature and duration of rewarming in direct proportion to a man's weight. He was not at liberty to provide data just yet, but the method his team had developed was so effective that the Luftwaffe air-sea rescue service had employed this very technique during the war. The experiments, said Weltz, had been conducted on large animals. Cows, horses, and "adult pigs."

Dr. Alexander was in Germany to investigate Nazi medical war crimes. He got straight to the point and asked Weltz whether human beings had ever been used in these Luftwaffe experiments.

"Weltz explicitly stated that no such [human experiments] had been done by him and that he did not know of any such work having been done," Dr. Alexander wrote in his classified report. But the way in which Weltz responded made Dr. Alexander deeply suspicious of him.

Dr. Alexander was in a conundrum. Should he have Weltz arrested? Or was it best to try to learn more? "I still felt it wiser for the purposes of this investigation not to resort to coercive measures such as an arrest," Alexander explained. He asked Weltz to take him to the laboratory where these experiments on large animals were performed.

Weltz claimed that because of heavy bomb damage in Munich the Luftwaffe's test facility for its rewarming techniques had been moved out to a dairy farm in the rural village of Weihenstephan. Alexander and Weltz drove there together in an army jeep. An inspection of the farm revealed a state-of-the-art low-pressure chamber concealed in a barn. This, Weltz explained, was where Luftwaffe pilots learned performance limits under medical supervision. Also called a high-altitude chamber, the apparatus allowed aviation doctors to simulate the effects of high altitude on the body. But the rewarming facilities were nowhere to be seen. Where were they? Dr. Alexander asked.

Weltz hesitated and then explained. They'd been moved, Weltz said—this time to an estate near Freising, at a government-owned experimental agricultural station. Dr. Alexander insisted on seeing the Freising facility, and the two men got back into the army jeep and drove on. In Freising, Alexander was shown yet another impressive medical research facility, also hidden in a barn, complete with a library and X-ray facilities, all meticulously preserved. But the laboratory was clearly designed to handle experiments on small animals, mice and guinea pigs, not larger animals like cows, horses, and adult pigs. There were records, drawings, and charts of the freezing experiments—all carefully preserved. But, again, they chronicled experiments on small animals, mostly mice. Where had the large animal experiments taken place? Weltz took Alexander to the rear of the barn, behind a stable and into a separate shed located far in the back of the property. There Weltz pointed to two dirty wooden tubs, both cracked.

It was an extraordinary moment, Dr. Alexander would later testify, horrifying in its clarity. Neither of the tubs could possibly fit a submerged cow, horse, or large pig. What these tubs "could fit was a human being," Dr. Alexander said.

The grim reality of Dr. Weltz's Luftwaffe research became painfully clear. "I came away from all these interviews with the distinct conviction that experimental studies on human beings, either by members of this group themselves, or by other workers well known to and affiliated [with] the members of this group, had been performed but were being concealed," Dr. Alexander wrote. Without an admission of guilt, he had only suspicion. To make an arrest, he needed evidence. He thanked Dr. Weltz for his assistance and told him that he would be returning sometime in the future for a follow-up visit.

Dr. Alexander had been on German soil for two weeks, and the deviance of Nazi science overwhelmed him. In a letter to his wife, Phyllis, he described what had become of German science under Nazi rule. "German science presents a grim spectacle," he wrote.

"Grim for many reasons. First it became incompetent and then it was drawn into the maelstrom of depravity of which this country reeks — the smell of the concentration camps, the smell of violent death, torture and suffering." German doctors were not practicing science, Alexander said, but "really depraved pseudoscientific criminality." In addition to investigating crimes committed in the name of aviation medical research, Dr. Alexander was the lead investigator looking at crimes committed in the name of neuropsychiatry and neuropathology. In this capacity, he came face-to-face with the odious and core Nazi belief that had informed the practice of medicine under Hitler's rule. Not only were all people not created equal in the eyes of the Third Reich, but some people were actually not humans at all. According to Nazi ideology, *Untermenschen* — subhumans, as they were called, a designation that included Jews, Gypsies, homosexuals, Poles, Slavs, Russian prisoners of war, the handicapped, the mentally ill, and others — were no different from white mice or lab rabbits whose bodies could thereby be experimented on to advance the Reich's medical goals. "The sub-human is a biological creature, crafted by nature," according to Heinrich Himmler, "which has hands, legs, eyes, and mouth, even the semblance of a brain. Nevertheless, this terrible creature is only a partial human being. . . . Not all of those who appear human are in fact so." German citizens were asked to believe this pseudo-science; millions did not protest. German scientists and physicians used this racial policy to justify torturous medical experiments resulting in maiming and death. In the case of the handicapped and the mentally ill, the *Untermenschen* theory was used by German doctors and technicians to justify genocide.

As a war crimes investigator, Dr. Alexander was one of the first American servicemen to learn that the Reich had first sterilized and then euthanized nearly its entire population of mentally ill persons, including tens of thousands of children, under the Law for the Prevention of Genetically Diseased Offspring. All across southern Germany, one German physician after the next admitted to Dr. Alexander his knowledge of the child euthanasia program. This included Dr. Alexander's own mentor and former professor, the neurologist Karl Kleist. During an interview in Frankfurt, Kleist confessed to Dr. Alexander that he had known of the policy of euthanasia, and he handed over military psychiatric reports that allowed him to circumvent personal responsibility and claim he was just following orders. Kleist was not arrested, but a few days later he was removed from his teaching job. The former teacher and student never spoke again, and it remains a mystery if Dr. Alexander requested that Kleist be fired. Within a few years Kleist's name would appear on a secret Paperclip recruiting list. It is not known if he came to the United States.

Each day brought atrocious new information. "It sometimes seems as if the Nazis had taken special pains in making practically every nightmare come true," Dr. Alexander later told his wife, comparing Reich medicine to something out of a dark German fairy tale. Whereas doctors who knew about the euthanasia program tended to be forthcoming with information — the program was "justified" by a German law kept secret from the general public — it struck Dr. Alexander that criminal human experiments by Luftwaffe doctors, like the freezing experiments Weltz was involved in, appeared to have been more skillfully concealed. If Dr. Alexander wanted to learn the facts about what Luftwaffe doctors had been up to during the war, he knew that he had to understand the bigger picture. And he also had to determine where else the crimes might have taken place. The best way to do this was to interview the man nearest the top, Dr. Hubertus Strughold. Strughold had directed the Aviation Medical Research Institute for the Luftwaffe for ten of the twelve years of Nazi Party rule. When Dr. Alexander learned that Dr. Strughold was in Göttingen, in the British zone, he headed there.

En route to Göttingen, Dr. Alexander had a fortuitous break. "A curious coincidence played into my hands," he wrote. "On my way to Göttingen...while having dinner in the Officers' mess of the 433rd A. A. Bn. [Army Battalion] in camp Rennerod, Westerwald, I happened to meet another casual guest, an army chaplain, Lieutenant Bigelow. In the course of our conversation, Lt. Bigelow told me he was quite eager to get my ideas about rather cruel experiments on human beings, which had been performed at Dachau concentration camp. He had learned of them from a broadcast a few days earlier when ex-prisoners of Dachau had talked about these grim experiments over the Allied radio in Germany." This was exactly the kind of lead Dr. Alexander was looking for, and he asked Lieutenant Bigelow to share with him anything else he remembered from the radio report.

Lieutenant Bigelow told Alexander that as a member of the clergy he had ministered to many war victims. He had heard frightful stories about what had gone on in the medical blocks at the concentration camps. But nothing compared to what he had heard in that radio report. Doctors at Dachau had frozen people to death, in tubs of ice-cold water, to see if they could be unfrozen and brought back to life. These experiments were apparently meant to simulate conditions that Luftwaffe pilots went through after they'd been shot down over the English Channel, Bigelow said. Dr. Alexander now had a solid new lead. To his mind, the experiments Bigelow was referring to sounded "strikingly similar to the animal experiments performed by Dr. Weltz and his group" at the Freising farm. Was the Luftwaffe involved in medical research at the concentration camps? Dr. Alexander asked the chaplain if he had caught any of the names of the doctors involved in the Dachau medical crimes. Bigelow said that he couldn't recall but that he was certain he had heard that the Luftwaffe was involved. More determined than ever to investigate, Dr. Alexander continued on to Göttingen to interview Dr. Hubertus Strughold.

At the Institute for Physiology in Göttingen, Dr. Alexander located Strughold and arranged to interview him, getting straight to the point. Dr. Alexander told Dr. Strughold about the radio report claiming that freezing experiments had been conducted by the Luftwaffe at Dachau. Did Strughold, as the physician in charge of aviation medical research for the Luftwaffe, know about these criminal experiments at Dachau? Dr. Strughold said that he had learned of the experiments at a medical symposium he attended in Nuremberg in October 1942. The conference, called "Medical Problems of Sea Distress and Winter Distress," took place at the Hotel Deutscher Hof and involved ninety Luftwaffe doctors. During that conference, Strughold said, a man named Dr. Sigmund Rascher presented findings that had been obtained from experiments performed on prisoners at the Dachau concentration camp. This was the same man "who had been mentioned over the allied radio the other day," Strughold said. He called Rascher a fringe doctor whose only assistant at Dachau was his wife, Nini. Both Raschers were now dead.

Did Strughold approve of these experiments? Strughold told Dr. Alexander that "even though Dr. Rascher used criminals in his experiments, he [Strughold] still disapproved of such experiments in non-consenting volunteers on principle." Dr. Strughold promised Dr. Alexander that within his institute in Berlin he had "always forbidden even the thought of such experiments...firstly on moral grounds and secondly on grounds of medical ethics." Alexander asked Strughold if he knew of any other Luftwaffe doctors who had been involved in human experiments at Dachau. Strughold said, "Any experiments on humans that we have carried out were performed only on our own staff and on students interested in our subject on a strictly volunteer basis." He did not reveal that a number of doctors on his staff had visited Dachau regularly and worked on research experiments there.

Also in Göttingen, Dr. Alexander interviewed several other doctors who had worked for Strughold, asking each of them specific questions about human experiments. Each doctor told a strikingly similar story. Dr. Sigmund Rascher was to blame for everything that went on at Dachau and now Rascher was dead. But one man, a physiologist named Dr. Friedrich Hermann Rein, let an important clue slip. Dr. Rascher had been an SS man, Rein said. This information gave Dr. Alexander a significant new piece of the puzzle he did not have before, namely, that the SS was also involved in the concentration camp freezing experiments. This was a revelation.

The day after this disclosure, Dr. Alexander received further extraordinary, related news. "I learned that the entire contents of Himmler's secret cave in Hallein, Germany [*sic*], containing a vast amount of miscellaneous specially secret S.S. records, had recently been discovered and taken" to the Seventh Army Document Center in Heidelberg. This huge trove of papers had been discovered by soldiers hidden away in yet another cave. The papers had been stamped with the unmistakable logo of the SS, and they bore Himmler's personal annotations, drawn in the margins in the green pencil he liked to use. Dr. Alexander set out for the Document Center to see what he could glean from the files. These papers would turn out to be among the war's most incriminating discoveries in a single document find.

In Heidelberg, Himmler's documents were being inventoried and sorted out when Dr. Alexander first arrived. One of the men tasked to the job was Hugh Iltis, the son of a Czech doctor, who, with his family, fled Europe in advance of the genocide. Iltis was a nineteen-year-old American soldier fighting on the front lines in France during the last months of the war when, he recalls, "a car showed up and an officer leapt out and pointed at me, then shouted 'You, come with me!'" Iltis climbed into the car and sped away from the battlefield with the officer. Someone had learned that Hugh Iltis was a fluent German speaker (and perhaps that his father was a leading geneticist and anti-Nazi). Iltis was needed in Paris to translate captured Nazi documents, and the work kept on coming. Now, six months later, here he was in Heidelberg documenting atrocities for the War Crimes Commission. His discovery of the Himmler papers—it was Iltis who identified how important they were—would also become the most important collection of documents on Nazi human experimentation to be presented at the doctors' trial.

Alexander told Iltis what it was that he was looking for: documents written by Dr. Sigmund Rascher that involved experiments on humans. Together, the two men broke the original seals on the innocuously named Case No. 707-Medical Experiments, papers that turned out to include years of correspondence between Rascher and Himmler.

"The idea to start the experiments with human beings in Dachau was obviously Dr. Rascher's," Alexander explained in his classified scientific intelligence report. But as Alexander learned from the papers, Rascher was far from the only Luftwaffe doctor involved. Nor were the human experiments limited to freezing experiments. Even more damning, Dr. Alexander learned that one of Dr. Strughold's closest colleagues and his coauthor, a physiologist named Dr. Siegfried Ruff, was in charge of overseeing Rascher's human experiments at Dachau. This was stunning news. "Dr. Ruff, and his assistant Dr. Romberg, joined forces [with Rascher] and arrived in Dachau with a low pressure chamber which they supplied," Alexander wrote in his report. This low-pressure chamber was used for a second set of deadly experiments involving high-altitude studies. Sitting inside the Seventh Army Documentation Center reading the Himmler papers, Dr. Alexander realized that Dr. Strughold had lied to him when he had said that the only Luftwaffe doctor involved in the Dachau experiments had been the "fringe doctor" Rascher. In fact, Strughold's friend and colleague Dr. Ruff was deeply implicated.

Most disturbing to Alexander were a group of photographs showing what happened in the course of the experiments as healthy young men — classified by the Nazis as *Untermenschen* — were strapped into a harness inside the low-pressure chamber and subjected to explosive decompression. These photographs, astonishing in their sadism, were essentially before, during, and after pictures of murder in the name of medicine.

Other photographs among the Himmler papers documented the freezing experiments as they were being conducted at Dachau. Rascher's experiments were by no means the solo act of one depraved man. There were photographs of yet another of Dr. Strughold's Luftwaffe colleagues, Dr. Ernst Holzlöhner, holding prisoners down in tubs of icy water while their body temperatures were recorded as they died. It is believed that Rascher's wife, Nini, took the photographs.

In a classified CIOS report, Dr. Alexander expressed doubts about the veracity of Dr. Strughold's earlier testimony from their first interview in Göttingen. The Dachau experiments were joint endeavors by the Luftwaffe and the SS, and, despite Strughold's denials, several aviation doctors on his staff, including individuals who reported to directly to him, were named in the Himmler papers. "Strughold at least must have been familiar with the parts played by his friend and co-worker Ruff," Dr. Alexander wrote. In his report, he advised SHAEF that, while he could not yet say if Dr. Strughold was directly involved in the death experiments, clearly Reich medical crimes "were still being covered up by" him.

On June 20, Alexander headed back to Munich to confront Dr. Weltz. Instead, he found a colleague of Weltz's, a Dr. Lutz, who broke down and confessed that he'd been aware of the human experiments, but that they'd been conducted by his team members, not him. Lutz claimed to have been offered the "human job," by Weltz, but declined to accept, on grounds that he was "too soft."

Before confronting Strughold, Dr. Alexander first returned to Dachau to locate eyewitnesses. There, he found three former prisoners who offered testimony. John Bauduin, Oscar Häusermann, and Dr. Paul Hussarek had managed to stay alive at the concentration camp by working as orderlies for the SS. After Dachau was liberated, the three men chose to stay behind so as to help investigators piece together medical crimes. They formed a group, calling it the Committee for the Investigation of SS Medical Crimes. From them, Dr. Alexander learned that the experiments had been conducted on Jews, Gypsies, homosexuals, and Catholic priests in the secret, freestanding barracks called Experimental Cell Block Five. "In general, the death of prisoners transferred to Block 5 was expected within 2–3 days," testified John Bauduin. The second witness, Dr. Hussarek, a Czech scholar sent to Dachau for committing "literary crimes," told Dr. Alexander that "only a few experimental subjects survived the low pressure experiments. Most were killed." All three men agreed that only one individual was known to have survived the experiments, a Polish priest named Leo Michalowski.

Father Michalowski's testimony provided a critical missing link in the medical murder experiments and how they were so skillfully concealed. Luftwaffe reports used the words "guinea pigs," "large pigs," and "adult pigs" as code words for their human subjects. In one of Weltz's papers confiscated by Dr. Alexander, entitled "Alcohol and Rewarming," Weltz wrote that "shipwreck experiments had been [simulated] in large pigs." The pigs were placed in tubs of water with blocks of ice and given alcohol to see if a rewarming effect occurred. The results, wrote Weltz, showed that "alcohol in pigs does not increase or accelerate the loss of warmth." In sworn testimony, Father Michalowski described what had been done to him at Dachau: "I was taken to room No. 4 on Block 5.... I was dropped in the water in which ice blocks were floating. I was conscious for one hour...then given some rum." In Weltz's paper the word "large pig" really meant "Catholic priest."

On June 22, Dr. Alexander returned to the Heidelberg Document Center to locate more information with the help of Hugh Iltis. Armed with new details and key words culled from survivor testimony, Dr. Alexander found what Dr. Rascher had called his "experiment reports." These charts, Alexander noted, were a scientific chronicle of medical murder. Rascher had also had bigger plans. He was working with the SS to have the aviation medicine experiments relocated from Dachau to Auschwitz. "Auschwitz is in every way more suitable for such a large serial experiment than Dachau because it is colder there and the greater extent of open country within the camp would make the experiments less conspicuous," Rascher wrote. Dr. Alexander also learned about a grotesque "motion picture of the record of the experiments" that had been shown at a private screening at the Air Ministry, at the behest of Himmler. The Luftwaffe doctor overseeing this event was yet another close colleague of Dr. Strughold's, a physiologist and government official named Dr. Theodor Benzinger. Dr. Alexander was unable to locate Benzinger but noted the name in his report. He then returned to Göttingen to interview Strughold a second time to see if he could confirm that Strughold had been lying to him.

In Göttingen, things had changed. Investigators working on scientific intelligence projects for the U.S. Army Air Forces (AAF) and the Royal Air Force (RAF) had interviewed many of the Luftwaffe doctors, including Dr. Strughold. Their conclusions were remarkably different from Dr. Alexander's. None of the RAF or AAF officers had traveled to the Document Center in Heidelberg to read the Himmler files. Instead, their reports were meant to serve and support Armstrong and Grow's secret, new research lab.

RAF Wing Commander R. H. Winfield wrote in his report, "Strughold was the mainspring of German Aviation Medical Research" and had a large staff of colleagues, including Dr. Siegfried Ruff, who all appeared to have "suffered tremendously from their isolation during the war years." Winfield, having no idea that Dr. Ruff had been the person in charge of overseeing Rascher's work at Dachau, stated that his "interrogations [of Ruff] revealed very little information not already known to the Allies." Winfield saw Dr. Strughold as a patrician figure, "considerably disturbed about the welfare of his staff who, unable to evacuate Berlin, were now threatened by the Russians."

Representing the U.S. Army Air Forces was Colonel W. R. Lovelace, an expert in high-altitude escape and parachute studies. The following decade, Lovelace would become famous as the physician for NASA's Cold War–era Project Mercury astronauts. For his confidential CIOS report, entitled "Research in Aviation Medicine for the German Air Force," Lovelace interviewed Dr. Strughold and many of his colleagues, including the freezing expert, Georg Weltz. Like Winfield, Lovelace was in the dark about the medical murder experiments going on inside the concentration camps. He saw Weltz's research as benign and dedicated five pages of his CIOS report to praising his studies on "rapid rewarming of the cooled animal." Lovelace was particularly impressed by the fact that Weltz had frozen a "guinea pig" to death and was still able to record a heartbeat after death. "[T]he heartbeat may continue for some time if the animal is left in the cold," Lovelace wrote in summation of Weltz's findings.

Unlike Dr. Alexander, Colonel Lovelace was able to interview Strughold and Ruff's colleague Dr. Theodor Benzinger, the high-altitude specialist who ran the Reich's Experimental Station of the Air Force Research Center, Rechlin, located north of Berlin. This was the same Dr. Benzinger who had overseen for Himmler the film screening at the Reich Air Ministry, in Berlin, of Dachau prisoners being murdered in medical experiments. And while Dr. Alexander had this information, Colonel Lovelace had no idea. Lovelace was particularly interested in Benzinger's work involving "high altitude parachute escapes," for which Benzinger had gathered much data and produced "studies in reversible and irreversible deaths." Benzinger told Lovelace that he performed his studies on rabbits.

Finally, Lovelace interviewed Dr. Konrad Schäfer, a chemist and physiologist whose wartime efforts to render salt water drinkable made him famous in Luftwaffe circles. With high praise from RAF and AAF officers, Doctors Ruff, Benzinger, and Schäfer were now each being considered for leading positions at the new research lab.

It was the end of June 1945, and Dr. Alexander's allotted time in the field as a war crimes investigator had drawn to a close. He was ordered back to London, where he would type up seven classified CIOS reports totaling more than fifteen hundred pages. Two weeks after Alexander left Germany, the chief of the Division of Aviation Medicine for the Army Air Forces, Detlev Bronk, and an AAF expert on the psychological and physiological stresses of flying named Howard Burchell, arrived in Germany to evaluate progress on the new research laboratory envisioned by Armstrong and Grow. Bronk and Burchell interviewed many of the same doctors and determined that they were all good candidates for the AAF center. Unlike Wing Commander Winfield and Colonel Lovelace, Bronk and Burchell had been made aware of some of the controversy surrounding Strughold and his Luftwaffe colleagues. In a joint report, they explained, "[N]o effort was made to assess [the doctors'] political and ethical viewpoints, or their responsibility for war crimes." They also concluded, "Strughold was not always quite honest in presenting the true significance of the work which he supported." But Bronk and Burchell stated that it was their position that army intelligence was better qualified to determine who was inadmissible "for political reasons" and who could be hired. As it turned out, military intelligence objected to hiring Dr. Benzinger and Dr. Ruff, on the grounds that both men had been hard-core Nazi ideologues. But in the following month, army intelligence determined that the doctors' work at Heidelberg would be "short term," and both men were cleared for U.S. Army employment.

A deal was made between the U.S. Army Air Forces and Dr. Strughold. He would serve with Armstrong as cochairman of a Top Secret research center that the AAF was quietly setting up at the former Kaiser Wilhelm Institute in Heidelberg, and to be called the Army Air Forces Aero Medical Center. No one outside a small group could know about this controversial project because, per JCS 1076, no foreign power was permitted to carry out military research of any kind in Germany, including in medicine.

Dr. Strughold handpicked fifty-eight Luftwaffe doctors for the research program, including Dr. Siegfried Ruff, Dr. Theodor Benzinger, and Dr. Konrad Schäfer — the first Nazi doctors to be hired by the U.S. Army Air Forces. In Munich, Dr. Georg Weltz was arrested and sent to an internment facility for processing. From there, he would be sent to the prison complex in Nuremberg to await trial.

In less than two years, many of the Nazi doctors chosen by Dr. Strughold would quietly begin their secret journeys to the United States.

CHAPTER EIGHT

Black, White, and Gray

In Washington, with policy now informally set, the debate over the Nazi scientist program became intense inside the State-War-Navy Coordinating Committee (SWNCC). Like its successor organization, the National Security Council, SWNCC acted as the president's principal forum for dealing with issues related to foreign policy and national security. The State Department was vocal in its opposition to the program. Exacerbating the situation for the State Department was a parallel issue it had recently become embroiled in. South American countries, Argentina and Uruguay in particular, were known to be giving safe haven to Nazi war criminals who had escaped from Germany at the end of the war. The State Department had been putting pressure on these countries to repatriate Nazis back to Europe to face war crimes charges. If it came out that the State Department was providing not only safe haven but employment opportunities for Nazi scientists in the United States, that would be cause for an international scandal. And while some generals and colonels in the War Department were decidedly for the Nazi science programs, others were fundamentally opposed to the idea. A secretly recorded conversation between two generals at the Pentagon summed up the conflict that the very idea of German scientists working for the U.S. military created.

"One of the ground rules for bringing them over is that it will be temporary and at the return of their exploitation they will be sent back to Germany," said one general, whose name was redacted.

The second general agreed. "I'm opposed and Pop Powers [a nickname for a Pentagon official] is opposed, the whole War Department is opposed," he said. To "open our arms and bring in German technicians and treat them as honored guests" was a very bad idea.

The Department of Justice was not happy about the voluminous workload that background checks on former enemy aliens would require. The Department of Labor was concerned about laws governing alien labor, and the Department of Commerce was concerned about patent rights. In an attempt to ease the contention, Undersecretary of War Robert Patterson sent a memorandum to the War Department General

Staff stating that the person to mediate these issues was John J. McCloy, the assistant secretary of war and chairman of SWNCC.

John J. McCloy would become an especially significant player in Operation Paperclip starting in 1949. But now, in the summer of 1945, he wore two hats related to the issue of Nazi scientists. On the one hand, Undersecretary of War Robert Patterson had put McCloy in charge of coordinating policy regarding Nazi scientists coming to the United States to work. On the other hand, Patterson's boss, Secretary of War Henry Stimson, had given McCloy the job of helping to develop the war crimes program. McCloy's position regarding the exploitation of Nazi science and scientists was clear. He believed that the program would help foster American military superiority while engendering economic prosperity. To McCloy, those ends justified any means. It was not that McCloy believed that the Nazis should go unpunished, at least not in the summer of 1945. For that, McCloy was a strong supporter of the International Military Tribunal (IMT) and the idea of a war crimes trial. But he was someone who saw these two categories as black and white. There were scientists and there were war criminals.

In McCloy's eyes, a war criminal was a Himmler, a Hess, a Göring, or a Bormann. Scientists, like industrialists, were the backbone of a healthy economy in this new, postwar world. In the summer of 1945, McCloy was regularly briefed on the capture and arrest of these war criminals as they were rounded up and taken to a Top Secret interrogation facility in Luxembourg, code-named Ashcan, where they would be squeezed for information before facing judgment at Nuremberg.

John Dolibois, an officer with Army Intelligence, G-2, the Collecting and Dissemination Division, spent a significant portion of the last eight months of the war watching and rewatching *Triumph of the Will,* the three-hour-long Nazi propaganda film by Hitler's favorite filmmaker, Leni Riefenstahl. Every Thursday night, inside a screening room at Camp Ritchie, America's Military Intelligence Training Center, located eighty miles north of Washington in the Catoctin Mountains, the twenty-six-year-old Dolibois used the film to teach German order of battle and Nazi Party hierarchy to colonels, generals, and intelligence officers preparing to go off to war.

The *Triumph of the Will* documentary was an ideal teaching tool and enabled Dolibois to point out to his students how individuals within the Nazi Party hierarchy spoke and gestured, what insignia they wore, who was subordinate to whom. Between the hateful speeches and the endless parades, the fawning inner circle and the Nuremberg rallies, John Dolibois had become so familiar with Hitler's inner circle that he could almost recite their speeches himself.

He enjoyed teaching, but, like so many dedicated Americans of his generation, Dolibois wished to see action overseas. There was a tinge of envy as well. He stayed in touch with his former colleagues from Officer Candidates School, most of whom had been sent to Europe months ago. Many had already been promoted to captains and majors. As the war in Europe drew to a close, John Dolibois had accepted that he was, in all likelihood, not going to be sent overseas as part of an interrogation team, called an IPW team, to interview newly captured prisoners of war. Then, on Easter Sunday, April 1, 1945, he received orders to ship out with the next detachment. Steaming out of the New York harbor only days later, he was standing on the deck of the *Île de France* when someone handed him a telegram. He'd been promoted to first lieutenant.

Things moved fast after he'd crossed the Atlantic. On April 13, Dolibois's ship landed in West Scotland. Every vessel in the harbor was at half-mast; President Roosevelt had died the day before. A quick train trip to London bore witness to "appalling devastation." Piles of rubble filled both sides of every street. Dolibois's channel crossing took place under a full moon, and he was grateful to arrive in the war-torn port at Le Havre, France, without incident. "Up until then our move from Camp Ritchie to Le Havre had been well orchestrated," explains Dolibois. "Now chaos set in." Driving into Munich, destroyed vehicles and weaponry littered the road. In the clearings in the woods sat small fleets of wrecked Luftwaffe airplanes, their wings torn off and their fuselages pockmarked with holes. Corpses rotted in ditches. "Suddenly the war was very real," Dolibois recalls.

His first assignment was at the Dachau concentration camp, just two days after its liberation. Dolibois had been sent to Dachau to look over groups of captured German soldiers to see if important generals, party officials, or scientists were hiding out among the crowd. "Primarily I was to watch for high ranking Nazis in disguise," remembers Dolibois. "We had reports that many of them were passing themselves off as ordinary German soldiers, thus hoping to be overlooked in the confusion and to disappear." His job was to intuit the meaning of certain manners of walk, greeting, and speech. Dolibois was on the lookout for anyone who might be useful to the Allies for a more detailed interrogation at a facility elsewhere.

At Dachau, John Dolibois scoured faces in the crowd for telltale marks, things that could not be hidden. The most obvious among them were the dueling scars of the Nazi elite. But at Camp Ritchie Dolibois had also become an expert in signs of concealment. Recently shaved facial hair or patches pulled off uniforms were indicators that a man had something to hide. True expertise, Dolibois knew, lay in recognizing nuance.

After a few days at Dachau, Dolibois received another assignment. He proceeded to Central Continental Prisoner of War Enclosure Number 32, or CCPWE No. 32. The mission he was now on was classified Top Secret. Everyone he asked about CCPWE No. 32 said that they had never heard of it before. When Dolibois's driver left the borders of Germany and began heading into Luxembourg, Dolibois became overwhelmed with memories. Luxembourg, of all places—how capricious to be on assignment here. John Dolibois was born in Luxembourg. He had moved to America when he was a twelve-year-old boy, with his father; his mother had died in the great influenza pandemic. Driving into Luxembourg in 1945, Dolibois was seeing his native country for the first time in fourteen years. As his army jeep made its way into a little spa town called Mondorf-les-Bains, images of his youth flooded his mind. He recalled Mondorf's "beautiful park, a quiet stream on which one could row a boat, lots of old trees, and acres of flowers." Mondorf was built a few miles from the Moselle River in antiquity, developed by the Romans as a health resort. It was known for its restorative qualities, its mineral baths and fresh air. How different it all looked now, another small city devastated by war. Most homes and shops had been plundered or destroyed. Driving along the main boulevard, Dolibois observed how the façades of many houses had been blown off. He could see people carrying on with their lives inside of what was left of their homes.

Only when his jeep pulled up to its destination did Dolibois realize that he'd arrived at the Palace Hotel. It was unrecognizable to him. A fifteen-foot-high fence ran around the main building, on top of which was a double-stringed curl of barbed wire. There was a second fence that appeared to be electrified. Camouflage netting hung from panels of fencing. Wide canvas sheets had been strung from tree to tree. Huge klieg lights illuminated the place. There were four guard towers, each manned by American soldiers holding powerful machine guns. Not even in photographs had John Dolibois seen an Allied prison facility in the European war theater as heavily fortified as this place was. At the front gate there was a jeep, parked and with its engine turned off. A stern-faced sergeant sat inside. His name tag read "Sergeant of the Guard, Robert Block." Block addressed Dolibois with a nod.

"Good afternoon, sergeant," Dolibois said. "I'm reporting for duty here."

Block just stared at him. Dolibois recalled asking what kind of place this was. What was going on inside?

Block said he had not been inside.

There was a long, uncomfortable pause. Finally

Block spoke. "To get in here you need a pass signed by God." He nodded at the prisoner-of-war facility behind him. "And have somebody verify the signature."

Dolibois handed over his papers. After Block looked at them, the gate swung open and Dolibois was waved inside. In spite of its fortifications, the Palace Hotel remained surprisingly unscathed by war. The boomerang-shaped building was five stories tall. The fountain at the front entrance lacked water, its stone-carved nymph rising up from an empty pool. Inside the hotel foyer Dolibois was greeted by two guards. A third soldier handed him a key and pointed up a flight of stairs. He told Dolibois to leave his things in room 30, on the second floor.

"I climbed up the stairs, located room 30 and let myself in with the key he had given me. It was an ordinary hotel room," remembers Dolibois, "with rather noisy wallpaper." Inside, the fancy light fixtures and plush furniture of a grand hotel had been replaced by a folding table, two chairs, and an army cot. Dolibois unpacked his duffel bag. There was a knock on the door.

Ashcan may have been heavily fortified on the exterior, but inside the facility, the prisoners were free to roam around. Dolibois opened the door and stood face-to-face with a large man dressed in a ratty pearl-gray uniform with gold braids on the collars and gold insignia on the shoulder pads. He held a pair of trousers draped over one arm. Clicking his heels, he nodded and introduced himself as if he were at a party, not in a prison. The man opened his mouth and barked, "Göring, Reichsmarschall!"

So this was Hermann Göring. Dolibois recognized him immediately from so many screenings of *Triumph of the Will*. Here was the man in flesh and blood. Göring was arguably the most notorious of Hitler's inner circle still alive. Former commander in chief of the Luftwaffe. Director of the Four Year Plan. Hitler's long-acknowledged successor until the perceived betrayal at the very end. It was Hermann Göring who ordered security police chief Reinhard Heydrich to organize and coordinate plans for a "solution to the Jewish question."

"At once I understood my assignment," recalled Dolibois. He was here in Luxembourg to interrogate the highest-ranking war criminals in the Nazi Party. This was not a Nazi propaganda film. The individuals who had so populated his mind and his teaching at Camp Ritchie for the past eight months were right here. And they were all prisoners now.

Göring stood before Dolibois, panting.

Göring said he had been unfairly tricked by his captors. "He had been told he was going to a palatial spa," Dolibois explained. When Göring arrived at Ashcan with his valet, Robert Kropp, he was expecting a vacation. He brought along eleven suitcases and twenty thousand Paracodin pills, and had made sure his toenails and fingernails had been varnished to a bright red shine for his stay. That the spa at Mondorf had lost its chandeliers and been turned into a maximum-security prison complex was not what Göring had in mind. His mattress was made of straw, Göring barked at Dolibois. He didn't have a pillow. A man of his rank deserved more.

Dolibois looked at Göring. Made a mental note.

"Are you by chance a welfare officer who will see to it that we are treated correctly, according to Articles of War?" Göring asked Dolibois.

In this question, Dolibois saw opportunity as an interrogator. "Yes," he said. He would be working "along those lines." Göring was pleased. "He made a great show again of heel-clicking, bowing and taking his 280 pounds out of my room."

Göring returned to his fellow prisoners. He told the other Nazis about the new officer's arrival and his responsibilities to see better treatment for all of them. Suddenly everyone wanted to speak with First Lieutenant John Dolibois.

CCPWE No. 32 was filled with Nazi "Bonzen," the big wheels, as Dolibois and the other interrogators called them. Hans Frank, the "Jew-Butcher of Cracow," arrived at Ashcan on a stretcher, in silk pajamas drenched in blood. He had tried to kill himself by slashing his own throat. Frank was captured with his thirty-eight-volume diary, written during the war, a damning confession of many crimes he was guilty of. "Dark-eyed and balding," noted Ashcan's commandant, Colonel Burton Andrus, Frank had "pale hairy hands." Other prisoners included members of the former German General Staff: Field Marshal Wilhelm Keitel, chief of the Oberkommando der Wehrmacht (OKW), or Armed Forces High Command; General Alfred Jodl, Keitel's chief of operations; Grand Admiral Karl Dönitz, commander of submarines and commander in chief of the German navy; Field Marshal Albert Kesselring, former chief of Armed Forces Italy and later Supreme Commander West; Joachim von Ribbentrop, foreign minister; and Albert Speer, minister of armaments and war production. These were the men who personally helped Hitler plan and execute World War II and the Holocaust—those who hadn't escaped, perished, or committed suicide.

"In a second circle, or clique, there were the real Nazi gangsters," Dolibois explained, "the old fighters—who had been with Hitler at the beginning of his rise to power." Among this group were Robert Ley, Labor Front leader; Julius Streicher, editor of the anti-Semitic newspaper and propaganda tool *Der Stürmer;* Alfred Rosenberg, Nazi philosopher; Arthur Seyss-Inquart, the man who betrayed Austria and became

Reichskommissar of Holland; and Wilhelm Frick, former minister of the interior and Reichsprotektor of Bohemia-Moravia.

Stripped of their power, small details spoke volumes to Dolibois. Göring was terrified of thunderstorms. Keitel was obsessed with sunbathing and staring at his reflection in Ashcan's only mirror, in its entrance hall. Robert Ley was repeatedly reprimanded for masturbating in the bathtub. Joachim von Ribbentrop, named by the Nazi Ministry of Propaganda as the best-dressed man in Germany for nine consecutive years, was a lazy slob. Day in and day out, John Dolibois interviewed them.

"Almost all the men at Ashcan were eager to talk," Dolibois recalls. "They felt neglected if they hadn't been interrogated by someone for several days.... Their favorite pastime was casting blame." The greatest challenge for Dolibois and his fellow interrogators was determining, or trying to determine, who was lying and who was telling the truth. "Cross-examination. Playing one prisoner off the other," according to Dolibois, was a tactic that worked best.

"Often, I was taken into their confidence when they needed a shoulder to cry on," Dolibois explains. "At Mondorf, they still couldn't believe they would be tried for their crimes."

CHAPTER NINE

Hitler's Chemists

At war's end, the staff of the U.S. Chemical Warfare Service had their sights set on bringing Hitler's chemists to the United States. The service saw unbridled potential in making the Nazis' nerve agent program its own and was willing to go to great lengths to obtain its secrets. Less than one month after British tanks rolled into the Robbers' Lair and found the enormous cache of tabun-filled bombs in the forests of Münster-Nord, the Chemical Warfare Service had obtained a sample of the nerve agent and was analyzing its properties in its Development Laboratory in the United States. Work began on May 15, 1945, and took two weeks to complete. The analysis revealed that tabun was a revolutionary killer that could decimate enemy armies. General William N. Porter, chief of the Chemical Warfare Service, requested that five

260-kg tabun-filled bombs be shipped from the Robbers' Lair to the United States "by air under highest priority" for field tests. Separately, General Porter asked the U.S. Army Air Forces and U.S. Army Ordnance to conduct their own feasibility studies to determine if tabun bombs could be used in combat by U.S. troops.

Most people looked upon chemical warfare as abhorrent. In a June 1943 speech, President Roosevelt himself had said that using chemicals to kill people was immoral and inhumane. The president had denied Chemical Warfare Service officials their request to change the service's name to the Chemical Corps because of the permanence the name change suggested. And yet here was the interesting news for the Chemical Service. When German nerve gas entered into the world of chemical warfare, it brought with it the assurance of a U.S. chemical warfare program in peacetime. According to chemical weapons expert Jonathan B. Tucker, "In 1945, in the aftermath of World War II, the U.S. Army Chemical Warfare Service decided to focus its research and development efforts on the German nerve agents, the technological challenges of which promised to ensure the organization's survival through the period of postwar demobilization and declining military budgets." Within several months of the German surrender, 530 tons of tabun nerve agent were shipped to the United States and used in Top Secret field tests.

Requests to bring German chemists to the United States for weapons work quickly followed. But, as had been the case with the V-2 rocket scientists, the notion of issuing visas to Hitler's chemists was met with hostility inside the State Department. When the chief of the State Department's Passport Division, Howard K. Travers, learned about this idea, he sent his colleagues an internal memo stating, "We should do everything we consistently can to prevent German chemists and others from entering this country."

In Germany, Alsos scientific director Samuel Goudsmit had been tracking Hitler's chemists ever since the Allies crossed the Rhine. Likewise, the CIOS chemical weapons team, led by Lieutenant Colonel Philip R. Tarr of the U.S. Chemical Warfare Service and his British counterpart Major Edmund Tilley, were continuing their relentless pursuit. When Alsos located the chemist Richard Kuhn at the Kaiser Wilhelm Institute for Medical Research in Heidelberg, they paid him a visit. Kuhn had once been an internationally revered organic chemist, but rumor had it that he had become an ardent Nazi during the war. Kuhn won the Nobel Prize in Chemistry in 1938 but turned down accepting the award at the request of Hitler, who called it a Jewish prize. Here now to interview Richard Kuhn, Samuel Goudsmit had·with him two American chemists, Louis Fieser of Harvard and Carl Baumann of the University of Wisconsin. Both men had actually worked with Kuhn in his laboratory before the war. After a cordial exchange of greetings, the interrogation began. Alsos sought information regarding the Third Reich's nerve agent program. What did Herr Kuhn know?

Kuhn, with his mop of straight reddish-brown hair, cunning smile, and schoolboy looks, swore that he had no connection whatsoever with Reich military research. He told his former colleagues that he was a pure scientist, an academic who spent the war working on the chemistry of modern drugs. Samuel Goudsmit had his doubts. "Richard Kuhn's record did not seem too clean to me," Goudsmit recalled after the war. "As president of the German Chemical Society he had followed the Nazi cult and rites quite faithfully. He never failed to give the Hitler salute when starting his classes and to shout 'Siegheil' like a true Nazi leader," Goudsmit recalled. But the Alsos leader did not have enough evidence to have Kuhn arrested, so he put him under surveillance instead.

Elsewhere in Germany, CIOS chemical weapons investigators Colonel Tarr and Major Tilley had been rounding up German chemists and sending them to prisoner-of-war facilities near where the individual arrests had taken place. Starting on June 1, 1945, these chemists would now be sent to a single location — a Top Secret interrogation facility outside Frankfurt. SHAEF was moving its headquarters from Versailles to Frankfurt and was to be dissolved in mid-July. The new organization in charge of all affairs, including scientific exploitation, was the Office of Military Government for Germany (OMGUS), whose commander was Eisenhower's deputy General Lucius D. Clay. The Allies were also reorganizing the way in which scientific intelligence was going to be collected moving forward. CIOS was transitioning into American and British components: FIAT (Field Information Agency, Technical) and BIOS (British Intelligence Objectives Sub-committee). CIOS teams would remain active while they completed open investigations.

The structure that housed this new interrogation center was none other than Schloss Kransberg, or Kransberg Castle, Hermann Göring's former Luftwaffe headquarters and the place where Albert Speer and his aide spent the night the last New Year's Eve of the war. The Allies gave it the code-name Dustbin. This medieval structure built high in the Taunus Mountains had grand rooms, hardwood floors, beautiful stone fireplaces, and shiny chandeliers. These were hardly gulag-type quarters. In terms of security classification, Dustbin was top-tier. The facility was the second-most classified interrogation center after

Ashcan. Dustbin was self-contained inside its centuries-old stone walls, and, as it went at Ashcan, the prisoners here were free to roam the grounds and chat among themselves. Karl Brandt, Hitler's doctor, organized morning gymnastics classes in the garden. Others played chess. Industrialists held lectures in the large banquet hall that Göring had once used as a casino. Speer took walks in the castle's apple orchard, almost always alone by choice. Whereas Ashcan housed the Nazi high command, Dustbin had many Nazi scientists, doctors, and industrialists under guard. This included more than twenty chemists with IG Farben and at least six members of its board.

Throughout the early summer of 1945, several key players in Farben's tabun gas program were still at large. For Major Tilley, the chronology regarding how Farben first began producing nerve gas and how it transformed into wide-scale production remained a mystery until a Farben chemist named Dr. Gerhard Schrader was captured and brought to Dustbin. Schrader was the man who created the nerve agent that had been found at Raubkammer, the Robbers' Lair. The information Schrader had was among the most sought-after classified military intelligence in the world. Tilley prepared for intense stonewalling from the Farben chemist. Instead Dr. Schrader spoke freely, offering up everything he knew, beginning with tabun's startling discovery in the fall of 1936.

*   *   *

Dr. Schrader had been working at an insecticide lab for IG Farben in Leverkusen, north of Cologne, for several years. By the fall of 1936, he had an important job on his hands. Weevils and leaf lice were destroying grain across Germany, and Schrader was tasked with creating a synthetic pesticide that could eradicate these tiny pests. The government had been spending thirty million reichsmarks a year on pesticides made by Farben as well as other companies. IG Farben wanted to develop an insect killer that could save money for the Reich and earn the company a monopoly on pesticides.

Synthesizing organic, carbon-based compounds was trial-and-error work, Dr. Schrader told Major Tilley. It was labor-intensive and dangerous. Schrader, a family man, took excellent precautions against exposure, always working under a fume hood. Even trace amounts of the chemicals he was using had cumulative, potentially lethal effects. Schrader suffered from frequent headaches and sometimes felt short of breath. One night, while driving home after working on a new product, Schrader could barely see the road in front of him. When he pulled over to examine his eyes in the mirror, he saw that his pupils had constricted to the size of a pinhead. Over the next several days his vision grew worse. He developed a throbbing pressure in his larynx. Finally, Schrader checked himself into a hospital, where he was monitored for two weeks before being sent home and told to rest.

Eight days after the respite, Schrader returned to work. He had been developing a cyanide-containing fumigant, which he had given the code name Preparation 9/91. Picking up where he'd left off with his work, he prepared a small amount of his new substance, diluting it to 1 in 200,000 units to see if it would kill lice clinging to leaves. He was stunned when his new creation killed 100 percent of the lice. Schrader repeated the experiment for his colleagues. They all agreed that Preparation 9/91 was a hundred times more lethal than anything anyone at the Leverkusen lab had ever worked with before.

Dr. Schrader sent a sample of this lethal new fumigant to Farben's director of industrial hygiene, a man named Professor Eberhard Gross (not to be confused with Dr. Karl Gross, the Waffen-SS bacteriologist connected with the Geraberg discovery). Gross tested the substance on apes and was duly shocked by the results. After a healthy ape was injected with a tiny amount of Preparation 9/91 — just 1/10th of a milligram per kilo of body weight — the ape died in less than an hour. Next, Gross tested the substance on an ape inside an inhalation chamber. He watched this healthy ape die in sixteen minutes. Professor Gross told Dr. Schrader that his Preparation 9/91 was being sent to Berlin and that he should wait for further instruction on what action to take next.

At Dustbin, Schrader told Major Tilley that when he learned his compound could kill a healthy ape through airborne contact in minutes, he became upset. His discovery was never going to be used as an insecticide, Schrader lamented. It was simply too dangerous for any warm-blooded animal or human to come into contact with. Schrader said his goal was to save money for the Reich. With the news of how powerful Preparation 9/91 was, Schrader felt he'd failed at his job. He got back to work, searching for a fumigant better suited for the task of killing weevils and leaf lice.

Meanwhile, Professor Gross brought the substance to his superiors. Starting in 1935, a Reich ordinance required all new discoveries with potential military application to be reported to the War Office. The Reich's Chemical Weapons Department began to evaluate Schrader's Preparation 9/91 for its potential use in chemical warfare. In May 1937 Schrader was invited to Berlin to demonstrate how he'd synthesized Preparation 9/91. "Everyone was astounded," Schrader told Tilley. This was the most promising chemical killer since the Germans invented mustard gas. Preparation 9/91 was classified Top Secret and given a code name: tabun gas. It came from the English word "taboo," something prohibited or forbidden.

Dr. Schrader was told to produce one kilogram for the German army, which would take over tabun production on a massive scale. Schrader got a bonus of 50,000 reichsmarks (the average German worker during this time period earned 3,100 reichsmarks a year) and was told to get back to work. Farben still needed him to develop a lice-killing insecticide.

With their new nerve agent tabun, Farben executives saw all kinds of business opportunities. Karl Krauch, the head of Farben's board of directors, began working with Hermann Göring on a longer-range plan to arm Germany with chemical weapons, ones that could eventually be dropped on the enemy from airplanes. In his report to Göring, Krauch called tabun "the weapon of superior intelligence and superior scientific-technological thinking." The beauty in the nerve agent, Krauch told Göring, was that it could be "used against the enemy's hinterland." Göring agreed, adding that what he liked most about chemical weapons was that they terrified people. He responded to Krauch in writing, noting that the deadly effects of nerve agents like tabun gas could wreak "psychological havoc on civilian populations, driving them crazy with fear."

On August 22, 1938, Göring named Karl Krauch his Plenipotentiary for Special Questions of Chemical Production. Farben was now positioned to build the Reich's chemical weapons industry from the ground up. The Treaty of Versailles had forced Germany to destroy all its chemical weapons factories after World War I, which meant factories had to be secretly built. This was an enormous undertaking, now an official part of the Nazis' secret Four Year Plan, and through Krauch IG Farben was made privy to the Reich's war plan before war was declared.

At the Dustbin interrogation center, Major Tilley asked Schrader about full-scale production. Based on the Allies' discovery of thousands of tons of tabun bombs in the forests outside Raubkammer, Farben must have had an enormous secret production facility somewhere. Dr. Schrader said that he was not involved in full-scale production. That was the job of his colleague, Dr. Otto Ambros.

Major Tilley asked Schrader to tell him more about Ambros. Schrader said that most of what Ambros did was classified but that if Major Tilley wanted to know more about what he actually did for Farben, Tilley should talk to individuals who sat on Farben's board of directors with Ambros, either Dr. Karl Krauch or Baron Georg von Schnitzler. Both men were interned here at Dustbin.

"Who is Mr. Ambros?" Major Tilley asked Baron Georg von Schnitzler, in an interview that would later be presented as evidence in a Nuremberg war crimes trial.

"He is one of our first, younger technicians," von

Schnitzler said. "He was in charge of Dyhernfurt [*sic*] as well as Auschwitz and Gendorg [*sic*]."

Where was Ambros now? Tilley asked von Schnitzler. The baron told Major Tilley to talk to Karl Krauch.

From Krauch, Major Tilley learned quite a bit more about Ambros. That he had been in charge of technical development of chemical weapons production at Gendorf and at Dyhernfurth. That Gendorf produced mustard gas on an industrial scale, and that Dyhernfurth produced tabun. Krauch also revealed a new piece of evidence. Dyhernfurth produced a second nerve agent, one that was even more potent than tabun, called sarin. Sarin was an acronym pieced together from the names of four key persons involved in its development: Schrader and Ambros from IG Farben and, from the German army, two officers named Rüdiger and Linde. Krauch told Major Tilley that the Dyhernfurth plant had fallen into Russian hands.

Karl Krauch said something else that caught Major Tilley by surprise. Before coming to Dustbin, Krauch said he had been in the hospital, where he'd been paid a visit by two American officers, one of whom was Lieutenant Colonel Tarr. "Judging from conversations I had a few months ago in the hospital with members of the USSBS [Colonel Snow] and Chemical Warfare [Colonel Tarr]," Krauch explained, "the gentlemen seemed mostly interested in sarin and tabun; they asked me for construction plans and details of fabrication. As far as I understood they intended to erect similar plants in the U.S.A. I told them to apply to Dr. Ambros and his staff at Gendorf."

Major Tilley was shocked. Lieutenant Colonel Tarr was his CIOS partner, and yet Tarr had neglected to share with him the story about visiting Krauch in the hospital. This was Tilley's first indication that Tarr was running a separate mission for the U.S. Chemical Warfare Service, one that apparently had a different objective than the CIOS mission. The full, dramatic story was about to unfold.

By June 1945, Otto Ambros had been questioned by soldiers with the Third Army numerous times. For reasons that remain obscure, no one from that division had been informed of the fact that Ambros was wanted for war crimes, or that he had served as Farben's chief of chemical weapons production throughout Hitler's rule. To the Third Army he was simply the "plain chemist" in the Bavarian village of Gendorf, the smiling, well-dressed businessman who supplied American soldiers with free bars of soap.

At Dustbin, Major Tilley relayed this critical new information about Dr. Otto Ambros to his FIAT superiors, who in turn sent an urgent message to the Sixth Army Group, also in Gendorf, ordering the immediate arrest of Dr. Ambros. The Sixth Army was to transport Ambros directly to Dustbin so that Major Tilley could interrogate him. A note card was placed in Ambros's dossier. Disparate bits of information were now coming into sharp focus. "Case #21877. Dr. Otto Ambros. Rumored to have been involved in use of concentration camp personnel for testing effectiveness of new poison gases developed at Gendorf."

CROWCASS notified SHAEF, insisting that Dr. Ambros be arrested. As the plant manager at Farben's Buna factory at Auschwitz, Otto Ambros had been linked to atrocities including mass murder and slavery. The Sixth Army Group swung into action. But when they arrived at Ambros's home in Gendorf, arrest orders in hand, Ambros was gone.

The first assumption was that Ambros had fled on his own. This proved incorrect. He had been taken away by Lieutenant Colonel Philip Tarr. Initially, the commanding officer at Dustbin found this impossible to comprehend. It was one thing for Tarr to try to interview Ambros before any other chemical warfare experts did. That kind of rivalry had been going on ever since the various scientific intelligence teams had crossed the Rhine. But why would Tarr defy orders from SHAEF to have Ambros arrested? While soldiers with the Sixth Army stood scratching their heads in Gendorf, Tarr and Ambros were actually headed to Heidelberg in a U.S. Army jeep. Their destination was an American interrogation center that was run by army intelligence officers with the Chemical Warfare Service. For days, no one at Dustbin had any idea where Tarr and Ambros had gone.

Otto Ambros had a razor-sharp mind. He was cunning and congenial, sly like a fox. He almost always wore a grin. The American war crimes prosecutor Josiah DuBois described him as having a "devilish friendliness" about him. He also had a distracting, rabbitlike habit of sniffing at the air. Ambros was short and heavyset, with white hair and flat feet. He was a brilliant scientist who studied chemistry and agronomy under Nobel Prize winner Richard Willstätter, a Jew. As a chemist, Ambros had a mind that was capable of pushing science into realms previously unexplored. Few men were as important to IG Farben during the war as Otto Ambros had been.

IG Farben first began producing synthetic rubber in 1935, naming it Buna after its primary component, butadiene. In 1937 Farben presented commercial Buna on the world stage and won the gold medal at the International Expo in Paris. When Germany invaded Poland in September 1939, the Reich's ability to import natural rubber diminished. Demand for a synthetic alternative skyrocketed, a fact Farben was well aware of in advance of Germany's attack. Tanks needed treads, aircraft needed tires, and Farben needed to produce rubber. Hitler directed Farben to increase its Buna production further. Dr. Ambros was put in charge and saw to it that Farben opened a second and then a third Buna plant so that supply could meet demand. As the invasion of the Soviet Union was secretly conceived by the German high command, Hitler again called upon Farben's board of directors to increase its synthetic rubber production. Farben needed to construct a massive new Buna factory. Otto Ambros was put in charge of masterminding this undertaking as well. The place chosen was Auschwitz.

Once, Auschwitz was a regular town. "Ordinary people lived there, and tourists visited to see the castle, the churches, the large medieval market square, and the synagogue," write historians Debórah Dwork and Robert Jan van Pelt. In the 1930s, visitors sent postcards from the area that read "Greetings from Auschwitz." When, in the fall of 1940, Otto Ambros pored over maps of this region, called Upper Silesia, in search of a Buna factory site, he found what he was looking for. The production of synthetic rubber required four things: water, flat land, good railway connections, and an abundance of laborers. Auschwitz had all four. Three rivers met in Auschwitz, the Sola, the Vistula, and the Przemsza, with a water flow of 525,000 cubic feet per hour. The land was flat and sixty-five feet above the waterline, making it safe from floodwaters. The railway connections were sound. But most important was the labor issue. The concentration camp next door could provide an endless labor supply because the men were cheap and could be worked to death.

For Farben, the use of slave labor could take the company to levels of economic prowess previously unexplored. First, a financial deal had to be made with the SS. Ambros was instrumental in this act. For months, before the building of the Buna factory got the go-ahead, the SS and Farben haggled over deal points. Some of the paperwork survived the war. On November 8, 1940, the Reich's minister of economics wrote to Farben's board of directors, requesting that they hurry up and "settle the question regarding the site." Otto Ambros lobbied hard for Auschwitz, and in December, IG Farben sent a busload of its rubber experts and construction workers to survey the work site. A Farben employee named Erich Santo was assigned to serve as Otto Ambros's construction foreman.

"The concentration camp already existing with approximately 7000 prisoners is to be expanded," Santo noted in his official company report. For Ambros, Farben's arrangement with the SS regarding slave laborers remained vague; Ambros sought clarity. "It is therefore necessary to open negotiations with the Reich Leader SS [Himmler] as soon as possible in order to discuss necessary measures with him," Ambros wrote in his official company report. The two men had a decades-old relationship; Heinrich Himmler and Otto Ambros had known one another since grade school. Ambros could make Himmler see eye-to-eye with him on the benefits that Auschwitz offered to both Farben and the SS.

In fact, the SS and IG Farben needed one another. Himmler wanted Farben's resources at Auschwitz and was eager to make the deal to supply the slaves, so SS officers hosted a dinner party for Farben's rubber and construction experts at the Auschwitz concentration camp, inside an SS banquet hall there. During the festivities, the remaining issues were finally agreed upon. Farben would pay the SS three reichsmarks a day for each laborer they supplied, which would go into the SS treasury, not to the slaves. "On the occasion of a dinner party given for us by the management of the concentration camp, we furthermore determined all the arrangements relating to the involvement of the really excellent concentration-camp operation in support of the Buna plants," Ambros wrote to his boss, Fritz Ter Meer, on April 12, 1941. "Our new friendship with the SS is proving very profitable," Ambros explained.

The SS agreed to provide Farben with one thousand slave workers immediately. That number, promised Himmler, could quickly rise to thirty thousand with demand. The relationship between Farben and the SS at Auschwitz was now cemented. Otto Ambros was the key to making the Buna factory a success. With his knowledge of synthetic rubber and his managerial experience—he also ran Farben's secret nerve gas production facilities—there was no better man than Otto Ambros for the Auschwitz job.

Major Tilley waited at Dustbin for the return of Tarr and Ambros. It was now clear to him that there was no single individual more important to Hitler's chemical weapons program than Otto Ambros had been. Ambros was in charge of chemical weapons at Gendorf and Dyhernfurth, and he was the manager of the Buna factory at Auschwitz. From interviewing various Farben chemists held at Dustbin, Tilley had also learned that the gas used to murder millions of people at Auschwitz and other concentration camps, Zyklon B, was a Farben product. Farben owned the patent on Zyklon B, and it was sold to the Reich by an IG Farben company. In one of these interviews, Tilley asked IG Farben board member Baron Georg von Schnitzler if Otto Ambros knew that Farben chemicals were being used to murder people.

"You said yesterday that a [Farben employee] 'alluded' to you that the poisonous gasses [sic] and the chemicals manufactured by IG Farben were being used for the murder of human beings held in concentration camps," Major Tilley reminded von Schnitzler in their interview.

"So I understood him," von Schnitzler replied.

"Didn't you question those employees of yours further in regard to the use of these gases?"

"They said they knew it was being used for this purpose," von Schnitzler said.

"What did you do when he told you that IG chemicals were being used to kill, to murder people held in concentration camps?" Major Tilley asked.

"I was horrified," said von Schnitzler.

"Did you do anything about it?"

"I kept it for [*sic*] me because it was too terrible," von Schnitzler confessed. "I asked [the Farben employee] is it known to you and Ambros and other directors in Auschwitz that the gases and chemicals are being used to murder people?"

"What did he say?" asked Major Tilley.

"Yes; it is known to all the IG directors in Auschwitz," von Schnitzler said.

For Lieutenant Colonel Philip R. Tarr, there was a mission at hand. Enemy Equipment Intelligence Service Team Number One, which he served on, needed information that only Dr. Otto Ambros had. Specifically, the team needed blueprints for equipment necessary for producing tabun nerve gas.

When Tarr and Ambros arrived in Heidelberg, the U.S. Chemical Warfare Service had another IG Farben chemist in custody whom they wanted Ambros to work with on a classified job. The man is referred to in documents only as Herr Stumpfi. Ambros and Stumpfi were told to drive to a special metals firm located in Hanau, where they were to locate "30 or 40 drawings of silver-lined equipment." The Chemical Warfare Service trusted Ambros to such a degree that they sent him and Stumpfi on this mission without an escort.

Manufacturing tabun gas was a precise and clandestine process. The United States desperately wanted to reproduce it, but attempting to do so without Farben's proprietary formula and its secret equipment was a potential death sentence for any chemists involved. Farben had spent millions of reichsmarks on research and development. Hundreds of concentration camp workers had died in this trial-and-error process. When the U.S. Chemical Warfare Service learned that the silver-lined equipment used to manufacture tabun gas on a large scale had been outsourced from a special metals firm called Heraus, they coveted those blueprints and plans. Dr. Ambros and Herr Stumpfi were to go to this engineering firm to locate these drawings and blueprints and bring them back to Heidelberg. The Chemical Warfare Service agents could not conduct this mission on their own because they had no idea what equipment to look for.

The two Farben chemists, Ambros and Stumpfi, set off on their secret assignment. "When they arrived at the factory in Hanau, personnel of a [U.S.] CIC [Counter Intelligence Corps] group with headquarters at that time in Hanau, arrested them," read a secret report. "When they explained their mission the CIC personnel concerned confirmed the German engineers' statement by communication with Heidelberg and the two Germans were released." Ambros and Stumpfi drove away. "The CIC personnel, concerned [after] having learned of the drawings through the two German engineers, then seized the drawings and took them to their own headquarters," read a classified Army report. The Chemical Warfare Service never obtained the drawings they were looking for. But at least Tarr had Dr. Ambros under his control, or so he believed.

Instead, somewhere between Ambros's release from the Heraus engineering firm in Hanau and his return to Heidelberg, he was able to communicate to his "network of spies and informants in Gendorf." From those sources, Ambros learned that soldiers with the U.S. Sixth Army had an order to arrest him. So instead of returning to Tarr's custody, Ambros drove to a fancy guesthouse that IG Farben maintained outside Heidelberg called Villa Kohlhof, where a staff of Farben employees tended to his every need.

CIOS and FIAT officials from Dustbin finally made contact with Tarr and ordered him to return with Ambros to the interrogation facility immediately.

But Tarr was no longer in control of Ambros. Major Tilley went looking for Ambros himself and found Hitler's chemist residing at Villa Kohlhof. Ambros told Major Tilley that he would agree to continue cooperating with the U.S. Chemical Warfare Service and the various Allied agencies that sought information from him, but only on one condition: that Tilley "secure the release of all chemical warfare personnel already detained at Dustbin." This was a preposterous request.

Tilley's superior, Major P. M. Wilson of FIAT's Enemy Personnel Exploitation Section, attempted to take control of the situation, ordering Ambros brought to Dustbin immediately. This was not a matter of cooperation, Wilson said. There were orders to arrest the man. Lieutenant Colonel Tarr intervened on Ambros's behalf. He lobbied the British Ministry of Supply (the agency responsible for British chemical warfare issues) for help getting Ambros's Dustbin colleagues released. To Tarr, extracting Ambros's esoteric knowledge outweighed the need to hold him accountable for his crimes. But the British also flatly refused to help Tarr. The matter stalled.

Lieutenant Colonel Tarr flew to Paris. That night, a telegram arrived at Dustbin, sent from Paris and purporting to be from the British Ministry of Supply. The telegram ordered the release of all Farben chemical warfare scientists at Dustbin, and was signed by a

British Ministry of Supply colonel named J. T. M. Childs. Officers at Dustbin suspected that something was amiss and contacted Colonel Childs about his outrageous request. Colonel Childs swore he had neither written the memo nor signed it and accused Lieutenant Colonel Tarr of forgery.

FIAT enhanced their efforts to have Dr. Ambros arrested in Heidelberg. The efforts failed. Ambros was able to evade capture by fleeing into the safety of the French zone. Double-crossing Lieutenant Colonel Tarr, Otto Ambros struck a deal with French chemical weapons experts. In exchange for information, he was given a job as plant manager at Farben's chemical factory in Ludwigshafen.

When FIAT officers at Dustbin learned what had happened, they were outraged. Ambros's escape had been entirely preventable. "It is evident that he was not kept in custody or under house arrest," noted Captain R. E. F. Edelsten, a British officer with the Ministry of Supply. Major P. M. Wilson saw the situation in much darker terms. Lieutenant Colonel Tarr had "taken steps to assist [Ambros] to evade arrest," he wrote in a scathing report. Wilson was appalled by "the friendly treatment being given to this man who is suspected of war criminality." But these were just words. Ambros was now a free man, living and working in the French zone.

The relationship among Tarr, Ambros, and the U.S. Chemical Warfare Service was far from over. It was only a matter of time before an American chemical company would learn of the army's interest in a whole new field of chemical weapons. An American chemist, Dr. Wilhelm Hirschkind, was in Germany at this same time. Dr. Hirschkind was conducting a survey of the German chemical industry for the U.S. Chemical Warfare Service while on temporary leave from the Dow Chemical Company. Dr. Hirschkind had spent several months inspecting IG Farben plants in the U.S. and British zones and now he was in Heidelberg, hoping to meet Ambros. Lieutenant Colonel Tarr reached out to Colonel Weiss, the French commander in charge of IG Farben's chemical plant in Ludwigshafen, and a meeting was arranged.

On July 28, 1945, Dr. Hirschkind met with Dr. Ambros and Lieutenant Colonel Tarr in Heidelberg. Ambros brought his wartime deputy with him to the meeting, the Farben chemist Jürgen von Klenck. It was von Klenck who, in the final months of the war, had helped Ambros destroy evidence, hide documents, and disguise the Farben factory in Gendorf so that it appeared to produce soap, not chemical weapons. Jürgen von Klenck was initially detained at Dustbin but later released. The Heidelberg meetings lasted several days. When Dr. Wilhelm Hirschkind left, he had these words for Ambros: "I would look forward after the conclusion of the peace treaty [to] continuing our relations [in my position] as a representative of Dow."

Only later did FIAT interrogators learn about this meeting. Major Tilley's suspicions were now confirmed. A group inside the U.S. Chemical Warfare Service, including his former partner, Lieutenant Colonel Tarr, did indeed have an ulterior motive that ran counter to the motives of CIOS, FIAT, and the United Nations War Crimes Commission. Tilley's superior at Dustbin, Major Wilson, confirmed this dark and disturbing truth in a classified military intelligence report on the Ambros affair. "It is believed that the conflict between FIAT...and Lt-Col Tarr was due to the latter's wish to use Ambros for industrial chemical purposes" back in the United States.

All documents regarding the Ambros affair would remain classified for the next forty years, until August of 1985. That an officer of the U.S. Chemical Warfare Service, Lieutenant Colonel Tarr, had sheltered a wanted war criminal from capture in the aftermath of the German surrender was damning. That this officer was also participating in meetings with the fugitive and a representative from the Dow Chemical Company was scandalous.

In 1945, the Chemical Warfare Service was also in charge of the U.S. biological weapons program, the existence of which remained secret from the American public. The program was robust; if the atomic bomb failed to end the war in Japan, there were plans in motion to wage biological warfare against Japanese crops. After the fall of the Reich, the staff of the Chemical Warfare Service began interrogating Hitler's biological weapons makers, many of whom were interned at Dustbin. The Chemical Warfare Service saw enormous potential in making the Nazis' biological weapons program its own and sought any scientific intelligence it could get. The man most wanted in this effort was Hitler's top biological weapons expert, Dr. Kurt Blome.

On June 29, 1945, Blome was sent to Dustbin. The officers assigned to interrogate him were Bill Cromartie and J. M. Barnes of Operation Alsos. Each man was uniquely familiar with Blome's background. Cromartie had been in Dr. Eugen Haagen's apartment in Strasbourg in November 1944 when he and Alsos scientific director Samuel Goudsmit made the awful discovery that the Reich had been experimenting on people during the war (Blome was named in the Haagen files). And it was Cromartie and Barnes who led the investigation of the Geraberg facility, the abandoned, curious-looking research outpost hidden in the Thuringian Forest. Both Cromartie and Barnes had concluded that Geraberg had been a laboratory for Reich biological weapons research and that Dr. Blome was in charge.

During his initial interview at Dustbin, Blome refused to cooperate. "When he was first interrogated, he was very evasive," Cromartie and Barnes wrote. But a few days later, when interrogated in more detail, Blome's "attitude changed completely and he seemed anxious to give a full account not only [of] what he actually did but what he had in mind for future work." Cromartie and Barnes were unsure if they should be enthused by Blome's seeming change of heart or suspicious of it. Blome had been observed in the Dustbin eating hall conversing at length with Dr. Heinrich Kliewe, the Reich's counterintelligence agent for bacterial warfare concerns. Perhaps the two men were concocting a misinformation scheme.

During the war, Dr. Kliewe's job had been to monitor bioweapons progress being made by Germany's enemies, most notably Russia. "Kliewe claims that he himself did all the evaluating of the reports received and determined what course of action his department should thenceforth follow," investigators wrote in Kliewe's Dustbin dossier. Kliewe told Blome that he would likely be taken to Heidelberg for a lengthy interrogation with Alsos agents, as Kliewe had been.

If Cromartie and Barnes were surprised by Blome's sudden willingness to talk, they were also aware that most of what he told them could not be independently verified. "It is quite impossible to check many of his statements and what follows is an account of what he related," read a note in Blome's Dustbin dossier. What Blome recounted was a dark tale of plans for biological warfare spearheaded by Heinrich Himmler.

Himmler had a layman's fascination with biological warfare. A former chicken farmer, the Reichsführer-SS had studied agriculture in school. According to Blome, it was Himmler who was the primary motivator behind the Reich's bioweapons program. Hitler, Blome said, did not approve of biological warfare and was kept in the dark as to specific plans. Himmler's area of greatest fascination, said Blome, was bubonic plague.

On April 30, 1943, Göring had created the cancer research post that was to be held by Blome. Over the next nineteen months, Blome explained, he met with Himmler five times.

At their first meeting, which occurred in the summer of 1943—Blome recalled it as being July or August—Himmler ordered Blome to study various dissemination methods of plague bacteria for offensive warfare. According to Blome, he shared with Himmler his fears regarding the dangerous boomerang effect a plague bomb would most likely have on Germany. Himmler told Blome that in that case, he should get to work immediately to produce a vaccine to prevent such a thing. To expedite vaccine research, Blome said, Himmler ordered him "to use human beings."

Himmler offered Blome a medical block at a concentration camp like Dachau where he could complete this work. Blome said he told Himmler he was aware of "strong objections in certain circles" to using humans in experimental vaccine trials. Himmler told Blome that experimenting on humans was necessary in the war effort. To refuse was "the equivalent of treason."

Very well, said Blome. He considered himself a loyal Nazi, and it was his intention to help Germany win the war. "History gives us examples of human disease affecting the outcome of wars," Blome told his Alsos interrogators, taking a moment to lecture them on history. "We know [that] from antiquity up till the time of [the] Napoleonic wars, victories and defeats were often determined by epidemics and starvation," Blome said. Spreading an infectious disease could bring about the demise of a marauding army, and Blome said that the failure of Napoleon's Russian campaign was "due in great part to the infection of his horses with Glanders," a highly contagious bacterial disease. History aside, Blome said he counseled Himmler on the fact that a concentration camp was a terrible place to experiment with bubonic plague because the population was too dense.

Blome then told Himmler that if he were to experiment with plague bacterium, he would need his own institute, an isolated facility far removed from population centers. Himmler and Blome agreed that Poland would be a good place, and they settled on Nesselst-edt, a small town outside the former Poznan' Univer-sity (by then operated by the Reich). Blome's research institute was to be called the Bacteriological Institute at Nesselstedt.

In the interim, in Berlin, Blome oversaw a field test using rats, history's traditional carrier of bubonic plague. A debate had been taking place inside the Hygiene Institute of the Waffen-SS as to whether or not rats were the best plague carriers. Himmler's idea was "to take infected rats on to U-boats and release them near the enemy shores so they could swim to land." Blome doubted that rats could swim great dis-tances. He believed they could swim only for as long as the air in their fur kept them afloat. To prove his point, Blome arranged for a test on a Berlin lake. "About thirty rats were taken out in a police boat and released at different distances from the shore to swim both with and against the wind." Blome said that the rats were dumber than he thought—that when placed in the water, "they had no idea where the shore was and swam around in different directions." A few of them drowned in ten minutes. The longest any of the rats swam for was thirty minutes. Of those released a little over a half-mile from shore, only a third reached land. As far as Blome was concerned, Himmler's U-boat dispersal idea was not practical.

Meeting Number Two took place a month or two

later, in September or October 1943, and was largely a repetition of the first, at least according to Blome. There was one significant development, however. Himmler asked Blome if he needed an assistant. Blome agreed that a bacteriologist would be helpful. Himmler assigned Dr. Karl Gross, formerly a staff member at the Waffen-SS Hygiene Institute.

The two doctors did not get along. Blome became convinced that Dr. Gross had been sent by Himmler to spy on him. He told his interrogators that he was under great pressure to work faster. Himmler repeatedly "reiterated that the methods of waging BW [biological warfare] must be studied in order to understand the defense against it." What this meant was that Himmler wanted Blome to infect human test subjects with plague to see what would happen to them.

The third meeting took place four or five months later, in February 1944. By this time, Blome said, the facility at Nesselstedt had been built. There was ample staff housing, a well-equipped laboratory, and an animal farm. The block for experimental work included a climate room, a cold room, disinfectant facilities, and rooms for "clean" and "dirty" experiments. There was an isolation hospital for sixteen people in the event that workers on Blome's staff contracted the disease. Work progressed slowly, Blome said, and Himmler became enraged. Rumors of an Allied invasion of the European continent had become a constant thorn in the side of the Reichsführer-SS. Why wasn't the Reich's bioweapons program more advanced, Himmler demanded to know. He asked Blome if it was possible to "do something now—for example disseminate influenza—that would delay the heralded Anglo-American invasion in the West." According to Blome, he told Himmler "it was impossible to do anything on these lines." Himmler proposed another idea: How about disseminating a virulent strain of hoof-and-mouth disease? Or tularemia, also called rabbit fever, which affected man in a manner similar to plague? Blome told Himmler that these were dangerous ideas, as any outbreak would surely affect Germany's troops. The Reich needed a massive stockpile of vaccinations before it could feasibly launch a biological attack.

Himmler stretched his thinking to target the Allies on their own soil. How about spreading cattle plague, also called rinderpest, in America or England? Himmler told Blome that infecting the enemy's food supply would have a sinister effect on enemy troops. Blome agreed and said he would investigate what it would take to start a plague epidemic among the enemies' cows. There was, however, a problem, Blome explained. An international agreement prohibited stocks of the rinderpest virus to be stored anywhere in Europe. Strains of cattle plague were available only in the third world.

Himmler said that he would get the cattle plague

himself. He sent Dr. Erich Traub, a veterinarian from the Reich's State Research Institute, located on the island of Riems, to Turkey. There, Dr. Traub acquired a strain of the lethal rinderpest virus. Under Blome's direction, trials to infect healthy cows with rinderpest began. Riems, in northern Germany in the Bay of Greifswald, was totally isolated and self-contained. It was the perfect place for these dangerous tests. The veterinary section used airplanes to spray the cattle plague virus on the island's grassy fields, where cows grazed. Blome said he didn't know much more about the program or its results—only that Dr. Traub, second in command at the research facility, was taken by the Russians when the Red Army captured Riems, in April 1945.

Blome's fourth meeting with Himmler took place in April or May of 1944. Himmler had become paranoid by now, Blome said. He believed that the Allies were plotting a biological weapons attack against the Reich. "Blome was summoned by telephone to see Himmler urgently. The latter had received a number of curious reports. Grass had come floating out of the sky over some part of Austria and a cow that had eaten some of it had died." Blome told Himmler he'd look into it. There were additional strange events, Himmler confided to Blome. "Some small balloons had been found near Salzberg [*sic*] and Berchtesgaden" not far from Hitler's mountain residence, the Berghof.

And potato beetles had been dropped in Normandy. Blome promised to study each incident.

Blome told Himmler he had a pressing issue of his own. Given the progress of the Red Army, he thought it was wise to move his plague research institute at Posen (Poznan′) somewhere inside Germany. The place Blome suggested was Geraberg, in the Thuringian forest, at the edge of the Harz Mountains. Himmler said that the Russians would never reach Posen. By early fall, he had changed his mind. In October a new biological weapons research facility was being built, concealed inside a pine forest in the village of Geraberg.

In the meantime, Blome told his interrogators, work on vaccines was moving forward—not at either of his research institutes but inside the army instead. Göring had moved epidemic control into the jurisdiction of a major general named Dr. Walter Schreiber, surgeon general of the Reich. Blome held the position of deputy surgeon general of the Reich, but the two men had equal positions under Göring, Blome explained. He, Blome, was in charge of creating the biological weapons; Dr. Schreiber was in charge of protecting Germans against biological weapons, should they be used—Major General Dr. Schreiber specialized in epidemic control. The sword and the shield.

Alsos was very interested in learning about these vaccines. Blome said that Major General Dr. Schreiber was the person to talk to about that. Where could he be found? Blome said Schreiber had last been seen in Berlin. Word was he had surrendered to the Red Army and was their prisoner now.

Blome said something else that alarmed interrogators. During the war, Dr. Kliewe told Blome that it was the Russians who had the single most extensive biological weapons program in the world—a program more advanced even than that of the Japanese. Further, the Russians had managed to capture everything from Dr. Traub's laboratory on Riems, including Dr. Traub's Turkish strain of cattle plague. Then, in the winter of 1945, the Russians had captured Blome's bacteriological institute at Nesselstedt, outside Posen. This meant the Russians now had the Reich's most advanced biological weapons research and development facility, its steam chambers, incubators, refrigerators, and deep-freeze apparatus. They also had all of the pathogens—viral and bacterial—that Blome had been working with when he had fled west.

Alsos had two key pieces of information now: The Russians had the laboratory from Posen and they had the doctor in charge of vaccines, Major General Dr. Walter Schreiber. They had both the science and the scientist.

CHAPTER TEN

Hired or Hanged

The future of the scientists at Dustbin was obscure. Would they be hired by the U.S. government for future work, or would they be prosecuted by the Allies for war crimes? At Ashcan the future facing the Nazi high command was almost certainly grim, although there was much work to do. War crimes prosecutors faced a monumental task. They had to build criminal cases from scratch, a conundrum described by Nuremberg trials prosecutor General Telford Taylor after the war. "Our task was to prepare to prosecute the leading Nazis on [specific] criminal charges.... The first question a prosecuting attorney asks in such a situation is, 'Where's the evidence?' The blunt fact was that, despite what 'everybody knew' about the Nazi leaders, virtually no judicially admissible evidence was at hand." For evidence, prosecutors like Telford Taylor were relying on interrogators like John Dolibois to glean facts from the Nazis interned at Ashcan.

Frustration was mounting at Ashcan. Prisoners were conspiring to withhold information. "They were whispering rather important secrets to each other," Ashcan's commander, Colonel Burton Andrus, recalled, "determined not to help on the question of hidden loot, the whereabouts of people like Martin Bormann, and the guilt for war atrocities." Colonel Andrus racked his brain for ideas. How to make guilty men talk? "To jog the prisoners' memories back to the reality of their grave situation we decided to show them atrocity films taken at Buchenwald."

Colonel Andrus assembled his fifty-two Nazi prisoners in one room. Before the film began, he addressed them with the following words: "You know about these things and I have no doubt many of you participated actively in them. We are showing them to you not to inform you of what you already know, but to impress on you the fact that we know of it, too."

The lights went down and the first frames of the documentary footage flickered across the screen. Colonel Andrus watched his prisoners. Hans Frank, governor general of occupied Poland, the lawyer and PhD largely responsible for the murder of the Jewish population there, put a handkerchief to his mouth and gagged. Joachim von Ribbentrop, the former champagne salesman who became Hitler's foreign minister and pressured foreign states to allow the deportation of Jews in their territories to extermination camps in the east, got up from his chair and walked out of the room. Albert Kesselring, commander of the Luftwaffe invasions into Poland, France, and the Soviet Union, who also led the Battle of Britain, turned white. Hermann Göring sighed as if bored. Julius Streicher, the schoolteacher who became Hitler's mouthpiece and encouraged readers to exterminate the Jews, sat on his chair "clasping and unclasping his hands." When the film was over, no one said a word. A little while later, Walter Funk, the fifty-five-year-old former minister of the war economy and president of the Reichsbank, asked to see Colonel Andrus alone.

Small, overweight, and suffering from venereal disease, this once supremely powerful Nazi Party official twisted his fingers anxiously, while telling Colonel Andrus he had a confession to make. Walter Funk "looked incapable of running a filling-station, let alone a bank," Andrus recalled. Funk fidgeted nervously. "I have something to tell you, sir," Andrus recalled Funk saying. "I have been a bad man, Colonel, and I want to tell about it." Funk started to cry. Then he told Colonel Andrus that it was he who gave the order that all gold in every prisoner's mouth — in every concentration camp across the Reich — be removed and collected for the Reichsbank's reserves. Funk confessed that at first he'd "had [the gold] knocked out of their mouths while they were alive, but [realized] if they were dead there was far less bother." Andrus was appalled. "I had never believed before that such a horrific order as the one Funk was confessing to had ever been given," Andrus said.

Colonel Andrus told Funk to leave. In his memoir, Andrus wondered what Walter Funk had hoped to gain from his confession; he never got his answer. Whatever motivated Funk that day at Ashcan eventually disappeared. Months later, at Nuremberg, Funk denied that the conversation with Colonel Andrus had ever taken place. Instead, he swore under oath that he had never been connected with orders involving acquiring gold from Jews. "If it was done," declared Funk, "it has been kept from me up till now."

Save Funk's tearful confession, the screening of the Buchenwald atrocity film had no apparent effect on the other prisoners. The men who gagged and left the room had nothing more to say, and the rest of the Nazi high command remained tight-lipped about their crimes. Colonel Andrus grew increasingly frustrated. The prisoners were allowed to sit outside in the Palace Hotel garden, in small groups out of earshot of prison guards. There, Andrus watched Hitler's closest confidants prattle on among themselves.

"They could sit out in the garden in the spring

sunshine and chatter to their hearts' content," recalled Andrus. "Undoubtedly they compared notes on the interrogation[s] which went on in sixteen booths every day." If only G-2 had wired the Ashcan garden for sound. Colonel Andrus got an idea. He made a list of the four men who he believed were the biggest "back-stabbers and gossips" in the group. "Those with outspoken views about their fellow Nazis — men ready to blame each other for their own crimes." Göring, "verbose," von Papen, "malicious," Kesselring, "vain," and Admiral Horthy, "the Prince of Austria who believed he was better than everyone else," were chosen. With permission from headquarters, Andrus devised an elaborate ruse intended to extract information from the Nazi prisoners.

The four men were told that they were being handed over to the British for interrogation; that soon they would be traveling to a villa in Germany to be questioned there. In reality, Army Intelligence, G-2, had rented a house located three and a half miles from Ashcan. It was a traditional German-style half-timber house with a high wall running around its perimeter. U.S. Army signal intelligence engineers spent several days running hundreds of yards of wires through the house, under floorboards, behind walls, inside furniture cushions and lighting fixtures. In a final touch, they wired the garden for sound. The engineers burrowed through the backyard garden and attached tiny microphones to a single tree above the sitting area. All wires connected back to a recording machine capable of laying audio tracks down onto gramophone records. This was high technology in the summer of 1945.

At the Ashcan detention center, Göring, von Papen, Kesselring, and Admiral Horthy were loaded into an ambulance, a common means of transport after the war, and driven away. Black curtains had been drawn across the vehicle windows so the prisoners couldn't see where they were being taken. For hours, the ambulance drove around in what Andrus called a "circuitous route," covering fifty miles of terrain but never leaving Luxembourg. When the ambulance pulled up to the Germanic-looking house and unloaded the Nazis, Göring was thrilled.

"We are at a house I know!" boasted Göring, assuring his colleagues they had arrived in beautiful Heidelberg. "I recognize the decor on the walls," Göring said once inside.

The bedrooms in the new quarters had fresh sheets, plush mattresses, and soft pillows. Strolling through the house, Göring pointed to a chandelier and warned his fellow Nazis to be wary of listening devices. When one of the prisoners asked if they could sit outside in the garden, a guard checked with a superior and said yes. Göring made note of a single patch of shade under a weeping willow tree and suggested it as a good spot.

The men dragged four garden chairs into the shade, sat down, and began to gossip.

"Heavy, guttural voices could be heard loud and clear," Andrus recalled. "They were being recorded onto the black gramophone disks."

It was a brilliant start. But only a start. Soon, it began to rain. The men moved inside. There, they sat around barely saying a word. The following day, it rained again. That evening, Colonel Andrus received an order from SHAEF. The eavesdropping project was being shut down. Andrus had twenty-four hours to pack up his prisoners and leave.

This time the ambulance took the direct route back to Mondorf, just three and a half miles away. "Göring was furious," Andrus recalled. "How could they have gotten back to the prison so quickly! They realized they'd been had."

Back at Ashcan, things moved in a whirlwind. John Dolibois received a "Letter of Authority" on August 10, 1945, stating that CCPWE No. 32 was going to be closed down. Dolibois was to be part of the transport team now taking prisoners to new locations. For reasons Dolibois was not privy to, thirty-three of the fifty-two Ashcan internees were going to a new prisoner of war interrogation facility, this one located in the small town of Oberursel in the Taunus Mountains. Only later would Dolibois learn that many of these Nazis would be hired by the U.S. Army to write intelligence reports on work they had done during the war. Oberursel was just a few miles from the Dustbin facility at Castle Kransberg. The transport would be a convoy of six ambulances, a command car, a jeep, a trailer, and a truck carrying the prisoners' suitcases. Dolibois was assigned to ride in the first ambulance. His prisoner list included Admiral Karl Dönitz, Field Marshal Albert Kesselring, General Walter Warlimont, cabinet minister Schwerin von Krosigk, OKW Foreign Office head Admiral Leopold Bürkner, and Admiral Gerhard Wagner.

The trip from Mondorf to Oberursel was a journey through ruined countryside. Crossing from Luxembourg into Germany, Dolibois watched the chatter among the Nazis in his backseat come to a "halting end" as they saw the destruction and despair everywhere. "This was their first look at the condition of their country" since the war, Dolibois explained. From churches to administrative buildings to shops, entire villages had been reduced to rubble. In three months of peace there were no resources to clean anything up. People were starving and trying to survive. "The destruction that was the aftermath of Hitler's determination to 'fight to the last man,'" said Dolibois.

The convoy arrived in Oberursel. Like many other army interrogation facilities across the American zone in postwar Germany, the one in Oberursel had been an important Third Reich military post during the war. Oberursel had a particularly storied past. This Dulag Luft, or Durchgangslager (terms for air force prisoner of war camp, or transit camp), had functioned as the sole interrogation and evaluation center for the Luftwaffe. It was here that Nazi interrogators had questioned every Allied pilot who had been shot down during the war. The Luftwaffe's lead interrogator, Hanns Scharff, kept a diary. "Every enemy aviator who is captured...will be brought to this place for questioning. It makes no difference whether he is taken prisoner at the front lines or whether he comes dangling down from the sky in the most remote location...he comes to Oberursel," wrote Scharff.

Physically, things had not changed much since Oberursel had changed hands. The interrogation facility centered around a large half-timber "mountain house" that served as an officers' club. Nearby there were fourteen buildings for officer housing. The prisoners' barracks, a large U-shaped building, contained 150 solitary jail cells. This building had been called "The Cooler" during the war, and it was the place where interrogator Hanns Scharff did most of his work with captured Allied pilots. Now it was the new home for thirty-three former Ashcan internees, at least for a while.

Dolibois turned his charges over to the guard detail at Oberursel. His orders said to return to Luxembourg and wait for new orders. Setting out for Ashcan, he rode in the lead jeep with an enlisted man. A little less than an hour south of Frankfurt, Dolibois came upon a row of U.S. Army trucks stopped along the side of the road. One of the soldiers stepped into the road and signaled for Dolibois's convoy to stop. John Dolibois climbed out of his jeep. He became overwhelmed by a horrific stench, "sickeningly sweet, nauseating," he later recalled. He heard retching. Several of the men in his convoy had gotten out of their vehicles and were now throwing up along the side of the road.

"What is that horrible smell?" Dolibois asked a soldier behind the wheel of one of the stalled vehicles. "What in God's name are you hauling?"

The captain climbed out of his jeep. He did not say a word but motioned for Dolibois to follow him behind one of the two-and-a-half-ton trucks. In silence, the captain untied a rope and flipped back a sheet of canvas that had obscured the cargo from view until now. Dolibois stared into the body of the truck. It took him a moment for him to realize what he was looking at: rotting corpses. "Putrefied," he recalled. "Most were naked. Some still wore the pajama-like striped pants, the concentration camp uniforms, now just rotting rags. It was the most horrible sight I had ever seen."

The army captain spoke in a flat, emotionless tone. "There are thousands of them, five truck loads," he told Dolibois. "We're hauling them from one mass grave to another. Don't ask me why." These were bodies from Dachau. Corpses found upon liberation. The army captain's convoy had come to a standstill after one of the vehicles had broken down. They'd been waiting at the roadside for an escort when Dolibois's convoy from Oberursel had arrived. The captain asked Dolibois if his group could escort them to the next military station down the road. Dolibois agreed. "I found myself leading a bizarre caravan: six empty ambulances, an empty weapons carrier, followed by five two-and-a-half-ton trucks loaded with the obscenity of the Nazi final solution," remembered Dolibois. The dead bodies were being taken to a proper burial spot on orders from the Office of Military Government, United States (OMGUS).

Back at Ashcan the world appeared different to him now. There, the rest of Hitler's inner circle remained, men "directly responsible for that ghastly transport," said Dolibois. If John Dolibois ever had a shred of doubt about the degree of barbarism and the collective guilt of the men he had spent three months interrogating at Ashcan, in that moment there was no hesitation anymore. At age ninety-three, John Dolibois says, "I still smell that foul odor of death."

But in August 1945, there was barely any time for the young interrogator to process what he had witnessed. Back at Ashcan he fell into bed and slept hard.

Come morning, John Dolibois received new orders from Colonel Andrus. There was still a group of Nazi Bonzen at Ashcan. Dolibois was to be on the team that would move them to the Nuremberg prison. Colonel Andrus had selected John Dolibois to fly with the remaining members of the Nazi high command on a transport plane.

Dawn, August 12, 1945. The sun had not yet risen in the sky. Ashcan's last group of Nazis were escorted out of the classified interrogation facility and driven in ambulances a short distance to the airport at Luxembourg City. A C-47 transport plane idled on the tarmac, stripped down to its bare bones. Inside the aircraft, a single row of seats ran down each side. There was a honey bucket toilet and a urinal attached to a door at the back of the airplane. To Dolibois, it seemed as if Colonel Andrus was nervous. "Something was cooking," Dolibois recalled. Security was always on the colonel's mind. "Aha!" Dolibois realized what it was. "Kaltenbrunner!"

Ernst Kaltenbrunner was considered the most dangerous Nazi among the high command. At 6'4", he was a giant man with a massive frame, a pockmarked face, and a huge head with seven or eight deep dueling scars running across both sides of his forehead, cheeks, and chin. He drank and smoked heavily and was missing teeth. Kaltenbrunner was described by the British journalist Rebecca West as looking like "a particularly vicious horse." He held a doctorate in law and specialized in secret police work. He was the head of the Reich Security Main Office and chief of the security police and the security services. He was as actively involved in concentration camp crimes as any Nazi had been. According to the OSS, even Heinrich Himmler was frightened of him.

If anyone was going to "cause trouble on the flight to Nuremberg," Dolibois explains, Colonel Andrus was worried that Ernst Kaltenbrunner might. "Special precautions needed to be taken." Kaltenbrunner would be handcuffed to Dolibois's left wrist. "If he starts to run or goes for the door," Colonel Andrus told Dolibois, "shoot!" Colonel Andrus wished John Dolibois "good luck" during the flight and said that he would see him once they landed in Nuremberg. Colonel Andrus would be traveling in a different airplane.

During the flight Kaltenbrunner did not run for the door. Instead, he wanted to talk. He told Dolibois that he desired for the young interrogator to understand that he, Kaltenbrunner, was not responsible for war crimes. "He felt the need to explain about the Jews," Dolibois recalled, with Kaltenbrunner saying, "admittedly, he hated them, but he said that he had not been involved in their treatment in concentration camps. In fact, he claimed to have remonstrated with

Hitler on the treatment of the Jews." As the airplane prepared to land in Nuremberg, Kaltenbrunner said what so many Nazis repeated after the war. "I am a soldier and I only obeyed orders," he told Dolibois.

"I didn't argue with him, I just listened," Dolibois said. "Kaltenbrunner was a ruthless killer determined to save his own skin. His soft talk did not change my mind about him, but it helped pass the time." When the aircraft finally touched the ground, Dolibois felt a huge wave of relief. The prisoners were unloaded, and Colonel Andrus assumed control of them once again. Without looking back, Dolibois hurried on to the transport plane. The plane taxied down the runway and took flight. Dolibois sat alone in the empty airplane. The weeks of interrogating the Nazi high command were over. The Nazis he had just flown with would now be tried for war crimes at the Palace of Justice in Nuremberg. A majority of them would be hanged.

Back at Ashcan, when Dolibois returned to his quarters to pack his bags, he came across a strange sight. Standing by the edge of the perimeter fence, not far from where Dolibois had first pulled up to this place in an army jeep three months before, he spotted a middle-aged man, apparently a local, with his hands in his pockets and a beret on his head. The man just stood there, shouting out in the direction of the Palace Hotel. Dolibois took a moment to pause and listen so he could make out what the man was saying.

"Hallo Meier! Hallo Meier!" the man shouted, again and again. "Wie gehts in Berlin?"

It took a moment for the significance of what the man was doing to register with Dolibois. The phrase translated into English as "Hello, Meier! Hello, Meier! So, how's it going in Berlin?" The beauty of the moment dawned on him. In the early days of the war, Hermann Göring was so confident the Third Reich would win the war that he'd famously bragged to the German people, "If the British and the Americans ever bomb Berlin, my name is Meier."

The image of the middle-aged Luxembourger taunting Göring was a perfect end to this chapter in John Dolibois's life. The man was enjoying himself so much Dolibois felt no need to tell him that Göring was gone. Besides, Hermann Göring was never coming back.

In Washington, on July 6, 1945, in a classified memorandum with the subject heading "Exploitation of German Specialists in Science and Technology in the United States," the Joint Chiefs of Staff finally approved—on paper—a Nazi scientist program. President Truman was not made aware of the initiative. The governing body that had been assigned to "exercise general supervision" over the program and to "formulate general policies for procurement, utilization and control of specialists" was the Military

Intelligence Division of the War Department, G-2. A five-page memo was sent out to eight agencies within the War Department outlining "principles and procedures" governing the classified program. The three most important points were that "certain German specialists...could be utilized to increase our war making capacity against Japan and aid our postwar military research," that "no known or alleged war criminals should be brought to the United States," and that "the purpose of this plan should be understood to be temporary military exploitation of the minimum number of German specialists necessary." According to this initiative, as soon as the jobs were completed "the specialists would be returned to Europe." Participants, it was noted, should be hereafter referred to as "eminent German specialists" as opposed to "German scientists," because not all the Nazis being requested for program approval had degrees in science. Included in the mix were Nazi bureaucrats, businessmen, accountants, and lawyers. The project also now had an official code name, Operation Overcast. The name Paperclip would not be used for another eight months.

Military agencies that were interested in hiring German specialists were to submit their requests to the assistant chief of staff, G-2. "Only the most compelling argument should bring a German specialist to this country," the initiative stated, and only the "chosen, rare minds whose continuing intellectual productivity we wish to use" would be approved. The British would be made aware of the program, in general terms. At some point after the first large group of scientists had arrived in the United States, a "suitable" press release would be generated by the War Department so as to "avoid possible resentment on the part of the American public."

A list of desired German scientists—"List I"—accompanied the memo. It included 115 rocket specialists. When the British learned about the U.S. Army's intentions to hire the German rocket scientists, they asked to first be allowed to conduct two rocket exploitation projects of their own. The Americans agreed and released into British custody a group of scientists, engineers, and technicians including Wernher von Braun, Walter Dornberger, and Arthur Rudolph.

The first British project was called Operation Backfire, a V-2 field test that took place on Germany's north coast, at a former Krupp naval gun range in Cuxhaven. Operation Backfire was designed to analyze technical data about the V-2 by having the Nazi rocket engineers fire four rockets, also taken piecemeal from Nordhausen, at a target in the North Sea. This would allow the British to evaluate various technical elements, from how the rocket was launched to its flight controls and fuels. Arthur Rudolph, the former Mittelwerk operations director, was considered an expert in launch techniques, and to his biographer, he later recalled a scene from Operation Backfire: "The V-2 ran on alcohol of the same chemistry as that appearing in say, Jack Daniels and Old Grandad [*sic*]. The people at the test site apparently knew that." One night, according to Rudolph, a group of British and German V-2 technicians got drunk together on the rocket fuel. A British officer came upon the group arm in arm, "apparently comrades now, and lustily singing, *Wir Fahren gegen England,* or 'We Will March Against England.'" General Dornberger was not part of the drinking and singing. The British kept him on a short leash, away from the test firing and always under a watchful eye. The British had alternative plans for Walter Dornberger. They were not interested in the knowledge Dornberger possessed. They wanted to try him for war crimes. After the test, he would not be returned to the Americans as the British had originally promised.

"The British pulled a sneaky on us," explained Major Staver, who attended Operation Backfire. The Americans were not permitted to take Dornberger back after the Operation; instead, Dornberger was declared "on loan" and was taken to England. There, he and von Braun were "interrogated for a week by the British and then kept behind barbed wire in Wimbledon for four and one-half weeks while waiting to be picked up by the Americans." Eventually, von Braun was returned but General Dornberger was not. Instead, he was issued a brown jumpsuit with the letters "PW" for Prisoner of War stenciled on the back. Under armed guard, he was taken to the London District Cage near the Windermere Bridge for interrogation. From there, General Dornberger was transferred first to a castle in Wales and then to Special Camp XI in Island Farm, South Wales, where he was an extremely unpopular prisoner.

"Walter Dornberger was definitely the most hated man in the camp," Sergeant Ron Williams, a prison guard, recalled. "Even his own people hated him. He never went out to the local farms to work like other prisoners." Wherever General Dornberger went while he was at Special Camp XI, he required an escort. The British feared that other prisoners might kill him.

On the morning of September 12, 1945, Wernher von Braun, Dr. Eberhard Rees, and five midlevel V-2 rocket engineers left their U.S. Army–sponsored housing, in the town of Witzenhausen, for the last time. The men climbed into two army jeeps, headed for France. The Germans knew they were heading to the United States to work. They were not aware that their driver, First Lieutenant Morris Sipser, spoke German. As First Lieutenant Sipser drove the group to their destination in Paris, he listened to von Braun crack crass anti-American jokes. The jeep crossed over the Saar River into France, and Sipser overheard von

Braun say to his colleagues, "Well, take a good look at Germany, fellows. You may not see it for a long time to come." In Washington, Operation Overcast had been approved as a "temporary" program, but von Braun, ever the visionary, had the foresight to see that many of the rocketeers and engineers were heading to America to stay. "We felt no moral scruples about the possible future use of our brainchild," von Braun later told *New Yorker* magazine writer Daniel Lang. "We were interested solely in exploring outer space. It was simply a question with us of how the golden cow could be milked most successfully."

After arriving in Paris, the Germans were taken to the officers' club at Orly Airport for dinner. Accompanying von Braun, Rees, and the five other V-2 engineers were four scientists from the Hermann Göring Research Institute, handpicked by Colonel Putt and headed for Wright Field. They were Theodor Zobel, Rudolph Edse, Wolfgang Noggerath and Gerhard Braun. Luftwaffe test pilot Karl Baur was also with the group; he had served as aircraft manufacturer Messerschmitt's chief Me-262 test pilot. Accompanying him he had his mechanic, Andreas Sebald. A little before 10:00 p.m., the Germans boarded a C-54 military transport plane waiting on the tarmac in the pouring rain.

"Quickly the plane moved through the clouds and a beautiful, clear sky with a moonlight night greeted us," Karl Baur wrote in his diary. "For the first time—I cannot recall the number of years—I enjoyed a flight as a passenger."

After a stop to refuel on the island of Santa Maria, in the Azores, the aircraft crossed the Atlantic, refueled in Newfoundland, and landed at New Castle Airport in Wilmington, Delaware, at 2:00 a.m. on September 20, 1945. Because the Germans were under military custody, they could not be traditionally processed by Customs. After a few hours, the sixteen Germans boarded a second, smaller airplane and were flown to the Naval Air Station at Squantum, in Quincy, Massachusetts.

At the naval base the Germans were loaded into sedans and driven out to the edge of dock, where a troopship waited for them. They boarded the vessel and made a short trip to a chain of small islands in Boston Harbor, the Harbor Islands, and out to the far end to a gravel shoal off Nixes Mate, where they were obscured from civilian view. There, a small Boston whaler idled on the sea. Its captain was named Corky, and the twenty-one-year-old intelligence officer who would take charge of the Germans was named Henry Kolm. Each German scientist was lowered down into the little Boston whaler by a harness hanging from a rope. "They were all seasick as can be," Kolm later recalled.

Fort Strong, on an island in the middle of Boston

Harbor, had been used as a training camp during the Civil War and remained in use on and off through World War I. The fortresslike nature of the coastal defense facility made it an ideal place for a secret military program like Operation Overcast. When the first German scientists arrived in September 1945, the fortress was still under army control but had not been used for nearly thirty years. Thick weeds grew in between the gun blocks and pedestal mounts. The administration and observation buildings had fallen into disrepair. But the army barracks was easily converted into what would become known as the Operation Overcast hotel. German prisoners of war were moved onto the island to work as staff, including as translators, cooks, bakers, and tailors. Kolm's job was to process the scientists, which included fingerprinting, medical examinations, and coordination with the FBI. This all took time, and the Germans were not known for patience. Before long an insidious unease settled in among the scientists, Kolm recalled.

When the weather was clear the Germans played volleyball games outside. Far off in the distance they could make out the Boston skyline and see the tall, shiny buildings on shore. But often the fog rolled in, and the island was soaked in a thick, dense mist for days at a time. To pass the time the scientists stayed indoors and played Monopoly, which they called the "capitalists' game." Still, it was impossible to deny that

Fort Strong took on a penal colony feel, as Kolm recalled. Soon the Germans started calling their new home Devil's Island. Finally, on September 29, 1945, Major James P. Hamill, an intelligence officer with the Army Ordnance Corps, was sent to the east coast of the island to escort a group of Germans off. Hamill had been in Nordhausen with Major Staver at war's end and had personally assisted in the mission to locate enough rocket parts to reconfigure one hundred of them at the White Sands Proving Ground. His presence on Devil's Island had to have been a welcome one among the rocket scientists. Hamill took six of them — Eberhard Rees, Erich Neubert, Theodor Poppel, August Schulze, Wilhelm Jungert, and Walter Schwidetzky — to Aberdeen Proving Ground, where they began translating, cataloguing, and evaluating the information from the Dörnten mine stash. Hamill's next mission was to escort Wernher von Braun by train to Fort Bliss, Texas.

The train ride began on October 6, 1945, and was a memorable event. Operation Overcast was a highly classified military matter, and Major Hamill was required to keep watch over von Braun twenty-four hours a day. Drawing any kind of attention to the German scientist was to be avoided at all cost. In St. Louis, Major Hamill and von Braun were assigned to a Pullman car filled with wounded war veterans from the 82nd Division, renowned for parachute assaults into Sicily and Salerno. Also in the train car were wounded war veterans from the 101st Airborne Division, men lauded for action in the Normandy invasion and the Battle of the Bulge.

Hamill quickly arranged for himself and von Braun to be moved into a different car. The train moved along to the Texas border. Hamill watched as the man sitting next to von Braun began engaging him in friendly conversation, asking von Braun where he was from and what business he was in. Von Braun, apparently well versed in lying on demand, replied that he was from Switzerland and that he was in "the steel business."

"Well it turned out that this particular gentleman knew Switzerland like the back of his hand," Major Hamill later recalled, "and was himself in the steel business." Von Braun quickly qualified "steel business" to mean "ball bearings." As it so happened the man was also an expert on ball bearings, Hamill explained. The train whistle blew. The approaching station stop was Texarkana, which was the businessman's destination. As the man prepared to exit the train he turned back to von Braun and waved good-bye to him.

"If it wasn't for the help that you Swiss gave us, there is no telling as to whom might have won the war," the businessman said.

Army Ordnance finally had their scientists on U.S. soil, and work commenced at Fort Bliss without delay.

In Germany, the drama between the U.S. Chemical Warfare Service and FIAT over the suspected war criminal Otto Ambros escalated. Despite orders from SHAEF to arrest him, Ambros remained a fugitive in plain sight. He lived and worked with impunity in the French zone. At Dustbin, FIAT officers continued to learn more about who Ambros was and the wartime role that he had played in chemical weapons research and development. In an interrogation with Albert Speer, FIAT learned that no single person had been as critical to the development of Hitler's vast arsenal of nerve agents and poison gases as Otto Ambros had been. "He is known to have spoken to Hitler at a high-level German conference on Chemical Warfare," a FIAT report read. Another stated, "Ambros' importance, from the Intelligence point of view, has been re-emphasized by the recent Chemical Warfare investigation at 'Dustbin.' Speer and the German Chemical Warfare experts agree that he is the key man in German Chemical Warfare production." Major Tilley was outraged by it all, but there was little he could do except put Ambros under surveillance.

Initial attempts to capture Dr. Ambros maintained the fiction of civility. "At the end of August or beginning of September 1945, an attempt was made to induce Ambros to return to the American zone," Tilley wrote in a FIAT report. In response, "Ambros claimed to be unable to return then as the French authorities would not permit him to leave the French zone." Major Tilley knew this was a lie. Ambros regularly traveled back and forth between Ludwigshafen, in the French zone, and the IG Farben guesthouse, Villa Kohlhof, outside Heidelberg, in the American zone. This made Tilley furious. "This man is thought to be far too dangerous and undesirable to be left at liberty, let alone be employed by the Allied authorities," Tilley wrote. FIAT authorized him to set up a sting operation in an attempt to capture and arrest Ambros. Captain Edelsten was assigned the job of tracking Ambros day and night. Through the Counter Intelligence Corps in Heidelberg, Captain Edelsten, working with a Colonel Mumford, set up a network of undercover agents who began to follow Ambros's every move. "Saw Ambros at LU [Ludwigshafen]," read one report. "Drives his own car, usually alone. Slept in car for two hours one night, on roadside," another set of field notes revealed. Ambros traveled frequently: to Freiburg, Rheinfelden, and Baden-Baden. He'd even been back to Gendorf—a brazen move, considering the Seventh Army had an outstanding arrest warrant for him there. But Ambros was quicker than the U.S. Army and had a better intelligence network in place as well. Whenever the U.S. Army showed up to arrest him, he was already gone.

Finally, Captain Edelsten reached Ambros on the telephone. "Edelsten told Ambros that Col. Mumford was anxious to see him again," and asked if a meeting could be arranged for the following Sunday, at the Farben villa. Ambros agreed. FIAT planned its take-down. Plain clothes CIC officers would wait outside Villa Kohlhof, out of eyesight, until after Captain Edelsten placed Ambros under arrest. Then they'd step in and transport the wanted war criminal back to Dustbin. Except Ambros was one step ahead. With his well established "private intelligence center complete with secretaries and errand boys," he learned the meeting was a ruse to arrest him. When Edelsten arrived at the villa, Ambros's secretary invited him inside. Smiling broadly, she apologized and said "Dr. Ambros is not able to come." Edelsten feigned understanding and sat down at a large banquet table elegantly set for eight. The secretary whispered in his ear, "You have been at Gendorf," as if to taunt him.

Ambros, it turned out, had his own people following the FIAT agents who had been tailing him, a posse of "various deputies [between] Ludwigshafen and Gendorf." Embarrassed and infuriated by the audacity of it all, Edelsten finished his tea and got up to leave. "Just as Edelsten was leaving, Ambros's car pulled in (Chevrolet)," and for a fleeting moment Edelsten believed he would capture him after all. Instead, a look-alike "emerged from [the] car and said [Ambros] was unable to come." Ambros had sent a double; Edelsten had been burned.

FIAT was being made fun of, and there was little

they could do. Captain Edelsten left the IG Farben villa red-faced and empty-handed. "Gave CIC description of Ambros's car," he noted in his report. "Light blue closed Saloon (Chevrolet)." Edelsten posted his men to watch the bridges around Heidelberg and promised to arrest Ambros if he ever arrived. But of course he never did. Adding insult to injury, the following day Dr. Otto Ambros sent Captain Edelsten a formal note, neatly typed on fancy stationery, with Farben's wartime address still embossed on the letterhead. "Sorry that I could not make the appointment," Ambros wrote. He signed his name in thick black ink.

At Dustbin, Major Tilley continued to interrogate Ambros's colleagues. In late August 1945, Tilley got a very lucky lead from Ambros's former deputy, the Farben chemist Jürgen von Klenck. Von Klenck was back at Dustbin now, and under intense scrutiny after having been in attendance alongside Ambros at the meeting with Tarr and the representative from Dow Chemical, Wilhelm Hirschkind. Jürgen von Klenck had amassed a sizable Dustbin dossier of his own. A Nazi ideologue, von Klenck had been a loyal party member since 1933 and an officer with the SS since 1936. One interrogator described him as "wily," "untrustworthy," and "not employable." Von Klenck was also an elitist and had made more than a few enemies because of this. Speaking in confidence with

Major Tilley, fellow chemist and Dustbin internee Wilhelm Horn candidly expressed his thought about Jürgen von Klenck. Von Klenck had a "magnetic presence, a brilliant mind, [was] handsome, polished and a wonderful talker, but lack[ed] the essential characteristics that would make him a truly great man," said Horn. This was because von Klenck was "an egoist… very proud of the nobility implied in his name, and an opportunist who knew how to make the best of his chances." Horn confirmed that von Klenck was a long-standing and avowed Nazi but that "it had always grated von Klenck's pride that such common people as Hitler and his minions were in the highest places," said Horn.

Tilley asked Horn how high up von Klenck was in the production of war gases. Horn revealed that von Klenck had been deputy chief of the ultra-secret Committee-C for chemical weapons. In other words, Jürgen von Klenck was Otto Ambros's right-hand man. Armed with this information, Major Tilley presented von Klenck with a piece of paper to sign. It was a declaration, "stating that he knew none of which had been concealed by others." Von Klenck refused to sign. Tilley explained that withholding information was a crime. Threatened with arrest, von Klenck admitted that there were a few things he hadn't been entirely truthful about.

He told Major Tilley that in the late fall of 1944,

Ambros had instructed him to destroy all paperwork regarding war gases, particularly the contracts between Farben and the Wehrmacht. Instead von Klenck had "carefully selected" a cache of important documents and secreted them away in a large steel drum. He had hired someone to bury the steel drum on a remote farm outside Gendorf. Where, exactly, von Klenck said he wasn't sure. He told Tilley that he had "deliberately refused to learn where the [documents] were buried in order to be able to deny that he knew of any documents concealed anywhere." He gave Major Tilley a list of possible hiding places.

For two months Major Tilley searched the countryside around Gendorf for the steel drum, interviewing locals and patiently waiting for a solid lead. Finally, on October 27, 1945, he found what he'd been looking for. During an interrogation with Gendorf's fire chief, a man called Brandmeister Keller, the location was revealed. There was more. Brandmeister Keller had also hidden documents for Otto Ambros. "At first, Keller denied that he had secreted any documents," read the FIAT report. "When he was told that his arrest order was in Major Tilley's pocket he remembered four boxes Ambros had asked him to fetch in 1945....Ambros gave him various barrels and boxes to hide with various farmers in Gendorf." But the most important barrel, the large steel drum from von

Klenck, had been "buried at the lonely farm of Lorenz Moser, near Burghausen."

Von Klenck's hidden barrel produced hard evidence, including a letter from Ambros stating that he had been in charge of document destruction—and why. "All papers which prove our cooperation with Tabun and Sarin in the low-works, the DL-plant in the upper-work, must of course be destroyed or placed in security," read one letter inside the drum, signed by Otto Ambros. It was attached to a stack of nerve agent contracts between Farben and Speer's ministry, papers that von Klenck had been ordered to destroy. These contracts chronicled "full details of TABUN production and other details of DYHERNFURTH (now in Russian hands), including a detailed plan of all buildings and much or most of the apparatus...photographs and drawings...and many other valuable data, covering the period from 1938 or earlier until March 1945." Now Major Tilley had two key pieces of evidence he had not had before.

There was more damning information pertaining specifically to Otto Ambros that had been hidden in the steel drum. One document, written by Albert Speer, described two meetings with himself, Ambros, and Hitler in June 1944 which not only confirmed the high position that Ambros held but that he was a war profiteer. "I (SPEER) reported to the Führer that Dr.

AMBROS of I.G. FARBEN had developed a new process by which Buna of the same quality as natural rubber can be produced. Some time in the future no further imports of natural rubber will be required... the Führer has ordered a donation of one million marks" to Ambros, Speer wrote. A final piece of evidence shined a light on Farben's long-term plans for its business venture at Auschwitz. In addition to its Buna factory, IG Farben planned to produce chemical weapons at the death camp. The company had "plans for further construction of CW plant at AUSCHWITZ... in February 1945," Tilley wrote in his FIAT report.

Tilley returned to Dustbin with the steel drum and the documents. Von Klenck appeared shocked when Tilley told him that the steel drum had been located. "There are indications that he did not expect them to be found," read Tilley's intelligence report. FIAT now had documentary evidence that Ambros was "guilty of contravening American Military Government laws by concealing documents connected with German military preparations." There were hundreds of documents, including "full data on Dyhernfurth, production sheets for Tabun and other war gases," as well as "many other matters which [Ambros] claimed to have burned in the furnaces at Gendorf in April 1945. It may well be that these papers, added to the formulas, production methods, and the secret CW contracts between IG and the Reich from 1935 to 1945, which were discovered in a Gendorf safe on that same day, may give us a more complete picture of German CW preparations than of most other fields of German armament and general war production." Major Tilley had pulled off a scientific intelligence coup d'état.

Tilley took the information to a colleague, a veteran intelligence officer, who reminded him that there was an alternative theory to be considered regarding von Klenck. "Basing oneself on experience with enemy agents who confess plots freely, one may come to the conclusion that a lesser secret has been admitted to deflect the investigation from a more important secret," the officer explained. Efforts to capture Ambros were redoubled. Two days after the steel drum find, on October 29, 1945, the British Intelligence Objectives Sub-committee issued a warrant for the "Arrest of Dr. Ambros." Now there were two out of three Allied nations lobbying for the immediate arrest of Hitler's favorite chemist.

Otto Ambros was "dangerous" and "undesirable" and should not be "left at liberty," read the arrest report. FIAT knew Ambros could remain protected as long as he stayed in the French zone. "He is wily and he will remain there as he knows the hunt for him is on in the U.S. Zone," the report explained. There was little to do but wait. But patience again paid off.

It took three months for Otto Ambros's hubris to get the better of him. On January 17, 1946, Ambros traveled outside the French zone and was arrested. He was sent to Dustbin, where Major Tilley was waiting to interrogate him. After FIAT squeezed him for information, Ambros was turned over to Colonel Burton Andrus, now commandant of the Nuremberg jail. Sometime in the foreseeable future Ambros would face judgment at Nuremberg.

Any suggestion that Otto Ambros would one day have a prominent and prosperous place in civilized society, and that the American government would be just one of the governments to employ him, would have seemed pure fantasy. Then again, the Cold War was coming.

PART III

"The past is a foreign country."

—L. P. Hartley

CHAPTER ELEVEN

The Ticking Clock

In the late summer of 1945, the Nazi scientist program underwent a significant organizational change. At the behest of the State-War-Navy Coordinating Committee, control over the program was removed from the Military Intelligence Division of the War Department, G-2, and given over to the Joint Chiefs of Staff. The newly created Joint Intelligence Objectives Agency (JIOA) would now be in charge of decision making for the rapidly expanding classified program. The JIOA was a subcommittee of the Joint Intelligence Committee, which provided national security information to the Joint Chiefs of Staff. To understand the JIOA's power, and how it ran the Nazi scientist program so secretively, is to first understand the nature of the Joint Intelligence Committee. According to national security historian Larry A.

Valero, who has written a monograph on the subject for the CIA, the JIC was and remains one of the most enigmatic of all the American intelligence agencies. "The JIC structure was always in motion, always morphing and changing, a flexible, ad-hoc system," Valero says. "Subcommittees came and went, so did staff officers, but JIC decisions always had to be by consensus and were always reported to the Joint Chiefs." Little has been written about the inner workings of the JIOA, but the stories of individual Nazi scientists, and the JIOA's trail of partially declassified papers, help to define this powerful postwar organization.

In the immediate aftermath of the German surrender, the Joint Intelligence Committee was focused on the emerging Soviet threat. Between June 15, 1945, and August 9, 1945, the JIC wrote and delivered sixteen major intelligence reports and twenty-seven policy papers to the Joint Chiefs. "The most important JIC estimates involved the military capabilities and future intentions of the USSR," says Valero. Those intelligence estimates determined that the Soviets were ideologically hostile to the West and would continue to seek global domination, an attitude they had managed to skillfully conceal during the war. In September 1945, the JIC advised the Joint Chiefs that the Soviet Union would postpone "open conflict" with the West in the immediate future but only so it could rebuild its military arsenal and by 1952 be back at fighting strength. After this date, said the JIC, the Soviets would be ready and able to engage the United States in "total war."

The following month, JIC intelligence report 250/4 (the fourth report in the JIC 250 series) warned the Joint Chiefs that "eight out of ten leading German scientists in the field of guided missiles" had recently gone missing from Germany, had most likely been captured by the Soviets, and were now at work in the USSR. Similarly threatening, noted the report, two German physics institutes had been seized by the Red Army and reassembled in the USSR—not just the laboratories and the libraries but the scientists as well. JIC 250/4 warned of "intensive Soviet scientific research programs" under way across Russia, all of which threatened the West. It was from within this environment of intense suspicion that the JIOA was created. The Nazi scientist program was an aggressive U.S. military program from the moment the JIOA took control, just a few weeks after two atomic bombs ended the war with Japan. The employment of German scientists was specifically and strategically aimed at achieving military supremacy over the Soviet Union before the Soviet Union was able to dominate the United States.

Attaining supreme military power meant marshaling all the cutting-edge science and technology that could be culled from the ruins of the Reich. In the eyes of military intelligence, the fact that the scientists happened to be Nazis was incidental—a troublesome detail. It had no bearing on the bigger plan. The clock was ticking and, according to the Joint Intelligence Committee, would likely run out sometime around 1952.

There was language in the existing Nazi scientist policy that now had to be dealt with by the JIOA. The phrase "no known or alleged war criminal" could not remain as part of policy nomenclature for long, nor could the phrase "no ardent Nazis." These words had been put there to appease a few generals in the Pentagon, certain individuals in the State Department and moralists like Dr. H. P. Robertson, General Eisenhower's chief of scientific intelligence. For the program to move forward according to this new strategy, the language needed to change.

With the Joint Intelligence Objectives Agency now in control, a new, aggressive recruitment process would begin. On its governing body, the JIOA had one representative of each member agency of the Joint Intelligence Committee: the army's director of intelligence, the chief of naval intelligence, the assistant chief of Air Staff-2 (air force intelligence), and a representative from the State Department. The diplomat in the group was outnumbered by the military officers three to one.

The State Department officer assigned to the JIOA

was Samuel Klaus, and he was perceived by his JIOA colleagues to be a troublemaker from day one. Samuel Klaus was a forty-two-year-old shining star in the State Department, a brilliant lawyer, avid horseman, and Hebrew scholar who also spoke Russian and German. Because Klaus was the man on the JIOA who was in charge of approving the visas for all incoming German scientists, it was important that he be on board with what the JIOA wanted to accomplish. But Samuel Klaus was fundamentally opposed to the Nazi scientist program, and this created intense conflict within the JIOA.

Klaus had hands-on experience with Nazi Party ideology, owing to his wartime work for the U.S. Foreign Economic Administration during the war. During the war, Klaus ran Operation Safe Haven, a program with international reach designed to capture Nazi assets, including stolen art and gold being smuggled out of Germany for safekeeping in neutral countries. During his years running Safe Haven, Klaus had interviewed hundreds of German civilians, and he had developed the belief that many "ordinary Germans" had profited from the Nazi Party and had had a tacit understanding of what was happening to the Jews. In his role as the State Department representative on the JIOA, Klaus argued that the Germans at issue were not brilliant scientists who had been unwittingly caught in a maelstrom of evil but rather that they were amoral opportunists of mediocre talent. JIOA records indicate that Klaus's sentiments were shared by at least two of his State Department colleagues, including Herbert Cummings and Howard Travers. But it was Samuel Klaus who was unabashedly vocal about how he felt and what he believed. At a JIOA meeting in the late fall of 1945, Klaus vowed that "less than a dozen [German scientists] would ever be permitted to enter the U.S." on his watch. For this, he was seen as a thorn in the side of military intelligence and he was also outnumbered. Per the JIOA's charter, it was required to share its plan with a cabinet-level advisory board, which included a representative from the Department of Commerce. As it so happened, the representative from the Commerce Department, John C. Green, was an advocate of the German scientist program, apparently without benefit of knowing who these German scientists really were. In the fall of 1945 Green came up with an idea that would undermine Samuel Klaus's resistance to Operation Overcast.

After the war in Europe ended, President Truman put the Department of Commerce in charge of a program designed to excite the nation about a unique form of reparations being culled from the defeated German state, namely, the acquisition of scientific and technical information. There would be no financial compensation coming from Nazi Germany, the Department of Commerce explained, but American industry could now benefit from a different kind of restitution: knowledge. Secretary of Commerce (and former vice president) Henry Wallace had been appointed by the president to supervise the release of thousands of what would become known as PB reports, named after the Commerce Department's Office of the Publication Board. These reports contained non-armaments-related information collected by CIOS officers in Germany after the war. The idea behind the PB reports was to get average Americans to start their own small businesses inspired by German technological advances. These new businesses would be a boon to America's postwar prosperity, the Commerce Department said.

Thanks to Reich scientists, the public was told, beverage manufactures could now sterilize fruit juice without heat. Women could enjoy run-proof hosiery. Butter could be churned at the rate of 1,500 pounds per hour. These lists seemed not to end. Yeast could be produced in unlimited quantities, and wool could be pulled from sheepskins without injuring the animal's hide—all because of brilliant German scientists. Hitler's wizards had reduced suitcase-sized electrical components to the size of a pinkie finger and pioneered electromagnetic tape.

Henry Wallace was one of the nation's greatest

champions of the idea that Americans could find prosperity thanks to science. Wallace had served under Roosevelt in 1944, when Roosevelt promised Americans sixty million jobs. The promise became the subject of a book by Wallace: *Sixty Million Jobs.* As secretary of commerce, he intended to make good on it. Business, industry, and government could work together to make the world prosperous in peace, Wallace said. German science was a jumping-off point.

The public was not made aware of a second list regarding captured German scientific and technical information, one marked classified. This list catalogued eighteen hundred reports on German technology with military potential. Subject headings included: "rockets," "chemical warfare," "medical practice," "aeronautics," "ordnance," "insecticides," and "physics, nuclear." The man in charge of both lists—the classified and the unclassified one—was Henry Wallace's executive secretary and his representative on the JIOA, John C. Green.

Regarding the classified list, Green got an idea. Peace and prosperity were, in principle, sound ideas. But there was big business in war. Green wanted to make the classified list available to certain groups in industry. "Specialized knowledge [should not be] locked up in the minds of German scientists and technicians," Green said. It needed to be shared. To help foster this sharing, in the fall of 1945 John C. Green traveled to Wright Field, in Dayton, Ohio, to meet with Colonel Donald Putt.

The first group of six Germans brought to Wright Field in the fall of 1945 lived in an isolated and secure housing area called the Hilltop, a cluster of five single-story wooden buildings and three small cottages that had once housed the National Youth Administration. Almost no one but the program's administrators knew that the German scientists were there. There was a single-lane dirt road that passed by the Hilltop, used only by locals who needed to visit the town dump. Trucks and station wagons filled with trash sped by the Hilltop's secret inhabitants, and when it rained the road turned into a sea of mud. This annoyed the Germans, and they began compiling a list of grievances to share with Colonel Putt at a later date. They knew better than to complain just yet. Considering the fate and circumstances of many of their colleagues back in Germany, theirs was a particularly good deal. But when the timing was right they would share this list of indignities, which would in turn affect the Nazi scientist program in the most unusual way.

The original scientists at Wright Field were listed as Dr. Gerhard Braun, motor research; Dr. Theodor Zobel, aerodynamics; and Dr. Rudolph Edse, rocket fuels; the specialists were Mr. Otto Bock, supersonics; Mr. Hans Rister, aerodynamics; and Mr. Albert Patin, a businessman. Their salaries averaged $12,480 a year, plus a $6.00 per diem—the equivalent of about $175,000 in 2013. Because of an "oversight," later caught and corrected, the Germans did not pay U.S. taxes for the first two years and twenty months of the program.

At the Hilltop, a husband-and-wife team of housekeepers looked after the Germans' domestic needs, washed their laundry, and made their beds. German prisoners of war who had already been brought to the United States and were not yet repatriated acted as cooks. The six scientists and specialists and the others who would soon follow carried military-issue identification cards that had a large green "S" stamped on the front, indicating that they were not allowed to leave the base on their own. A gate running around the perimeter of the Hilltop was to be locked from 5:00 p.m. to 7:00 a.m. each night. On weekends, U.S. Army intelligence officers escorted the Germans into Dayton, where they could exercise at the local YMCA. A priest was brought in from Cincinnati to deliver Sunday mass in German. "We would like you to know and to appreciate that you are here in the interest of science and we hope that you will work with us in close harmony to further develop and expand your various subjects of interest," read the introductory pamphlet issued to each specialist at Wright Field.

"We have tried to make you comfortable in the quarters assigned to you."

What the Germans craved most was respect, and this eluded them. During the war Hitler's scientists and industrialists had been treated with great admiration by the Reich. Most scientists enjoyed financial reward. But here at Wright Field, many of the Germans' American counterparts looked down on them with disdain. "The mere mention of a German scientist is enough to precipitate emotions in Air Corp personnel ranging from vehemence to frustration," one manager stated in an official classified report.

As commanding general for intelligence at Wright Field, Colonel Donald Putt was in charge of the German specialist program. Putt had great admiration for each of the scientists, having handpicked almost every one of them in Germany, at the Hermann Göring Aeronautical Research Center at Völkenrode, and elsewhere. He could not fathom why the specialists were looked upon with contempt. "[A]ll they wanted was an opportunity to work," Putt said. Colonel Putt's vision for the Germans' workload in America was threefold. Initially, he planned for the men to write reports on their past and future work. Next, those reports would be translated and circulated among American engineers at Wright Field. Then Air Technical Service Command would hold research and development seminars at Wright Field, with invitations sent out to defense contractors, university laboratories, and any other interested parties with Top Secret clearance and a contract with the Army Air Forces. But Putt's idea came to a grinding halt after the War Department weighed in on his proposal, responding to what it called "calculated risk." The German scientist program was a highly classified military program and needed to remain secret. A War Department memo required that the Germans remain "properly segregated from persons not directly concerned with their exploitation." There was to be no fraternizing with American scientists. Collaboration with defense contractors and others was impossible at this point in the program.

The German specialists were offended by the way they were treated. Word from Dr. Herbert Wagner, inventor of the Hs 293 missile, was that Gould Castle on Long Island, where Wagner resided and worked, had marble bathtubs. Naval Intelligence allowed their German scientists to take field trips into Manhattan. The Germans at Wright Field told Putt they felt like "caged animals," and they demanded that something be done about it. Putt saw opportunity here. He wrote to Army Intelligence, G-2, to say that the Germans' overall malaise was "critically affecting" their ability to work. When the Pentagon ignored Putt's concerns, he appealed to Major Hugh Knerr, his commanding general at Air Technical Service Command. Knerr wrote to the Pentagon. "Intangibles of a scientist's daily life directly affect the quality of his product," he said, but this too had little effect. In Washington the general feeling was that Operation Overcast was temporary and that the Germans should be happy to have jobs. Besides, the Nuremberg trials were about to begin.

On October 18, 1945, an indictment was lodged by the International Military Tribunal against the defendants named as major war criminals. The trial would take place inside the east wing of the Nuremberg Palace of Justice, with opening statements beginning November 20. Because this was a military tribunal, sentences would be passed by judges, not jurors. Nuremberg as a city had played a unique role in the rise of the Nazi Party. It had been the site of Hitler's Nazi Party rallies—colossal military parades supported by as many as 400,000 Nazi loyalists—and home to the Nuremberg race laws. Now the leaders in the regime would be tried in this location for conspiracy, crimes against peace, war crimes, and crimes against humanity.

As the trial began, twenty-one defendants sat crammed onto two benches, inside Courtroom 600, headphones over their ears. (There were twenty-two defendants; Martin Bormann was tried in absentia.)

Behind them on the wall, symbolically positioned over their heads, was a large marble statue of the hideous monster Medusa. The twenty-one present faced the death penalty if convicted. "The wrongs which we seek to condemn and punish have been so calculated, so malignant, and so devastating," Chief U.S. Prosecutor Justice Robert H. Jackson famously declared, "that civilization cannot tolerate their being ignored, because it cannot survive their being repeated." The trial would last almost a year. With stories about Nuremberg and the Nazi war crimes dominating world news, complaints about comfort from the Germans at Wright Field meant very little to the War Department General Staff.

The same month the Nuremberg trial opened, in October 1945, the Army Air Forces hosted a grand two-day-long fair at Wright Field. On display were captured German and Japanese aircraft and rockets seen by the public for the first time since war's end. Over half a million people from twenty-six countries came to marvel at the confiscated enemy equipment, said to be worth $150 million. Among the items on display were the V-2 rocket, the Focke-Wulf Fw 190 G3 fighter aircraft, and the Messerschmitt Me 262. Particularly fascinating to the public was that some of the airplanes still had swastikas painted on their tails. The fair was so popular that the Army Air Forces extended it for five days. There was no mention made of the fact that several of the men who had designed and engineered these weapons were living a stone's throw away, at the Hilltop.

Among the half-million visitors at the fair was John C. Green, the Commerce Department's representative on the JIOA advisory board and its executive secretary overseeing the PB reports. The board had just changed its name to the Office of Technical Services, underscoring its transformation from a passive "board" to a more active "service" that would make use of cutting-edge science and technology. As planned, Green tracked down Colonel Putt at the Wright Field Fair. He had a myriad of questions for Putt, all of which centered around one idea: How could all this science and technology on display benefit American industry moving forward? Initially, Putt was uneasy about Green's attention, but in the end he decided to take a gamble on him. After all, John C. Green had access to the classified list. Putt shared with him some information about the German scientists on the Hilltop. How they were like men kept in an ivory tower, how their talents were squandered by policy and prejudice in some circles in Washington, D.C. They needed employment opportunities, Putt lamented. Perhaps Green could help?

Green seemed amenable, and Putt took note. "During his visit to the Air Forces Fair, Wright Field, [John C. Green] evidenced keen interest and inquired as to the reaction of industry toward the possible employment of German scientists," Putt wrote in a memo. He was not yet clear if Green's "influence is favorable or unfavorable." But Putt decided to take the risk. He forwarded to Green several "letters of interest" from defense contractors regarding potential employment of the German scientists. These documents had already been received by Air Material Command. They included letters from Dow Chemical Company, the AiResearch Manufacturing Company, and the Aircraft Industries Association. Defense contracts meant that there was business in the wings waiting to be transacted. It was Washington, D.C., that stood in the way. Putt explained to Green that these private businesses did not have a high enough security clearance to deal directly with the German scientists themselves.

John C. Green wrote to JIOA explaining what he had in mind: German scientists of "international repute" should be allowed, with their families, to enter the United States for long-term work, argued Green. This was good for American businesses. The letters from the defense contractors indicated that there was a great demand for this kind of work. The Commerce Department would set up a board to weed out the Nazis and bring the good Germans in. The German scientists' knowledge and know-how would be "fully and freely" available to all Americans, said Green.

This boom to industry would help create tens of thousands of American jobs.

Inside the JIOA, reactions were mixed, particularly among advisory board members. The assistant secretary of the interior was skeptical as to how Commerce could guarantee to keep old Nazis out of the program. The War Department did not like the idea of having to bring the families to America. Army Intelligence felt Green's proposal had validity from an economic perspective. If Commerce got involved in the German scientist program, the army would not necessarily have to shoulder so much of the financial burden. The State Department continued to voice objections, saying that regardless of who footed the bill, visas were not going to be granted to former enemies of the state without thorough and individual investigations. The Nazi scientist program was a temporary military program, State said. Nothing more.

John C. Green had an alternative plan. Instead of arguing his case further to the JIOA advisory board, he appealed to his boss, Henry Wallace. In turn, Henry Wallace wrote directly to President Truman, requesting that the president support the German science program. Science would help create those sixty million jobs, Wallace said, and nothing had a higher national priority in peacetime than American jobs. It was "wise and logical" to bring to America "scientists of outstanding attainments who can make a positive contribution to our scientific and industrial efforts," Wallace wrote to President Truman on December 4, 1945. The knowledge these men possessed, Wallace said, "if added to our own would advance the frontiers of scientific knowledge for national benefit." To illustrate his point, Wallace used one of the most benign scientists in all of Germany, a concrete and road construction expert named Dr. O. Graff, who had helped design the autobahn. "If you agree that the importance of a selected few (approximately 50 in number) would be an asset to our economy, I suggest you declare that this to be U.S. policy," Wallace urged the president.

For Colonel Donald Putt at Wright Field and the military intelligence members of the JIOA, Henry Wallace's endorsement of the program was like a shot in the arm. Before Wallace's letter to the president, Samuel Klaus of the State Department had suggested that the public would be outraged by the program once they found out about it. It could not stay secret forever, nor was it meant to. Klaus had said that bringing Hitler's former scientists to America for weapons research and development gave the impression that the army and the navy were willing to make deals with the devil for national security gains. Henry Wallace's economically minded endorsement changed all that. It gave the German scientists program an air of democracy, offering counterbalance to what could be perceived as an aggressive military program.

Henry Wallace had been staunchly anti-Nazi during the war. Preceding Truman in the vice presidency, Wallace had publicly called Hitler a "supreme devil operating through a human form." In another famous speech, he had likened Hitler to Satan seven times. That Henry Wallace was encouraging President Truman to endorse the German scientist program in the name of economic prosperity gave Operation Overcast a future. Henry Wallace was exactly what the JIOA had been waiting for.

On November 4, 1945, a headline in the *Washington Post* caught the nation's attention: "Army Uncovers Lurid Nazi 'Science' of Freezing Men." The article, written by reporter George Connery, was a major news scoop. In an effort to garner support for subsequent military trials in Nuremberg, the War Department had leaked to Connery the secret CIOS report written by war crimes investigator Dr. Leo Alexander. The report chronicled the freezing experiments conducted at Dachau inside Experimental Cell Block Five. That human beings had been tortured to death by German physicians in the name of medical science was both horrifying and incomprehensible to most Americans. The *Post* article revealed that the only man believed to have survived the freezing experiments had been located by Dr. Leo Alexander. Most of the other victims—the so-called *Untermenschen* whom the Nazi doctors had experimented on—died in the process or were killed. It was likely that this sole surviving victim, a Catholic priest, would provide witness testimony in the Nuremberg courts. Americans were rapt.

Kept secret from the public was an astonishing hypocrisy. Less than 150 miles from the Nuremberg courtroom, several of the physicians who had participated in, and many others who were accessory to, these criminal medical experiments were now being employed by the U.S. Army at the Army Air Forces Aero Medical Center, the classified research facility in Heidelberg. This laboratory, dreamed up by Colonel Harry Armstrong and Major General Malcolm Grow at a meeting in France in the spring of 1945, would remain one of the best-kept secrets of Operation Paperclip for decades to come. Here, starting on September 20, 1945, fifty-eight doctors handpicked by Dr. Strughold had been working on medical research projects begun for the Third Reich. Some of the data the Nazi doctors were using in their new Army research had been obtained in experiments in which test subjects had been murdered.

For Grow and Armstrong, the plan was to have these Luftwaffe doctors reconfigure the results of their war work in Heidelberg under army supervision. The follow-on plan was for these doctors to come to the United States under Paperclip contracts. Because conducting military research inside Germany was a violation of Allied Control Council Law 25 of the Potsdam Accord, the Aero Medical Center's classified nature shielded the Nazi doctors from chance exposure.

The codirectors of the secret research facility, Colonel Harry Armstrong and Dr. Hubertus Strughold, were alike in many ways, so much so that some saw the two men as mirror images of one another. The growing success of the Aero Medical Center would prove to be a launching point for each man's meteoric postwar career. Armstrong would eventually be promoted to U.S. surgeon general of the air force. Strughold would become the father of U.S. space medicine.

Harry Armstrong, born in 1899, entered into the U.S. military when horses were still being ridden into battle. During World War I Armstrong learned how to drive a six-mule ambulance and decided to become a doctor. After receiving his medical degree from the University of Louisville, he opened a private practice in Minneapolis. He might have become a country doctor, but he was preoccupied by airplanes and dirigibles. Appointed first lieutenant in the Medical Corps Reserve, Armstrong entered the School of Aviation Medicine (SAM) at Brooks Field, in San Antonio,

Texas, in 1924. In 1925 he decided to specialize in a field of medicine few had ever heard of, and soon he would become a flight surgeon. He had never flown in an airplane before.

It was a master sergeant named Erwin H. Nickles who inspired Armstrong to make his first parachute jump. In a lecture that took place in a parachute hangar, Nickles presented the idea that one day entire troops of infantrymen just might jump out of airplanes into combat situations as a group. After the class was over, Armstrong got in a long conversation with Nickles. "He told me that he was puzzled by the fact that people who he supervised in practice jumps almost invariably failed to follow his instruction which was to count ten after leaving the airplane before pulling the rip cord," Armstrong explained. He said that Nickles feared jumpers would "black out or get into a panic and pull the rip cord too quickly." When Nickles "hinted that he would be very happy if some doctor would make a jump to see if they could solve his problem," Armstrong's mind was made up. "I decided I would make a practice jump and delay my opening as long as possible."

A few weeks later Armstrong was standing in the cockpit of a biplane, wearing a flying suit and a gabardine helmet and getting ready to jump. "I had a feeling of panic," Armstrong explained, but he hurled himself out of the aircraft anyway. As he fell through the air he kept his eyes closed and paid attention to what his body felt like as he descended. The feeling of panic disappeared, he later recalled. He did not lose consciousness or black out. Armstrong allowed himself to free-fall for approximately twelve hundred feet before he finally pulled the ripcord. His parachute opened and he floated the last one thousand feet to earth, where he landed in a grassy Texas field. Harry Armstrong had set a U.S. Army record. He was the first flight surgeon to make a free fall from an aircraft.

Armstrong finished school and returned to Minnesota, but with an insatiable love of flying. On March 21, 1930, he closed his practice in Minneapolis and joined the army for good. His life as one of the most important figures in the history of American aviation and aerospace medicine had begun.

When Armstrong arrived with his family at Wright Field, in 1934, the world was enamored with airplanes, which were not yet associated with war but with peacetime progress and the spirit of adventure. Jimmy Doolittle set a transcontinental record flying from California to New Jersey in eleven hours, sixteen minutes. Wiley Post and Harold Gatty circled the globe in eight days. At Wright Field, the primary task of the flight surgeon was determining who was physically fit to fly in the airplanes of the day. Armstrong was a man with a vision and he was also a soldier. He envisioned a future where wars would be fought in the air. The Army Air Corps' most advanced fighter aircraft was a biplane with a speed of around 200 miles per hour and a flight ceiling of 18,000 feet. Armstrong's work centered around resolving problems related to oxygen deprivation and exposure to cold.

One day Armstrong spotted a trapdoor in the floor of his office inside Building 16 at Wright Field. He opened the door, saw a staircase, and climbed down. He found himself in a basement filled with old machinery and drafting tables. An unusual-looking chamber, like something out of a novel by Jules Verne, caught his eye. It was shaped like a globe, made of iron, and had windows like submarine portholes. This was the army's first and only low-pressure chamber, built decades earlier for its World War I flight surgeon school. The school, located in Mineola, Long Island, had closed down after the war and the chamber had wound up here, at Wright Field.

Armstrong got an idea. Next door to his office, Army Air Forces engineers were designing airplanes that could fly faster, higher, and farther than ever before. Armstrong wanted to begin research and development on the medical effects that flying these new airplanes would have on pilots. He wrote a letter to the engineering division at Wright Field requesting permission. The letter was forwarded to Washington.

In no time, Armstrong was appointed director of the Physiological Research Unit (later called the Aero Medical Research Laboratory and other variations on the name) at Wright Field.

His lab took off. Shop mechanics built test chambers from old airplane parts. He hired a scientist from Harvard, a PhD named John "Bill" Heim. Using volunteer test subjects, Heim and Armstrong gathered data on how the body responds to speed, lower oxygen levels, and decompression sickness and extreme temperatures. But it was an experiment on himself, with a rabbit on his lap, for which Armstrong became legendary.

He had been wondering what really happened to the human body above 10,000 feet. Why, and at what specific height, would a man die? Armstrong climbed into the low-pressure chamber with the rabbit on his lap. His technician adjusted the pressure to simulate high altitude. Armstrong's chest began to tighten and his joints hurt. When he rubbed his hands, he felt tiny bubbles along his tendons, ones that he could move around under his skin. He surmised that these were nitrogen bubbles forming in his blood and tissues, and that death at high altitude was caused by blood clotting. Armstrong indicated to the technician that he should simulate an even higher altitude inside the chamber. He was wearing an oxygen mask, but the rabbit on his lap was not. Soon the rabbit would be dead. As the lab technician raised the pressure, the rabbit convulsed and died. When Armstrong got out of the chamber, he dissected the rabbit and found nitrogen bubbles, proving that his hypothesis was correct.

Armstrong's discovery gave way to a major milestone in aviation medicine. Working with Heim on more tests, he inserted a viewing tube into the artery of a test animal. The two men took data on what happens to a mammal's body at forty thousand, fifty thousand, and finally sixty-five thousand feet. They were the first to witness that body fluids boil at sixty-three thousand feet. This point would become known as the Armstrong line. This is the altitude beyond which humans cannot survive without a pressure suit.

In 1937, Captain Harry Armstrong was considered one of America's aviation medicine pioneers. On October 2 of that year he attended the Aero Medical Association's first international convention, which took place in the Astor Gallery of the Waldorf-Astoria hotel. There, he and Heim reported the results of their recent studies at Wright Field. One of the doctors most interested in these studies was the Luftwaffe physician representing Germany, Dr. Hubertus Strughold. The two men, Armstrong and Strughold, were pioneers in the same field. "We hit it off immediately," remembered Harry Armstrong decades later. That fortuitous meeting would profoundly shape Strughold's post-Nazi career.

*   *   *

Some men claim to be shaped by a single event. For Hubertus Strughold, it was watching Halley's Comet streak across the sky from his backyard tree house in Westtünnen, Germany, in 1910. Forever after, said Strughold, he became preoccupied with what lies above. That same year, a second event shaped the rest of his life. Strughold watched a solar eclipse through a viewing glass and nearly went blind. The lens wasn't as dark as he thought it was and he burned the retina of his right eye, causing permanent damage. "When I looked at somebody with the right eye, at his nose, he had no nose.... When I looked at somebody at the street, at a distance of about a hundred meters at somebody's head [with the right eye], he did not have any head. It was always clear with both eyes," Strughold later explained. Hubertus Strughold had learned the hard way that experiments using one's own body could be dangerous. Still, as a young man, he pursued auto-experimentation with vigor and imagination. In college he studied physics, anatomy, and zoology, but it was physiology that interested him most, the functions of living organisms and their parts.

At the University of Würzburg Strughold taught the world's first college course on the effects high altitude had on the human body. His experimental test data came from himself. On weekends he flew hot air balloons, recording everything from vision to ear pressure to muscle effects. Inside the flying balloon, Professor Strughold recorded how his body responded to rapid acceleration and descent, which in turn made him curious to know how he would feel during radical banking turns. For that he needed an airplane. Strughold found the perfect mentor in a World War I flying ace by the name of Robert Ritter von Greim.

Von Greim was a legend. Renowned for his fearlessness in battle, in World War I he had recorded twenty-eight kills. In the 1920s and 1930s he was considered one of the top pilots in Germany and performed a variety of flight-related jobs, including exhibition dogfights against fellow World War I flying ace Ernst Udet. When Adolf Hitler needed a pilot to fly him from Munich to Berlin for the Kapp Putsch coup attempt in 1920, he chose Robert Ritter von Greim for the job. In 1926, von Greim was hired by Chiang Kai-shek to set up the Chinese air force in Canton, China. Returning to Germany, von Greim opened a flight school, located at the top of a mountain in Galgenberg, two miles from where Hubertus Strughold taught aviation medicine to college students.

Strughold hired von Greim to teach him how to fly, paying him six marks per lesson. The two men became fast friends. In Robert Ritter von Greim, Hubertus Strughold found a brilliant match—another man willing to push pilot performance to the edge of unconsciousness. The men would strap themselves into harnesses in von Greim's open cockpit airplane and fly loops and rolls in the skies over Galgenberg. Strughold kept track of their physiological reactions to extreme flight, seeking answers to questions. Can a man draw a straight line while flying upside down? Can a pilot mark a bull's-eye on a piece of paper immediately after a barrel roll? With how little oxygen could a man legibly write his name? How far up can a man fly before his vision fades? Von Greim was challenged by Strughold's strange requests, and he was willing to fly faster and higher as his new young physician friend recorded data on von Greim's pilot performance and physical capabilities in the air. Strughold knew his tests were original and hoped they would attract interest from the United States. In 1928 his wish came true when he received a prestigious fellowship from the Rockefeller Foundation. Strughold packed his bags, boarded the SS *Dresden,* and headed to New York.

Hubertus Strughold took to America like a fish to water, he later explained. As a Rockefeller Foundation fellow at the University of Chicago, he was at the center of the music scene in the roaring twenties. Listening to jazz music was his favorite pastime after flying. In Chicago he attended vaudeville shows, parties, and dances and became fluent in English. He loved to drink and almost always smoked. His thick German accent distinguished him from everyone else around and made most people remember him. His first scientific paper in English was on oxygen deficiency and how to revive a heart using electric shock. For research he used dogs as test subjects, importing them from Canada at a time when experimenting on dogs from the United States was illegal. Strughold attended conferences in Boston and visited the laboratories at Harvard, at the Mayo Clinic in Minnesota, and at Columbia University in New York. And he met and became friendly with American aviation medicine pioneers like Harry Armstrong.

The Rockefeller fellowship lasted only a year. Back in Germany, Strughold and von Greim took up where they left off. Von Greim was now flying a double-decker Udet Flamingo aircraft, an aerobatic sports biplane made of wood. To determine how many g-forces a man could take before his eyeballs suffered damage, the two men would climb high in the air, then dive down toward the ground until one of them blacked out.

Von Greim's longtime friend and colleague Adolf Hitler took power in 1933. In secret, von Greim was called upon by Hermann Göring to rebuild the German Air Force, in violation of the Treaty of Versailles. Through his personal friendship with von Greim, Strughold ingratiated himself into this inner circle of Nazi power. In 1935 he was offered a job that would shape the rest of his life. As director of the Aviation

Medical Research Institute of the Reich Air Ministry in Berlin, he was now in the uppermost echelon of Luftwaffe medical research. The lab, located in the suburb of Charlottenburg, featured a state-of-the-art low-pressure chamber and a ten-foot centrifuge in which test subjects could be exposed to varying degrees of gravitational pull. The chamber could take both apes and humans to between fifteen and twenty Gs.

The job required time, commitment, and most of all dedication. The Reich needed the Luftwaffe to help conquer all of Europe. This was why Strughold had packed his bags and moved to Berlin. He would report directly to Erich Hippke, chief of the Luftwaffe's medical corps, who reported to Hermann Göring. This was a vertical career move for Strughold. He'd gone from a university teaching post to the top of the Reich's aviation medicine chain of command. The Reich had vast resources and a desire to conduct groundbreaking experiments for the benefit of its pilots. There were risks but with risks came rewards. For Strughold, the reward was monumental. For ten years he enjoyed a career as one of the most powerful physicians working for the Third Reich.

In Berlin, Strughold treated his expansive new laboratory as a haven for risk takers. Colleagues, including officials from the Nazi Party and the SS, would stop by to marvel over his work with the low-pressure chamber and the centrifuge. Experiments were almost always in progress. Strughold's medical assistants were forever allowing themselves to be hooked up to these odd-looking contraptions with pipes, valves, and hoses projecting from all sides. Assistants, one with and one without an oxygen mask, would allow themselves to be locked inside the low-pressure chamber in order to determine how high up a man could go before becoming unconscious. During one experiment, two officials with the Reich Air Ministry were on hand to observe. The man without the oxygen mask began to lose consciousness. First his eyes closed, then his head fell to his chest. The second man inside the chamber, wearing a mask, administered first aid. It did not take long for the man to quickly recover.

"Our studies are all very risky," Strughold told the Nazi Party officials. "They require great ability on the part of the assistants and great responsibility. If the man did not get oxygen...he might be dead in five minutes."

As the Luftwaffe prepared its pilots for war, Strughold continued to use his staff as test subjects. He also experimented on himself. He was said to have "ridden the centrifuge" for a full two minutes, simulating what it would be like to experience fifteen times the force of gravity while flying an airplane.

After Germany invaded Poland in September 1939, the war spread into areas of extreme climates, from Norway to North Africa to the Russian front. These new combat theaters created urgent new medical problems for the Reich, most notably for foot soldiers but also for the Luftwaffe. As the war progressed and the Luftwaffe unveiled one new airplane after the next, the pilot physiology challenges grew. By 1940 new engine systems were being developed, including turbo and jet engines, with countless pilot parameters to explore, including the effects of speed, lower oxygen levels, decompression sickness, and extreme temperatures on the body. A web of institutions sprang up across Germany and its newly conquered lands, all financed by the deep pockets of the Reich Research Council, and including Strughold's Aviation Medical Research Institute in Berlin. The institute worked hand in hand with two Luftwaffe facilities close by, and Strughold developed strong relationships with the director of each institute: Dr. Theodor Benzinger, of the Experimental Station of the Air Force Research Center at Rechlin, and Dr. Siegfried Ruff, of the German Experimental Station for Aviation Medicine, Aero Medical division, in Berlin.

This is why, after the war, when Strughold was asked by Colonel Armstrong to be the codirector of the classified AAF Aero Medical facility in Heidelberg, Strughold asked Benzinger and Ruff to come along. He put each man in charge of one of the four areas of aviation research at the new facility. They trusted one another. They all had the same secrets to protect.

*   *   *

Dr. Theodor Benzinger was tall, thin as a rail, 5'11" and just 138 pounds. He had dark blue eyes, sharp, angular features, and kept his black hair slicked back, with a pencil part. Born in 1905, Benzinger was described, in his army intelligence dossier, as "an old school Prussian, willful, self-serving and willing to get what he wants by any means." At Heidelberg he was put in charge of a department that developed oxygen equipment for airplanes. Benzinger was a committed Nazi and had been from the earliest days of National Socialism. He joined the Nazi Party the year Hitler took power, in 1933. He was also a member of the SA, holding the position of medical sergeant major. He and his wife, Ilse Benzinger, were members of the NSV, the Nazi Party's so-called social welfare organization, which was overseen by Reich propaganda minister Joseph Goebbels. Ilse was active in NSV-sponsored programs like Mother and Child, whereby unwed German mothers could birth Aryan children on bucolic baby farms.

In 1934, the twenty-nine-year-old Dr. Benzinger was made department chief of the Experimental Station of the Air Force Research Center. Like Harry Armstrong, Benzinger predicted that pilots would fly high-altitude missions to sixty thousand feet sometime in the near future. In service of this idea Benzinger and his staff at Rechlin researched high-altitude durability and explosive decompression. They took great risks experimenting on themselves. On one occasion, one of Benzinger's technicians died as a result of complications from oxygen deprivation experienced inside a low-pressure chamber. In addition to researching aviation medicine, Benzinger became a pilot and served as a colonel in the Luftwaffe. He flew reconnaissance and combat missions over the British Isles. In 1939, showing "bravery before the enemy" Benzinger was awarded the Iron Cross, Class I and Class II.

In Heidelberg, at the Army Air Forces Aero Medical Center, Strughold put Dr. Siegfried Ruff in charge of work involving the effect g-forces have on human beings. This was work that Ruff had begun at the test center in Rechlin with Dr. Benzinger during the war. Ruff did not have the same striking looks as his colleague, Dr. Benzinger. Ruff's smiling, professorial posture made it hard to imagine he had spent so much time supervising medical experiments inside the Dachau concentration camp, including Rascher's murderous high-altitude studies in Experimental Cell Block Five. Like Benzinger, Ruff was an avowed and dedicated Nazi. He joined the party in 1938. The facility Dr. Ruff was in charge of for the Third Reich was located just ten miles across town from the institute that Dr. Strughold oversaw. As the directors of the two most important Luftwaffe medical facilities in Berlin, Ruff and Strughold collaborated closely on a number of projects during the war.

Ruff and Strughold coauthored several papers together and coedited *Aviation Medicine (Luftfahrt-medizin)*. One of the articles they cowrote so fascinated the U.S. Army Air Corps that in 1942 intelligence officers had it translated and circulated among flight surgeons at Wright Field. The two men also coauthored a book called the *Compendium on Aviation Medicine,* which served as a kind of handbook for Luftwaffe flight surgeons and included articles on explosive decompression and oxygen deficiency. At Heidelberg, Dr. Ruff was in charge of this work again, only now it was paid for by the U.S. Army.

Working directly under Dr. Ruff at the Aero Medical Center was Dr. Konrad Schäfer, listed in declassified documents as also researching the effects of g-forces on the body. This was not Schäfer's primary area of expertise. His wartime research work, which had been supported by both the Nazi Party's Reich Research Council and the Luftwaffe, was the pathology of thirst. Schäfer was a tall man, slightly overweight with a receding hairline and thick-lensed glasses that made him appear slightly cross-eyed. Unlike most of his colleagues, Schäfer avoided joining the Nazi Party, which he later said cost him jobs. In 1941 he was drafted and sent to a Luftwaffe air base at Frankfurt on the Oder. When his talents as a chemist came to light—he'd worked as chief physiological chemist for the firm Schering AG—Schäfer was transferred to Berlin and given an assignment in Luftwaffe sea emergencies. "This included research on various methods to render seawater potable," Schäfer later explained under oath.

Sea emergencies were an area of great concern. As the man in charge of aviation medical research for the Luftwaffe, Dr. Strughold had solutions to sea emergencies high on his priority list. During the air war, every pilot knew that drinking ocean water destroyed the kidneys and brought death faster than suffering indomitable thirst. But German pilots shot down over the sea and awaiting rescue were known to break down and drink seawater anyway. The Luftwaffe announced a contest. Any doctor or chemist who could develop a method to separate the salt from seawater would be greatly rewarded. Konrad Schäfer, one of Strughold's protégés in Berlin, aimed to solve that conundrum. Schäfer worked "in co-operation with IG Farben to create Wolfen, a mixture from barium and silver zeolith," he later explained, which he synthesized into "a tablet named Wolfatit [which] was developed to separate the salt in a residue." The results produced drinkable water, which was a remarkable achievement. Schäfer had succeeded where so many other doctors and chemists had failed.

Dr. Oskar Schröder, head of the Luftwaffe Medical Corps, was thrilled. Konrad Schäfer had "developed a process which actually precipitated the salts from the sea water," Schröder later testified. But another group of Luftwaffe doctors were already backing a different process, called the Berka method, which was bad news for the Schäfer process. "It was thought by the Chief of the Luftwaffe Medical Service to be too bulky and expensive," Schröder explained.

A second contest was proposed; this one to see which desalination method was superior. The effectiveness of both the Schäfer process and the Berka method would be tested on the *Untermenschen* at Dachau. A Luftwaffe physician named Hermann Becker-Freyseng was assigned to assist Dr. Schäfer, and to coauthor with him a paper documenting the results of the contest. The senior doctor advising Becker-Freyseng and Schäfer in their work was Dr. Siegfried Ruff. The resultant paper, called "Thirst and Thirst Quenching in Emergency Situations at Sea," described saltwater medical experiments conducted on prisoners inside Experimental Cell Block Five.

Dr. Hermann Becker-Freyseng had been a member of the Nazi Party since 1938. His specialty was oxygen poisoning in the human body. An odd-looking man, Becker-Freyseng's unusually large ears gave the appearance of handles on either side of his head. During the war Becker-Freyseng served as chief of the Department for Aviation Medicine and Medical Services in the Luftwaffe, another branch under an umbrella of medical facilities and laboratories overseen by Dr. Strughold. Becker-Freyseng was held in great esteem by his colleagues, many of whom, under interrogation, described him as "heroic" for the masochistic extremes he was willing to go in auto-experimentation. Becker-Freyseng conducted over one hundred experiments on himself, many of which rendered him unconscious. At least one took him to the brink of death. The story repeated most often about Becker-Freyseng was of a self-experiment he did in a chamber, also with a rabbit. Determined to learn how much oxygen would poison a man, Becker-Freyseng went into a low-pressure chamber with a rabbit with the goal of staying inside for three days. A few hours shy of his goal, Becker-Freyseng began to show symptoms of paralysis. "The rabbit died, Becker-Freyseng recuperated," Strughold later explained under oath. That was all during the war. Now Ruff, Benzinger, Schäfer, Schröder, and Becker-Freyseng, with the approval of Strughold and Armstrong, continued their work on secret aviation medical projects initially conceived for Hitler's war machine.

The Army Air Forces Aero Medical Center in Heidelberg was a squat, brick, two-story facility facing the Neckar River. Only a few months prior it had been the Kaiser Wilhelm Institute for Medical Research, a bastion of Nazi science where chemists and physicists worked on projects for the Reich's war machine. At its front entrance, the Reich's flag came down and the

U.S. flag went up. Photographs of Hitler were pulled from the walls and replaced by framed photographs of Army Air Forces generals in military pose. Most of the furniture stayed the same. In the dining room German waiters in white servers' coats provided table service at mealtimes. A single 5" x 8" requisition receipt, dated September 14, 1945, made the transition official: "This property is needed by U.S. Forces, and the requisition is in proportion to the resources of the country." The mission statement of the project, classified Top Secret, was succinct: "the exploitation of certain uncompleted German aviation medical research projects." Dr. Strughold was put in charge of hiring doctors, "all of whom are considered authorities in a particular field of medicine."

Across the American zone of Germany, entire laboratories were dismantled and reassembled here at the secret facility in Heidelberg. More than twenty tons of medical research equipment was salvaged from the Tempelhof Airport, in Berlin, including a "huge human centrifuge...and a low pressure chamber the length of two, ordinary Pullman cars." There was equipment here that American physicians had never seen before: esoteric items including a Nagel Anomaloscope, a Zeiss-made interferometer, an Engelking-Hartung adaptometer, a Schmidt-Haensch photometer, and a precision-built Siemens electron microscope—with which to study night vision, blood circulation, g-forces, and the bends. Even the low-pressure chamber from Georg Weltz's research facility at Freising, near the Munich dairy farm, was brought to Heidelberg. This was the laboratory where Dr. Leo Alexander had experienced his revelation that Nazi doctors had been freezing people to death.

It was a precarious time for doctors who had previously worked for the Reich. With the Nuremberg trial under way, the international press had its attention focused on war crimes. German doctors were looked at with suspicion. Articles about Nazi doctors, including the November 1945 piece in the *Washington Post* about the "science" of freezing humans, put a spotlight on German medicine. Many doctors fled the country to South America through escape routes called ratlines. Others tried to blend in by offering their services in displaced-persons camps. Some killed themselves. Maximilian de Crinis, chief of the psychiatric department at the University Charité in Berlin, swallowed a cyanide capsule in the last days of the war. Ernst-Robert Grawitz, physician for the SS and president of the German Red Cross, killed himself and his family, including his young children, by detonating a small bomb inside his house outside Berlin. The Reich Health Leader, Leonardo Conti, hanged himself in his cell at Nuremberg. Ernst Holzlöhner, the senior doctor at the University of Berlin who conducted the freezing experiments at Dachau with Sigmund Rascher, committed suicide in June 1945 after being interrogated by British investigators.

The list of suicides was long, but the number of German doctors believed to have been involved in war crimes was even longer. The U.S. war crimes office for the chief counsel wrote up a list of doctors involved in medical research that resulted in "mercy killings," a euphemism used by the Reich for its medical murder programs. The list was classified with a strict caveat that access to it remain "restricted for 80 years from the date of creation." This meant that, by the time the world would know who was on this list, it would be the year 2025, and everyone named would be dead.

A copy of the list was given to the commander of the Army Air Forces Aero Medical Center, Robert J. Benford. Five doctors working at the center starting in the fall of 1945 were on the list: Theodor Benzinger, Siegfried Ruff, Konrad Schäfer, Hermann Becker-Freyseng, and Oskar Schröder. Instead of firing these physicians suspected of heinous war crimes, the center kept the doctors in its employ and the list was classified. The list remained secret from the public until 2012, when the Department of Defense (DoD) agreed to declassify it for this book.

CHAPTER TWELVE

Total War of Apocalyptic Proportions

By the end of January 1946, 160 Nazi scientists had been secreted into America. The single largest group was comprised of the 115 rocket specialists at Fort Bliss, Texas, led by Wernher von Braun. The men resided in a two-story barracks on the Fort Bliss reservation and worked in a laboratory that was formerly the William Beaumont General Hospital. They ate in a mess hall shared with Native American Indians, which only enhanced von Braun's perception that he was living life inside an adventure novel. "It is such a romantic Karl May affair," von Braun wrote in a letter to his parents in Germany. Karl May was a German novelist famous for his cowboy and Indian westerns. Soon, von Braun would begin writing a novel of his own, in the science fiction genre, about space travel to Mars.

Von Braun loved the desert landscape, the cactus, the vast gypsum dune fields, and the long drives in open army jeeps. Rocket work was not perfect, but it progressed. "Frankly we were disappointed with what we found in this country during our first year or so," von Braun later recalled. "At Peenemünde, we'd been coddled. Here they were counting pennies," he said of the U.S. Army. V-2 launchings would take place about eighty miles away, on the White Sands Proving Ground, and getting there meant a long and beautiful ride. An army bus took scientists around the Franklin Mountains, through El Paso, and along the Rio Grande to Las Cruces. Next came the rugged journey over the San Andreas pass and into the Tularosa Basin, where the army's proving grounds began. Twelve to fifteen Germans were sent at a time to White Sands, where they lived in barracks alongside men from the General Electric Company and a technical army unit. The actual rocket firings took place inside a single forty-foot-deep pit, with the Germans watching the launches from a massive but rudimentary concrete blockhouse nearby. When the first V-2 was launched, in April 1946, it climbed to three miles. Although one of the fins fell off, von Braun felt inspired to draft a memo to Robert Oppenheimer, director of Los Alamos, proposing the idea of merging his missile with the atomic bomb. The memo turned into a proposal, "Use of Atomic Warheads in Projected Missiles," submitted to the army. In it, von Braun discussed building a rocket that could carry a two-thousand-plus-pound nuclear payload a distance of one thousand miles.

Two personal changes in von Braun were afoot. The first was that he joined an Evangelical Christian church and became "born again," something he rarely discussed in public. The second was that he decided to marry his first cousin, Maria von Quistorp, the daughter of his mother's brother, Alexander von Quistorp. Von Braun was nearly twice her age—she had just turned eighteen in the summer of 1946—and she lived in Germany. From Texas, von Braun began making plans to bring his future bride to the United States.

Descriptions of life out west in America varied from scientist to scientist. "The conditions of employment were considered to be fair and generous by all," said Dieter Huzel, the engineer who'd stashed the V-2 documents in the Dörnten Mine. Arthur Rudolph liked the fact that the swimming pool and the bowling alley were made available to the Germans exclusively one afternoon a week. He told his biographer, who did not want to be publicly identified and wrote using the pseudonym Thomas Franklin, that he missed his family and his Bible. Von Braun's brother, Magnus, was being investigated by the FBI for selling a platinum bar he had illegally smuggled into the United States. Interrogators with the Department of

Justice found Magnus von Braun to be "snobby" and "conceited" and said that he seemed to pose "a worse threat to security than a half a dozen discredited SS Generals."

For Army Ordnance there were many problems to overcome. Funding was scarce. Shrinking military budgets offered very little room for missile development immediately after the end of a world war. Also at Fort Bliss the army discovered that not all the so-called rocket scientists had the talents they allegedly possessed. Karl Otto Fleischer, Major Staver's original lead for the Dörnten mine and the man who led him on the wild goose chase around the Harz and to the Inn of the Three Lime Trees, claimed to have been the Wehrmacht's business manager, when in reality he had been in charge of food services. In Texas, Fleischer was assigned the job of club manager until he was finally "repatriated" to Germany. Von Braun had also sold the army on hiring Walter Weisemann, a Nazi public relations officer who had done some work in the Peenemünde valve shop. Von Braun called him an "eminent scientist." In reality, Weisemann learned engineering in America working for the army.

Fifteen hundred miles across the country, in the winter of 1946, there were now thirty German scientists at Wright Field. Colonel Putt considered this sum to be offensively low. At least once a month, he wrote to

Army Air Forces headquarters in Washington requesting more German scientists and inquiring why the importation of these "rare minds" was happening at a snail's pace. In fact, there was very little for the Germans to do at Wright Field, and many of them were restless. The Air Documents Research Center, formerly in London, had also moved to the Air Material Command headquarters, at Wright Field. There, five hundred employees sorted, catalogued, indexed, and put on microfiche some 1,500 tons of German documents captured by Alsos, CIOS, and T-Forces after the war. So abundant was the material that more than one hundred thousand technical words had been added to the Air Material Command's English-language dictionary. The plethora of information provided some work opportunities for a predominantly idle group of German specialists who in turn resented this kind of work. The Germans perceived themselves as inventors and visionaries, not librarians or bureaucrats.

One of the Germans, a Nazi businessman named Albert Patin, had been keeping track of the groups' complaints, which now made their way to Colonel Putt's desk. It was not just the lack of challenging work, said the Germans; it was the whole package deal. The Hilltop was a dump. Payments were slow. The mail to Germany was even slower. Dayton had no civilized culture. The laboratory facilities at Wright Field were nothing compared to the grand laboratories of the Third Reich. In general, the Germans told Putt, they were beginning to "distrust" their American hosts "based on promises broken by USA officers."

Colonel Putt's next move was a controversial one. He appealed to Albert Patin for help. At fifty-eight, Patin was one of the more senior Germans at the Hilltop. He was a wartime armaments contractor whose numerous factories produced equipment for the Luftwaffe under the Speer ministry. When Patin's facilities were first captured by the U.S. Army, one of the American technical investigators, Captain H. Freiberger, was so amazed by Patin's industrial vision that he called "the soundness of his principles a revelation." For Colonel Putt, Albert Patin's wartime innovations represented the best and the brightest of Reich science. Putt coveted the scientific inventions that Patin's factories mass-produced, which included navigation aids, in-flight steering mechanisms, and automatic control devices. This kind of technology would give the Army Air Forces a ten-year jump on anything the Russians had, Putt believed.

Hiring Patin for a U.S. Army Air Forces contract meant ignoring his past. His armaments factories used slave labor, which was a war crime. In an autobiographical report for Putt, Albert Patin admitted that many of the people in his six-thousand-person workforce were slave laborers supplied by Heinrich

Himmler's SS. Patin stated that he was not ashamed of this; he explained to Colonel Putt that he had been one of the better bosses in the Third Reich. He didn't encircle his factories in electric fencing like other industrialists did. Patin acknowledged that his wartime access to Hitler's inner circle benefited his businesses, but he did not see how this made him a war profiteer. He was just following orders. Patin took summer holidays with the Göring family and winter trips with Albert Speer's munitions procurement chief, Dieter Stahl, but so did a lot of people. He was no better and no worse.

During this slow period at Wright Field, Colonel Putt had Albert Patin survey the other Germans. He told Patin to be alert to grudges so that Patin could formalize his list of complaints. Putt would in turn forward this summary to his superiors at Air Materiel Command. Patin's job, Putt counseled, was to emphasize how the Germans had become depressed, even suicidal, without their families and without the promise of long-term work. Colonel Putt sent Patin's summary of complaints to Air Force Headquarters in Washington, to the attention of Brigadier General John A. Samford. In his own cover letter, Putt requested that immediate action be taken to "improve the morale [of the Germans] and save the existing situation."

The name Albert Patin had already caught General

Samford's eye. All mail sent to the scientists at Wright Field was screened by army intelligence first. Albert Patin regularly received letters from his staff back in Germany, many of whom also sought work in the United States. A letter had recently been sent to Patin in which lucrative offers from French and Russian intelligence agents were discussed. Brigadier General Samford's office was made aware of this unwelcome development. Coupled with the summary of German scientists' dissatisfaction, General Samford took action. He sent the complaint list as well as Patin's intercepted mail to the War Department. "Immediate action in this situation is imperative if we are to divert the services of valuable scientists from France and Russia to the United States," General Samford warned his colleagues.

The timing created a perfect storm. The Joint Intelligence Committee was in the process of implementing a major policy change. It had just warned the Joint Chiefs that the existing idea of using restraint when dealing with the Soviets needed to be reconsidered. "Unless the migration of important German scientists and technicians into the Soviet zone is stopped," read a JIC memo to the Joint Chiefs, "we believe that the Soviet Union within a relatively short time may equal United States developments in the fields of atomic research and guided missiles and may be ahead of U.S. developments in other fields of great military importance, including infra-red, television and jet propulsion." The JIC also stated, incorrectly, that German nuclear physicists were helping the Russians develop a nuclear bomb and that "their assistance had already cut substantially, probably by several years, the time needed for the USSR to achieve practical results." In reality the Soviets had gotten to where they were in atomic bomb development not because of any German rare minds but by stealing information from American scientists at Los Alamos. Not until 1949 would the CIA learn that the Russian mole was a British scientist named Karl Fuchs, who worked on the Manhattan Project.

In response to the perception that the Soviets were getting all the "important German scientists," the Joint Intelligence Committee proposed to the JIOA that three changes be implemented in the Nazi scientist program, effective immediately. The first was to do everything possible in Germany to prevent more scientists from working for the Russians. The second was that the U.S. Army was to make sure that German scientists and their families were given whatever it was they were asking for, including American visas. Third, a list was drawn up proposing that as many as one thousand additional Germans be brought to America for weapons-related research.

For Samuel Klaus, the second proposed policy change was untenable. The Nazi scientist program was originally defined as "temporary," with scientists working under military custody. That was how the War Department was able to circumvent immigration law for all the scientists already here. Now the JIOA was demanding that immigration visas be issued to scientists and their families. Even if the policy change were approved, Klaus argued, the visa process was a slow-going one. The State Department was legally required to approve each scientist's visa application individually. This was not an overnight task but a lengthy investigative process. The person requesting a visa was required to list on his or her application contacts who would in turn be interviewed by a representative from State. The Office of the Military Government in Germany needed to compile a security report on each individual scientist. Nazi Party records would have to be pulled from the Berlin Document Center. If the scientist had won an honorary award from the National Socialist German Workers Party (NSDAP)—the Nazi Party—or was a member of the SS or the SA, that needed to be explained. This was the law, Klaus said.

With the new information about the Soviets, Robert Patterson, now secretary of war, shifted from being weary of the Nazi scientist program to becoming its champion. Only a year earlier, Patterson had called the German scientists "enemies…capable of sabotaging our war effort," and had warned the Joint Chiefs of

Staff that "[b]ringing them to this country raises delicate questions." Now he stated in a memorandum that
"the War Department should do everything possible to
clear away obstacles that may be raised in the State
Department." This in turn caused Secretary of State
James F. Byrnes, Samuel Klaus's boss, to soften his
opposition to Operation Overcast. Due to the emerging Soviet threat, Secretary of State Byrnes and Secretary of War Patterson agreed informally that leaving
German scientists unsupervised inside Germany, where
they could be bought by the Russians, was too dangerous. If the State Department required individual investigations, so be it, Byrnes said. German scientists and
their families should be allowed to enter the country
under temporary military custody with an interim State
Department blessing, Patterson wrote. The logic was
simple. If we don't get them, the Russians will.

The State-War-Navy Coordinating Committee,
now acting as an advisory body to JIOA, confirmed
agreement with the positions of the Secretaries of War
and State but added another consideration to the argument. German scientists left to their own devices presented "serious military implications to the future of
United States Security," according to SWNCC. In
other words, Samuel Klaus's argument could now be
used against him in the military's attempt to speed up
the visa application process. Yes, the German scientists
were inherently untrustworthy—so much so that they

could not be trusted if they were left unsupervised, let alone left available to competing powers.

On March 4, 1946, SWNCC Paper No. 275/5 went into effect. German scientists could now be admitted to the United States in a classified program that was in the "national interest." This shifted the focus from whether or not someone was a Nazi to whether they were someone the Russians would be interested in. The commander in chief of U.S. Forces of Occupation in Germany and commander of U.S. Forces, European Theater (USFET), General Joseph T. McNarney, was told to draft a list of one thousand top scientists in Germany who were to be brought to the United States at once so the Russians couldn't get them. A military intelligence officer named Colonel R. D. Wentworth was assigned to provide General McNarney with material support on behalf of Army Intelligence, G-2. The scientists' families were to be given food and clothing and were to be housed in a secret military facility northeast of Munich called Landshut until their visa applications were approved. It was a radical revision of the initial terms of the German scientist program, and it was exactly what the JIOA had envisioned all along.

The following month, the members of JIOA were called together to spend an entire day hammering out new program protocols. Expert consultants like Alsos scientific director Samuel Goudsmit were invited to attend. Expedite the German scientist program, said the Joint Chiefs of Staff. There were now 175 German scientists in America under military custody, none of whom had visas. The consensus, save Klaus, was that the application process needed to be sped up. The thorniest issue had to do with getting the State Department to approve certain individuals who had clearly been Nazi ideologues, including members of the SS and SA. Also at issue were those men who received high awards for their important contributions to the Nazi Party. These were people that by regulation were entirely ineligible for citizenship.

The meeting resulted in a clever workaround. Army Intelligence officers reviewing the OMGUS security reports of certain scientists could discreetly attach a paperclip to the files of the more troublesome cases. Those files would not be presented to the State Department right away. Instead, those men would remain under military custody in America, most likely for a longer period of time than some of their fellows. As a result, the Nazi scientist program got a new code name. Operation Overcast had apparently been compromised after the families of the German scientists starting calling their U.S. military housing Camp Overcast. So from now on, the Nazi scientist program would be called Operation Paperclip.

* * *

Not everyone understood the discreet paperclip-attached-to-the-file protocol. The first major setback came just a few months later, on July 17, 1946. General Joseph McNarney wrote to JIOA stating that he had worked with Colonel Wentworth to identify 869 German scientists who were ready to sign Paperclip contracts. But there was an obstacle. "There is a large number of former Nazis and mandatory unemployables among those shown on the list," General McNarney wrote. "These [men] cannot now or later be employed in the United States zone of Germany except in the labor category." McNarney was following USFET rules that said all members of the SS and the SA had to go through mandatory denazification trials.

Citing America's "national interest," the JIOA would now change the language of the core principle guiding Paperclip's original charter. "No known or alleged war criminals" and "no active Nazis" would become no persons who might try and "plan for the resurgence of German military potential." Assistant Secretary of War Howard Petersen felt this new language would allow the JIOA to "bypass the visa people," as stated in a memo dated July 24, 1946. But this was meant to be temporary. Eventually, State Department officials like Samuel Klaus would take umbrage at this language. What JIOA really needed was an endorsement from President Truman.

By the summer of 1946 the relationship between the United States and the Soviet Union was shattering. The legendary Long Telegram, written by George F. Keenan, America's diplomat in Moscow, had been received at the State Department, reviewed by the president and his advisers, and sent to every U.S. embassy around the world. After analyzing the Soviet's "neurotic view of world affairs," Keenan warned his bosses at the State Department that "in the long run there can be no permanent peaceful coexistence" with the Soviet Union. The two nations were destined to become steadfast enemies, Keenan said.

Influenced by Keenan's insights, President Truman asked White House counsel Clark Clifford to prepare a study of the current state of affairs and the future prospects regarding Soviet-American relations from a military standpoint. To do so Clifford culled reports and briefings from the Secretaries of War, State, and Navy as well as the attorney general, the Joint Chiefs of Staff, various directors of military and civilian intelligence, and George Keenan. The result was an alarming Top Secret analysis. The report's conclusion was made clear in the introduction: "Soviet leaders believe that a conflict is inevitable between the U.S.S.R. and the capitalist states, and their duty is to prepare the Soviet Union for this conflict." Clifford warned that Soviet leaders were on a path "designed to lead to eventual world domination." The Russians were developing atomic weapons, guided missiles, a strategic air force, and biological and chemical weapons programs. The idea of "peaceful coexistence of communist and capitalist nations is impossible," Clifford wrote. The only way to counter this threat was to use the "language of military power." Not military force, but military threat.

On August 30, 1946, the undersecretary of the State Department, Dean Acheson, asked President Truman to make a decision on Paperclip. If the president did not act quickly, Acheson wrote, many of the German scientists "may be lost to us." After four days of deliberation Truman gave his official approval of the program and agreed that Operation Paperclip should be expanded to include one thousand German scientists and technicians and allow for their eventual immigration to the United States. With presidential approval official, the attorney general was able to expedite the proposed changes to the program. A new JIOA contract was drawn up, allowing scientists who had been in the United States for six months to sign on for another year, and with the government maintaining the right to renew the contract for another five years. Operation Paperclip was transitioning from a temporary program to a long-term one. Former enemies of the state would now be eligible for coveted U.S. citizenship.

In response to the Clifford Report, the Joint

Intelligence Committee conducted its own classified assessment of the Soviet threat, JCS 1696. The Soviet Union, wrote the JIC, sought world domination and would begin by bringing other nations into Soviet control to isolate the capitalist world. JIC saw a future war with the Soviet Union as being of apocalyptic proportion. In a war "with the Soviet Union we must envisage complete and total hostilities unrestricted in any way on the Soviet part by adherence to any international convention or humanitarian principals," noted JCS 1696. "Preparations envisaged on our part and our plans must be on this basis." In other words, for the United States to prepare for "total war" with the Soviets, America had to maintain military supremacy in all areas of war fighting, including chemical warfare, biological warfare, atomic warfare, and any other kind of warfare the other side dreamed up.

Copies of the classified report were sent out to thirty-seven or thirty-eight people, says CIA historian Larry A. Valero, including the Joint Chiefs of Staff. It was not known if President Truman received a copy of JCS 1696, as he was not on the distribution list.

One of the scientists on the JIOA list of one thousand was Dr. Kurt Blome. The Allies were unsure what to do with Hitler's biological weapons maker. Clearly, no discreet paperclip attached to Blome's file would be able to whitewash the reality of his inner-circle role as deputy surgeon general of the Third Reich. But if the United States were to go to war with the Soviet Union it would mean "total war" and, according to JCS 1696, would likely include biological warfare. America needed to "envisage" such a scenario and to plan for it, with both sword and shield. Dr. Blome had spent months at the Dustbin interrogation facility, Castle Kransberg, but had recently been transferred to the U.S. Army Military Intelligence Service Center at Darmstadt, located eighteen miles south of Frankfurt. In the summer of 1946, Dr. Blome was employed there by the U.S. Army "in the capacity of a doctor."

Dr. Kurt Blome's expertise was in great demand, but his future was as yet undecided. In his Posen laboratory, Blome had made considerable progress with live plague pathogens, including bubonic and pneumonic plague. How far that research progressed remained vague, likely because it would put an unwanted spotlight on human experiments many believed had taken place there. Blome repeatedly told investigators that he had intended to conduct human trials but never actually did.

Blome's American counterpart in wartime plague-weapon research was a left-leaning bacteriologist named Dr. Theodor Rosebury. During the war, the biological weapons work Rosebury conducted was so highly classified that it was considered as secret as atomic research. He had worked at a research facility outside Washington, D.C., called Camp Detrick. It was like Posen, only bigger. Detrick had 2,273 personnel working on Top Secret biological warfare programs. Like Blome, Rosebury worked on bubonic plague. Rosebury's colleagues worked on 199 other germ bomb projects, including anthrax spore production, plant and animal diseases, and insect research, in an effort to determine which bugs were the most effective carriers of certain diseases.

Almost no one in America had any idea that the U.S. Army had been developing biological weapons until January 3, 1946, when the War Department released a slim, sanitized government monograph called the Merck Report. That is when the American public learned for the first time that the government's Top Secret program had been "cloaked in the deepest wartime secrecy, matched only by the Manhattan Project for developing the Atomic Bomb." The rationale behind developing these kinds of weapons, the public was told, was the same as it had been with America's wartime chemical weapons program. If the Nazis had used biological agents to kill Allied soldiers, the U.S. military would have been prepared to retaliate in kind. Yes, the war was over, Americans were now told, but unfortunately there was a new and emerging threat out there, the Merck Report warned, an invisible and insidious evil capable of killing millions on a vast, unknowable scale. America's bioweapons program needed to continue, the Merck Report made clear. America may have won the war with the mighty atomic bomb, but biological weapons were the poor man's nuclear weapon. Biological weapons could be made by just about any country "without vast expenditures of money or the construction of huge production facilities." A bioweapon could be hidden "under the guise of legitimate medical or bacteriological research," the report said.

The Merck Report was written by George W. Merck, a forty-eight-year-old chemist and the owner of Merck & Co., a pharmaceutical manufacturer in New Jersey. Merck had served Presidents Roosevelt and Truman as civilian head of the U.S. biological warfare effort during the war. Merck & Co. made and sold vaccines, notably the first commercial U.S. smallpox vaccine, in 1898, and, in 1942, it manufactured penicillin G, among the first general antibiotics. During World War II, U.S. soldiers received smallpox vaccines. The man diagnosing the bioweapons threat, George Merck, was also the man whose company might sell the government the solution to combat the threat. In 1946 this was not looked upon with the same kind of scrutiny as it might have been decades later, because America's military-industrial complex had yet to be broadly revealed.

The Merck Report did not specify what kind of germ warfare had been researched and developed by the United States, only that it took place at a Top Secret facility "in Maryland." Camp Detrick was a 154-acre land parcel surrounded by cow fields about an hour's drive north of Washington, and under the jurisdiction of the former Chemical Warfare Service, then the Chemical Corps. After the release of the Merck Report, and coupled with the ominous "total war" prospects as outlined in the Clifford Report and the JCS 1696, Congress would grant vast sums of money to the Chemical Corps for biological weapons research and Detrick would expand exponentially.

Dr. Kurt Blome had information that was coveted by the bacteriologists at Camp Detrick, and plans were being drawn up to interview him. And then, in the summer of 1946, a totally unexpected event occurred inside the Palace of Justice in Nuremberg that would render hiring Dr. Kurt Blome for Operation Paperclip an impossibility, at least for now. In the tenth month of the trial, the Soviets presented a surprise witness, putting an unforeseen and unwelcomed focus on Dr. Kurt Blome. The witness was Major General Dr. Walter Schreiber—the shield to Blome's sword.

On August 12, 1946, prosecutors for the Soviet Union stunned the tribunal by announcing that a missing Nazi general and the former surgeon general of the Third Reich, Major General Walter P. Schreiber, was going to testify against his colleagues at Nuremberg.

Schreiber was brought forth as a witness to show that, after the Nazis' crushing defeat at Stalingrad, the Third Reich was planning to retaliate by conducting a major biological warfare offensive against Soviet troops. This was the first time information about biological warfare was being presented at the trial. The Allies were not informed that Schreiber was going to be a witness. U.S. prosecutors asked to interview him in advance of his testimony, but the Soviets denied the request. The medical war crimes investigator, Dr. Leopold Alexander, appealed to speak with Schreiber himself, to no avail.

During the war, Schreiber held the position of wartime chief of medical services, Supreme Command, Wehrmacht. He was the Third Reich's highest-ranking major general who was also a physician, and he held the title Commanding Officer of the Scientific Section of the Military Medical Academy in Berlin. Most important, he was the physician in charge of vaccines. Schreiber had been in Soviet custody for sixteen months, since April 30, 1945, when he was captured by the Red Army in Berlin. According to Schreiber, he had opened a large military hospital in a subway tunnel around the corner from the Führerbunker and had been tending to "several hundred wounded" soldiers when the Soviets captured him. After being taken by train to the Soviet Union, he was moved around various interrogation facilities, he said, until he ended up in Lubyanka Prison, the notorious penitentiary located inside KGB headquarters in Moscow. Nuremberg was Schreiber's first public appearance since war's end. No one, including his family members, had any idea where he had been.

That the former surgeon general for the Third Reich was now going to help Russian prosecutors send his former Nazi colleagues to the gallows for their crimes was as ironic as it was outrageous. Dr. Schreiber was on the U.S. Army's Central Registry of War Criminals and Security Suspects list. Along with many of those colleagues, and since war's end, Major General Walter Schreiber had been sought by the Allied forces for possible war crimes. If the Americans had located him and had a chance to interrogate him, he might well have been in the dock at Nuremberg facing the hangman's noose alongside his colleagues. Instead, here he was, testifying against them.

It was Monday morning, August 26, 1946, when Dr. Schreiber took the stand: Day 211 of the trial. Colonel Y. V. Pokrovsky, deputy chief prosecutor of the Soviet Union, presented Schreiber to the court as a witness. German defense lawyer Dr. Hans Laternser, counsel for the General Staff and Army High Command (Oberkommando des Heeres, or OKH), objected, on the grounds that the evidence was submitted too late. "The Tribunal is not inclined to admit any evidence so late as this, or to reopen questions which have been gone into fully before the Tribunal," said the tribunal's president, Lord Justice Sir Geoffrey Lawrence, "but, on the other hand, in view of the importance of the statement of Major General Schreiber and its particular relevance, not only to the case of certain of the individual defendants but also, to the case of the High Command, the Tribunal will allow Major General Schreiber to be heard as a witness." In other words, Schreiber was a high-ranking Nazi general and the judges wanted to hear what he had to say. With that, Schreiber was brought to the witness stand.

He stood five foot six and weighed 156 pounds. A long-sleeve shirt covered the saber scars on his right forearm. At fifty-three years old, Schreiber had been an active military doctor since 1921. He was an expert in bacteriology and epidemiology and had traveled the globe studying infectious diseases, from a plague outbreak in West Africa to a malaria epidemic in Tunisia. He claimed to understand medical aspects of desert warfare and winter warfare better than anyone else in the Third Reich. He was also an expert in biological and chemical weapons, in typhus and malaria epidemics, and in the causes and conditions of jaundice and gangrene. When war came Schreiber, an affable and ambitious son of a postal worker, was catapulted to the top of the Wehrmacht's medical chain of command. This was in part due to the Reich's zealously germ-phobic core. Schreiber's vast knowledge of and experience with hygiene-related epidemics made his expertise highly valuable to the Nazi Party. He was put in charge of the research to fight infectious disease, and also in remedial means to defend against outbreaks. In this way, he became privy to Reich medical policy from the top down. In 1942, Hermann Göring also put General Schreiber in charge of protection against gas and bacteriological warfare, which is how he came to be in charge of the Reich's program to produce vaccines.

"I swear by God the Almighty and Omniscient that I will speak the pure truth and will withhold and add nothing," Schreiber promised when taking the oath. Major General G. A. Alexandrov, the Russian assistant prosecutor, asked Schreiber what event had compelled him to testify at Nuremberg.

"In the second World War things occurred on the German side which were against the unchangeable laws of medical ethics," Schreiber said from the stand. "In the interests of the German people, of medical science in Germany, and the training of the younger generation of physicians in the future, I consider it necessary that these things should be thoroughly cleared up. The matters in question are the preparations for bacteriological warfare, and they give rise to epidemics and experiments on human beings." Schreiber was saying that the

Reich had been preparing for offensive biological warfare and had used the *Untermenschen*—the subhumans—as guinea pigs.

General Alexandrov asked General Schreiber why he had waited so long to come forth—if he had been coerced into making statements or if he had taken the initiative himself.

"I myself took the initiative," Schreiber declared. "When I heard the report of Dr. Kramer and Professor Holzlehner [Holzlöhner] here in Nuremberg I was deeply shocked at the obviously perverted conceptions of some of the German doctors," he said. Holzlöhner and Schreiber had been close friends and colleagues. After Schreiber learned about Holzlöhner's murderous freezing experiments—at the Nuremberg conference of 1942, "Medical Problems of Sea Distress and Winter Distress"—he invited Holzlöhner to come give the same lecture at the Military Medical Academy in Berlin.

But how had Dr. Schreiber heard these revelatory reports if he was in prison in the Soviet Union? Alexandrov asked. This had to have been a question on many people's minds.

"In the prison camp German newspapers were available in the club room," Schreiber claimed, making the notorious Lubyanka Prison sound like an aristocratic men's club as opposed to the draconian penal institution that it was. "I had to wait and see whether

352

this Court itself might not raise the question of bacteriological warfare," Schreiber said. "When I saw that it did not raise this question I decided in April to make this statement."

"Witness," said General Alexandrov, "will you kindly tell us what you know about the preparations by the German High Command for bacteriological warfare?"

"In July 1943, the High Command of the Wehrmacht called a secret conference, in which I took part as representative of the Army Medical Inspectorate," said Schreiber. "A bacteriological warfare group was formed at this meeting. As a result of the war situation the High Command authorities now had to take a different view of the question of the use of bacteria as a weapon in warfare from the one held up till now by the Army Medical Inspectorate," Schreiber testified. "Consequently, the Führer, Adolf Hitler, had charged Reich Marshal Hermann Goering to direct the carrying out of all preparations for bacteriological warfare, and had given him the necessary powers," Schreiber said. In this statement, Schreiber was contradicting the generally accepted notion that Hitler had never authorized his generals to use chemical or biological weapons against Allied troops. In fact, no chemical or biological weapons were ever used in World War II, which made it strange that Schreiber had been brought all the way to Nuremberg to testify to something that was ultimately irrelevant to the war crimes trial. Why, then, was Schreiber really there?

"At [this] secret conference it was decided that an institute should be created for the production of bacterial cultures on a large scale," Schreiber said, "and the carrying out of scientific experiments to examine the possibilities of using bacteria [in warfare]. The institute was also to be used for experimenting with pests which could be used against domestic animals and crops, and which were to be made available if they were found practicable."

"And what was done after that?" Major General Alexandrov asked rather pointedly.

"A few days later, I learned…that Reich Marshal Goering had appointed the Deputy Chief of the Reich Physicians' League, [Dr. Kurt] Blome, to carry out the work, and had told him to found the institute as quickly as possible in or near Posen."

"And what do you know about the experiments which were being carried out for the purpose of bacteriological warfare?" General Alexandrov asked.

"Experiments were carried out at the institute in Posen," Schreiber said ominously, referring to Blome's institute for plague research. "I do not know any details about them. I only know that aircraft were used for spraying tests with bacteria emulsion, and that insects harmful to plants, such as beetles, were experimented with, but I cannot give any details. I did not make experiments myself."

Alexandrov asked if the army high command knew about these experiments; Schreiber replied, "I assume so."

"Will you kindly tell us precisely what the reason was for the decision of the OKW to prepare for bacteriological warfare?" Alexandrov asked.

"The defeat at Stalingrad," Schreiber said, "led to a reassessment of the situation, and consequently to new decisions. It was no doubt considered whether new weapons could be used which might still turn the tide of war in our favor."

"So why didn't the Reich use biological weapons?" Alexandrov asked.

Instead of answering the question, General Schreiber went into minute detail regarding a meeting in March 1945 with Dr. Blome. "In March 1945, Professor Blome visited me at my office at the Military Medical Academy," Schreiber recalled. "He had come from Posen and was very excited. He asked me whether I could accommodate him and his men in the laboratories at Sachsenburg so that they could continue their work there; he had been forced out of his institute at Posen by the advance of the Red Army. He had had to flee from the institute and he had not even been able to blow it up. He was very worried at the fact that the installations for experiments on human beings at this institute, the purpose of which was obvious, might be easily recognized by the Russians for what they were. He had tried to have the institute destroyed by a Stuka bomb but that, too, was not possible. Therefore, he asked me to see to it that he be permitted to continue work at Sachsenburg on his plague cultures, which he had saved," Schreiber said.

Dr. Blome wasn't on trial. Why was Schreiber spending so much of his testimony talking about Dr. Blome? "During his visit Blome told me that he could continue his work at an alternative laboratory in [Geraberg,] Thuringia," Schreiber said, "but that this was not yet completed. It would take a few days or even a few weeks to complete it, and that he had to have accommodation until then. He added that if the plague bacteria were to be used when the military operations were so near to the borders of Germany, when units of the Red Army were already on German soil, it would, of course, be necessary to provide special protection for the troops and the civilian population. A serum had to be produced. Here again time had been lost, and as a result of all these delays it had never been possible to put the idea into effect."

Was Schreiber's testimony focused against Dr. Kurt Blome out of some kind of personal rivalry or vendetta? On the witness stand, Schreiber also fingered a number of other Reich medical doctors, none of whom was on trial. In addition to naming Kramer and Holzlöhner as organizers of the freezing experiments, Schreiber said that a man called Dr. Ding "had artificially infected [KZ prisoners] with typhus using typhus-infected lice" and that the "talented surgeon" Dr. Karl Gebhardt had "carried out cranium operations on Russian prisoners of war and had killed the prisoners at certain intervals in order to observe the pathological changes." Schreiber testified that the "Defendant Goering had ordered these experiments," and that the "Reichsführer-SS Himmler had kindly made available the subjects for the experiments." But in a disproportionate amount of his testimony Schreiber circled back to Dr. Blome's plague research for the Reich.

Dr. Hans Laternser was given an opportunity to cross-examine the witness. Laternser asked Schreiber if his testimony for the Russian assistant prosecutor was prepared. Schreiber said no.

"Was any advantage promised to you for making this report?" Laternser asked.

"No, nothing was promised me. I would refuse to allow anybody to hold out advantages to me," Schreiber said.

"Well, let us assume that such a devilish idea as actually to use bacteria did exist. Would that not have involved your troops in serious danger?" Laternser asked.

"Not only our troops, but the whole German people; for the refugees were moving from East to West. The plague would have spread very swiftly to Germany."

"I have one more question, Witness. Did you ever write down your objections to this bacteriological warfare?" Dr. Laternser asked.

Schreiber said, "Yes, in the memorandum which I mentioned before."

Dr. Laternser asked, "When did you submit that memorandum?"

"In 1942; may I now—"

"That is enough," Laternser interrupted. He'd caught Schreiber in a lie. "The conference took place in July 1943!"

Laternser had no further questions. The tribunal adjourned. Perhaps embarrassed by the fact that their star witness had been caught in a lie, the Russians did not call Schreiber back to the stand. Dr. Alexander made yet another attempt to interview him, again without success. The Russians said they were sorry, but Dr. Schreiber had already been transported back to Moscow. It was a curious event, but something did result from Schreiber's bizarre testimony. Two days later, a military vehicle pulled into the U.S. Army Military Intelligence Service Center at Darmstadt, where Blome had been employed by the army as a post doctor. Dr. Blome was arrested and taken to the prison complex at the Nuremberg Palace of Justice. A "confidential change of status report" now listed him as a prisoner in the custody of the 6850 Internal Security Detachment, Nuremberg, where Colonel Burton Andrus served as prison commandant.

Circumstance had altered Blome's future. He was off the Paperclip list and instead placed on a list of defendants who would face prosecution at the upcoming Nuremberg doctors' trial.

One hundred and fifty miles from Nuremberg, at the Army Air Forces classified research facility in Heidelberg, the massive undertaking forged ahead. For an entire year now, day in and day out, fifty-eight German physicians in white lab coats had been working on an array of research projects in state-of-the-art laboratories studying human endurance, night vision, blood dynamics, exposure to bomb blast, acoustic physiology, and more. They all reported to Dr. Strughold, who reported to the facility's commanding officer, Colonel Robert J. Benford. High-ranking military officers regularly visited the facility, including its two founders, General Malcolm Grow and Colonel Harry Armstrong. Grow was working in Washington, D.C., as the air surgeon (soon to be the first surgeon general of the U.S. Air Force). Harry Armstrong had returned to Texas where he was now commandant at the School of Aviation Medicine (SAM) at Randolph Field.

Working alongside the Nazi doctors in Heidelberg were dozens of army translators preparing English-language versions of the physicians' reports. By September 1946 there were over a thousand pages of documents completed. Soon everything would be compiled into a two-volume monograph for the Army Air Forces entitled *German Aviation Medicine, World War II.*

Work progressed well for Strughold's staff of doctors in Heidelberg until the institute was thrown into psychological chaos. On September 17, 1946, military security officers with the Counter Intelligence Corps, 303 Detachment, arrived at the center with five arrest warrants in hand. Doctors Theodor Benzinger, Siegfried Ruff, Konrad Schäfer, Hermann Becker-Freyseng, and Oskar Schröder were wanted by the International Military Tribunal, Nuremberg, for "War Crimes as suspect." The men were arrested and taken to the prison complex at the Nuremberg Palace of Justice, to the same wing where Dr. Blome was already incarcerated. If any of this were to come to light — that the U.S. Army Air Forces had been employing war crimes suspects and had them conducting military research at a facility inside Germany, expressly prohibited by Allied peace agreements — Harry Armstrong's institute would be shut down, Operation Paperclip would be exposed, and the U.S. Army would have an international scandal on its hands.

*   *   *

The Nazi doctors' trial was the first of the so-called subsequent trials to take place after the trial of the major war criminals at Nuremberg. It began on December 9, 1946. Unlike with the first trial, the twenty-three defendants at the doctors' trial — twenty doctors and three SS bureaucrats — were virtually unknown figures in the eyes of the American public. What was known, from earlier press coverage, was that these proceedings would put lurid Nazi science on trial. In the words of chief prosecutor General Telford Taylor, Nazi doctors had become proficient in the "macabre science" of killing. Torturous medical experiments conducted on concentration camp prisoners included freezing experiments, high-altitude tests, mustard gas research, seawater drinkability tests, malaria research, mass sterilization, and euthanasia. The *New York Times* called the doctors' crimes "beyond the pale of even the most perverted medicine" and cautioned that some details were difficult to report because they were impossible to comprehend. The *New York Times* cited one particularly grotesque example. Perfectly healthy people, "Jews and Slavs," had been murdered at the request of SS physician Dr. August Hirt for a university skeleton collection of the *Untermenschen*. This was the same anatomist named by Dr. Eugen Haagen in papers discovered by Alsos officers in Strasbourg in November 1944. Hirt, an expert in dinosaur anatomy, had committed suicide before the trial. The defendants ran the gamut from "the dregs of the German medical profession" to doctors who had once been internationally esteemed, like Dr. Kurt Blome.

On October 12, 1946, the *Stars and Stripes* newspaper, which operated from inside the Pentagon, listed the individual names of the doctors charged—a list that included the five Luftwaffe doctors who had been arrested at the U.S. Army Air Forces Aero Medical Center: Theodor Benzinger, Siegfried Ruff, Konrad Schäfer, Hermann Becker-Freyseng, and Oskar Schröder. In a matter of weeks, these physicians had gone from being employed by the U.S. Army to being tried by the U.S. military for war crimes. The ultimate judicial punishment was on the line: Each doctor faced a possible death sentence.

The following week, inside the prison complex at the Nuremberg Palace of Justice, there was a strange occurrence involving Dr. Benzinger. In Benzinger's pretrial investigation he admitted being aware of the fact that medical experiments were taking place at concentration camps and that nonconsenting test subjects had been murdered in the process. Benzinger also admitted that he had attended the October 1942 conference in Nuremberg, "Medical Problems of Sea Distress and Winter Distress," where data from murdered people was openly discussed among ninety

Luftwaffe doctors. During Benzinger's Nuremberg incarceration, prosecutors revealed to him that they had a new detail regarding his accessory to medical crimes, namely, the "motion picture of the record of the [medical murder] experiments" that had been shown at a private screening at the Air Ministry. Benzinger did not deny that he had been one of a select group of doctors invited by Himmler to attend this film screening; nor did he deny that he was one of nine persons chosen by Himmler to host the event. Benzinger was part of an elite inner circle of Luftwaffe doctors favored by the Reichsführer-SS, he conceded, but that was not a crime. But prosecutors also had a document that suggested Benzinger was far more implicated in the crimes than he let on. "After the showing of the film, most of the spectators withdrew and a small group of doctors remained behind [including] Benzinger. They asked Rascher... for a verbal report on the experiments," the document read. Benzinger insisted that all he did was listen. That there was no evidence of his having participated in any of the medical murders and that there were no documents and no eyewitnesses that could prove otherwise.

On October 12, 1946, Dr. Theodor Benzinger was announced as a defendant in the forthcoming doctors' trial. And then, just eleven days later, on October 23, 1946, Benzinger was released from the Nuremberg prison without further explanation. He was returned directly to the custody of the Army Air Forces, as stated in his declassified Nuremberg prisoner file. After spending a little over a month in the Nuremberg jail, Benzinger was back in Heidelberg continuing his U.S. Army research work. There was no explanation as to why Benzinger was dropped from the list of defendants in the upcoming doctors' trial. It would take decades for an important clue to be revealed, by Nuremberg trial expert and medical history professor Paul Weindling. As it turned out, in the fall of 1946, Benzinger had recently completed a paper on pilot physiology concerning stratospheric, or extremely high-altitude, aircraft. "The US [Army] Air Forces, Wright Field circulated his report on this topic in October 1946—just weeks after his detention and release" from the prison complex at the Nuremberg Palace of Justice, Weindling explains. The suggestion is that perhaps someone in the Army Air Forces felt Benzinger's expertise was more in the "national interest" than was trying him for war crimes. One of the Nuremberg prosecutors, a Boston attorney named Alexander G. Hardy, was outraged when he learned of Benzinger's release and insisted that the "interrogations were sloppy."

Back at the AAF Aero Medical Center in Heidelberg, the arrests of the five doctors on war crimes charges had everyone on edge. The classified programs taking place there began to quietly wind down. While

Doctors Ruff, Schäfer, Hermann Becker-Freyseng, and Schröder faced judgment at Nuremberg, thirty-four of the doctors remaining at the center prepared for shipment to the United States under Operation Paperclip. One of the first doctors in the group to head to America for work, in February 1947, was Dr. Theodor Benzinger.

At Nuremberg, while the doctors from Heidelberg remained incarcerated in one wing of the prison awaiting trial, preparations of another kind were also going on. The trial of the major war criminals was over. On the morning of October 1, the judges took turns reading the verdicts: nineteen convictions and three acquittals of the twenty-two accused major war criminals (one, Bormann, in absentia). That same afternoon, the tribunal pronounced what sentences would be imposed: twelve death sentences, three life sentences, and four lengthy prison terms. Albert Speer, the only defendant who pled guilty, was sentenced to twenty years.

The doctors from Heidelberg were housed in a separate wing of the prison where pretrial interviews would continue for another two months. Also in custody inside the prison complex at the Palace of Justice was Dr. Otto Ambros. Ambros would be tried in a subsequent trial, Case VI or the Farben trial, which was scheduled to begin in the summer of 1947.

As commandant of the facility, Colonel Burton Andrus was in charge of all prisoners, including those who were soon to be hanged. The condemned had roughly two weeks to live. Andrus described the surreal atmosphere of the last days at Nuremberg for the Nazis who had been sentenced to death. Ribbentrop, Kaltenbrunner, Frank, and Seyss-Inquart took communion and made a last confession to Father O'Conner, the Nuremberg priest. Göring bequeathed his shaving brush and razor to the prison barber. Streicher penned six letters. Ribbentrop read a book. Keitel requested that a German folk song be played on the organ as he was hanged. "On the night of 14th October, I made arrangements for the tightest security to be put around the prison," Colonel Andrus recalled. "I wanted the condemned men to know only at the very last minute that their time had come to be hanged."

The Nuremberg gymnasium was chosen as the location where the executions would take place. Each night, the U.S. Army prison guards played basketball there to blow off steam. The night before the gallows were constructed Andrus allowed the usual game to go on. "Late at night," remembered Andrus, "when the sweating players had trotted off to their showers, the grim-faced execution team passed through a door specially cut into the catwalk wall, and began their tasks in the gymnasium. A doorway had been cut into the blind side of the building so that no prisoner would see scaffolding being carried in." The condemned prisoners were also shielded from seeing the stretchers that would soon carry their corpses away.

While the gallows were being built, drama unfolded in the prison. Göring had requested death by firing squad instead of hanging—to be hanged was something he considered beneath him. His plea was considered by the Allied Control Commission and rejected. The night before he was to be hanged, Göring swallowed a brass-and-glass vial of potassium cyanide that he had skillfully managed to keep hidden for eighteen months. In a suicide note, he explained how he had managed to keep the vial hidden from guards by alternating its hiding place, from his anus to his flabby navel. War crimes investigator Dr. Leopold Alexander would later learn that it was Dr. Rascher of the notorious Dachau experiments who had originally prepared Göring's suicide vial for him.

Shortly after midnight in the early morning hours of October 16, 1946, three sets of gallows had been built and painted black. Each had thirteen stairs leading up to a platform and crossbeam from which a noose with thirteen coils hung. The executioner was Master Sergeant John C. Woods, a man whose credentials included hanging 347 U.S. soldiers over a period of fifteen years for capitol crimes including desertion. At 1:00 a.m. Colonel Andrus read the names of the condemned out loud. After each name, a bilingual assistant said, *"Tod durch den Strang,"* or death by the rope.

One by one the Nazis were hanged. At 4:00 a.m. the bodies were loaded onto two trucks and driven to a secret location just outside Munich. Here, at what was later revealed to be the Dachau concentration camp, the bodies of these perpetrators of World War II and the Holocaust were cremated in the camp's ovens. Their ashes were raked out, scooped up, and thrown into a river.

When asked by *Time* magazine to comment on the hangings, executioner John C. Woods had this to say: "I hanged those ten Nazis...and I am proud of it...I wasn't nervous....A fellow can't afford to have nerves in this business....The way I look at this hanging job, somebody has to do it."

So it went, just one year and a few months after the end of World War II.

Some Nazis were hanged. Others now had lucrative new jobs. Many, like the four German doctors from the Army Air Forces Aero Medical Center at Heidelberg, now awaiting trial at Nuremberg, were in the gray area in between. Was the old German proverb true? *Jedem das Seine.* Does everyone get what he deserves?

CHAPTER THIRTEEN

Science at Any Price

The same week that the major war criminals convicted at Nuremberg were hanged and their ashes thrown into a river, Undersecretary of State Dean Acheson called Samuel Klaus into his office at the State Department to discuss Operation Paperclip. At issue was the fact that JIOA had circulated a new Top Secret directive, JIOA 257/22. The way in which Paperclip participants' would now receive visas had officially been changed. Instead of allowing State Department representatives to conduct preexaminations in Europe prior to visa issuance, as was required by law, that process would be completed here in America by the commissioner of the Immigration and Naturalization Service (INS). "The Department of State would accept as final, the investigation and security reports prepared by JIOA, for insuring final clearance of individuals concerned," wrote JIOA director Colonel Thomas Ford. Acheson and Klaus were both aware that JIOA had wrested control of how visas were issued and had done so in defiance of U.S. law. But the president had signed off on the directive. Operation Paperclip was now officially a "denial program," meaning that any German scientist of potential interest to the Russians needed to be denied to the Russians, at whatever cost.

There were now 233 Paperclip scientists in the United States in military custody. The State Department was told to expect to receive, in the coming months, their visa applications and those of their family members. The information contained within the scientists' OMGUS security dossiers promised to be the "best information available." Samuel Klaus knew this vague new language meant military intelligence officers could withhold damaging information about certain scientists from State Department officials. The pipeline to bring ardent Nazis and their families into the United States was open wide.

Three weeks later the *New York Times* reported for the first time that there were Nazi scientists living in America under a secret military program. Its sources were the Russian army's Berlin-based newspaper, *Tägliche Rundschau,* and the Russian-licensed East German newspaper *Berliner Zeitung.* In a follow-up article, an anonymous source told the newspaper that "one thousand additional German scientists" were on the way. "All were described as volunteers and under contract," the article reported. "Their trial periods are generally six months, after which they can apply for citizenship and have their dependents brought to the United States." *Newsweek* magazine revealed that the secret military program's classified name was Project Paperclip.

Rather than deny the story, the War Department decided to go public with a sanitized version of its program. They would also make several scientists at Wright Field "available to press, radio and pictorial services." An open house was organized with army censors releasing details and photographs that would foster the appearance that all of the German scientists in the United States were benign. At Wright Field the "dirigible expert" Theodor Knacke gave a demonstration with a parachute. The eighty-year-old Hugo Eckener, former chairman of the Zeppelin Company, explained to reporters that thanks to his army contract he was now working with Goodyear on a new blimp design. Alexander Lippisch, inventor of the Messerschmitt Me 163 jet fighter, was photographed in a suit holding up a scale model of a sleek, futuristic, delta-wing jet, his hawkline nose staring down the end of the airplane. The emphasis on Lippisch was not that his jet fighter held records for Allied shoot-downs in the war but that his aircraft set international records for speed. Ernst Eckert, an expert in jet fuels, discussed high-speed gas turbines in his thick German accent. It was by mistake that the War Department allowed Eckert to chat with reporters, considering that his JIOA file listed him as a Nazi ideologue and former member of the SS and the SA. The program was becoming unwieldy, and no matter how hard the JIOA tried to maintain control, they could not keep an eye on all things. One American officer, assigned as a spokesman for the Germans, told reporters he so enjoyed working with German scientists, "I wish we had more of them."

Other German scientists at Wright Field were kept away from reporters, particularly those men who had been members of Nazi Party paramilitary squads like the SA and the SS. In aerodynamicist Rudolf Hermann's intelligence file, it was written that during the war, while working inside the wind tunnels in Kochel, Bavaria, Hermann had held morning roll calls in his brown SA uniform, and that he often gave speeches in support of Hitler. The information in engineer Emil Salmon's OMGUS security report was even more incriminating. At the aircraft factory where he had worked, Salmon had been known to carry a rifle and wear an SS uniform. "He also belonged to the Storm Troops [sic] (SA) from 1933–1945 and held the position of Troop Leader (Truppfuehrer)," read one memo. When bringing him to America, the army stated,

"This Command is cognizant of Mr. Salmon's Nazi activities and certain allegations made by some of his associates in Europe," namely, that during the war Emil Salmon had been involved in burning down a synagogue in his hometown of Ludwigshafen. But Emil Salmon was now at Wright Field because the Army Air Forces found his knowledge and expertise "difficult, if not impossible, to duplicate." Emil Salmon built aircraft engine test stands.

For various press events, the army provided photographs of some of the more wholesome-looking German scientists—definitely no one with a dueling scar. There were pictures of white-haired men playing chess, window-shopping outside a Dayton, Ohio, toy store, smoking cigarettes, and sunning themselves on army grounds. To be invited to the open house, a reporter had to agree in advance to clear his or her story with army censors before going to press. The military placed its own article in the *Stars and Stripes* purporting to tell the official story: None of the Germans had ever been Nazis; the men were under strict supervision here in the United States; they were all outstanding scientists and technicians "vital to national security"; they were moral family men.

The news stories about the scientists at Wright Field generated a flurry of response, including newspaper editorials and letters to congressmen. A Gallup Poll the following week revealed that most Americans believed that bringing one thousand more German scientists to America was a "bad idea." Journalist and foreign affairs correspondent Joachim Joesten was outraged by the very idea of Paperclip, writing in the *Nation*, "If you enjoy mass murder, but also treasure your skin, be a scientist, son. It's the only way, nowadays, of getting away with murder." Rabbi Steven S. Wise, president of the American Jewish Congress, penned a scathing letter to Secretary of War Patterson that was made public. "As long as we reward former servants of Hitler, while leaving his victims in D. P. [displaced-persons] camps, we cannot even pretend that we are making any real effort to achieve the aims we fought for." Eleanor Roosevelt became personally involved in protesting Operation Paperclip, organizing a conference at the Waldorf-Astoria hotel with Albert Einstein as honored guest. The former First Lady urged the United States government to suspend visas for all Germans for twelve years. When professors at Syracuse University learned that a new colleague, Dr. Heinz Fischer, an expert in infrared technology and a former member of the Nazi Party, had been sent by the army to work in one of their university laboratories under a secret military contract, they wrote an editorial for the *New York Times.* "We object not because they are citizens of an enemy nation, but because they were and probably still are Nazis."

The Society for the Prevention of World War

III—a group of several thousand writers, artists, scholars, and journalists—did not mince words in their December journal. The group had been set up during the war to advocate for harsh punitive measures against a nation they perceived as inherently aggressive and militaristic, and against individuals they believed had substantially profited from the Nazi regime. "These German 'experts' performed wonders for the German war effort. Can one forget their gas chambers, their skill in cremation, their meticulous methods used to extract gold from the teeth of their victims, their wizardry in looting and thieving?" The society, which counted William L. Shirer and Darryl Zanuck among its members, urged all fellow Americans to contact the War Department and demand that Hitler's scientists be sent home.

One engineer at Wright Field actually was about to be sent home. But tunnel engineer Georg Rickhey's downfall came not because of the demands made by the public but because of the actions by a fellow German.

In the fall of 1946, of the 233 Nazi scientists in America, 140 were at Wright Field. With that many single men living together in isolation, the Hilltop became divided into social cliques. The Nazi businessman Albert Patin continued to serve as Colonel Putt's ears and eyes among the Germans, reporting to Putt about the Germans' needs and complaints. In this arrangement Patin wielded power at the Hilltop. With the arrival of former Mittelwerk general manager Georg Rickhey, in the summer of 1946, Albert Patin saw a business opportunity. Rickhey was a former employee of the Speer ministry. He served as the general manager of the Mittelwerk slave labor facility in Nordhausen where the V-2 rockets were built. Colonel Beasley of the United States Strategic Bombing Survey had evacuated Rickhey to London, and after that work was completed, Rickhey was hired to work at Wright Field. When asked by military intelligence officers what his job was, Rickhey said, "Giving my knowledge and experience in regard to planning, construction and operating underground factories."

Georg Rickhey was a tunnel engineer, and his value to the U.S. military was the knowledge he had gained while overseeing vast underground building projects for the Reich. A memo in Rickhey's dossier, written by the army's Office of the Deputy Director of Intelligence, kept his most powerful negotiating secret a secret: "He was in charge of all tunnel operations directly under Hitler's headquarters in Berlin." Georg Rickhey had overseen the building of the Führerbunker, where Hitler lived the last three months and two weeks of his life. With its more than thirty rooms, cryptlike corridors, multiple emergency exits, and hundreds of stairs, all located more than thirty feet under Berlin, the Führerbunker was considered an engineering tour de force. The army was impressed by how well Hitler's Führerbunker had withstood years of heavy Allied bombing and was interested in learning from Georg Rickhey how to build similar underground command centers of its own.

As the Cold War progressed, the U.S. Army would begin the secret construction of such facilities, notably ones that could continue to function in the aftermath of a chemical, biological, or nuclear attack. It would take decades for journalists to reveal that, starting in the early 1950s, several sprawling, multifloor, underground command centers had been secretly built for this purpose, including one in the Catoctin Mountains, called Raven Rock Mountain Complex, or Site R, and another in the Blue Ridge Mountains, called Mount Weather.

Georg Rickhey's expertise in underground engineering was not limited to the construction of the Führerbunker. During the war, he had also served as a director of the Reich's Demag motorcar company, where he oversaw the construction of a massive underground facility where tanks had been assembled. And as general manager of the Mittelwerk, he oversaw the construction of the rocket assembly facility near Nordhausen. In his army intelligence dossier, it was noted that Rickhey had overseen the underground construction of more than 1,500,000 square feet of space. At Wright Field, Rickhey was meant to start consulting with American engineers on underground engineering projects for the army. But that work was slow and Rickhey was given a second job. Despite poor English skills, Rickhey was put in charge of examining V-2 documents captured at Nordhausen and, in his words, the "rendering of opinions on reports."

In the late summer of 1946 Albert Patin and Georg Rickhey started running a black market operation at the Hilltop, selling booze and cigarettes to their colleagues at premium prices. Rickhey had years of experience in wartime black market operations. Military intelligence would later learn that during the war, what little rations the slave laborers were allotted by the Speer ministry, Rickhey would sometimes sell off at a price. At Wright Field, Rickhey and Patin's black market business quickly expanded, and the men enlisted outside assistance from Rickhey's sister, Adelheid Rickhey, who was living in a hotel in New Jersey at the time. Adelheid Rickhey agreed to move to Ohio to help the men expand their black market business into Dayton and beyond. Because the Germans' mail was monitored, it did not take long for the higher-ups at Wright Field to learn what was going on, but the Army Air Forces took no action against either man. The business continued into the fall of 1946, when it came to a head.

In their private time Rickhey and Patin liked to gamble. They also liked to drink. The two men regularly hosted parties at the Hilltop, staying up late drinking and playing cards. One night, in the fall of 1946, a sixty-three-year-old German aircraft engineer named Hermann Nehlsen decided he had had enough. It was a little after midnight during the second week of October when Rickhey, Albert Patin, and a third man were playing cards. The noise woke up Hermann Nehlsen, who knocked on the door and told the three cardplayers to quiet down. After Nehlsen's second request was ignored, he opened the door, walked into the room, and turned off the lamp on the card table. Rickhey, by Nehlsen's account, was drunk. As Nehlsen was leaving, Rickhey lit a candle and laughed at his German colleague. A man "could still play cards with a good kosher candle," Rickhey joked.

The incident was a tipping point for Hermann Nehlsen, and he went to Colonel Putt to file a formal complaint against the two men. There was more to the story, Nehlsen told Colonel Putt. Rickhey was a war criminal. He had been the primary person behind a mass hanging at Nordhausen, the hanging by crane of a dozen prisoners, and had bragged about it on the ship ride coming over to the United States. As for Albert Patin, Hermann Nehlsen told Colonel Putt he was an ardent Nazi as well, a member of the SA. Patin's companies used slave laborers from concentration camps. What Hermann Nehlsen did not know was that Colonel Putt was not interested in the past histories of the German scientists in the Army Air Forces' employ. Or that Putt had a gentleman's agreement with Albert Patin and had been using him to keep an eye on the other Germans at the Hilltop. Colonel Putt said he would look into the allegations. Instead, Putt had Nehlsen watched for future security violations. Hermann Nehlsen remained angry about the incident and wrote so in a letter to a friend in New York City named Erwin Loewy. Perhaps Nehlsen knew that all of his mail was read by military screeners, or maybe he did not. Either way, what Nehlsen wrote to Loewy caught the eye of Wright Field mail censors. Nehlsen's letter was turned over to the intelligence division to be analyzed. Several weeks later, Hermann Nehlsen violated security and left Wright Field for a weekend visit with a relative in Michigan. Colonel Putt reported the violation to Air Material Command headquarters and arranged to have Nehlsen transferred to Mitchel Field, an air force base in New York.

But Hermann Nehlsen's letter to Erwin Loewy about Georg Rickhey had a life of its own. It made its way to Army Air Forces Headquarters, in Washington, D.C., where it was read by Colonel Millard Lewis, Executive Assistant Chief of Air Staff-2. The allegations were serious, Colonel Lewis decided. They involved alleged war crimes. Lewis sent a memorandum to the director of intelligence for the War Department General Staff summarizing the situation and advising that an investigating officer be assigned to look into the matter. "Mr. Nehlsen stated that Dr. Georg Rickhey, age 47, specialist in the field of engineering and production of guided missiles, was employed in the underground plant at Nordhausen as a strong Nazi Party member, where in 1944, twelve foreign workers were strung up on a cross beam and raised by a crane in the presence of the group of workers," and killed, read Colonel Lewis's memorandum. "One of the group who acted as an observer asserted that Dr. Rickhey was the chief instigator for the execution." The Pentagon assigned an Air Corps Major named Eugene Smith to look into the Georg Rickhey case.

Back at Wright Field, Colonel Putt likely believed that the allegations against Rickhey had fallen by the wayside. Meanwhile, Hermann Nehlsen had been banished to Mitchel Field. In January 1947 Putt recommended that Rickhey be given a long-term military contract for employment at Wright Field. That request was authorized, and on April 12, 1947, Georg Rickhey signed a new, five-year contract with the War Department. Separately, and without Putt's knowledge, the Rickhey investigation was moving forward. At Army Air Forces Headquarters in Washington, Major Eugene Smith made preparations to travel to various military bases for interviews. It was Smith's job to talk to Rickhey's former colleagues from Nordhausen, to write up his findings, and to file a report.

Colonel Putt did not learn about this investigation until Major Smith arrived at Wright Field. Putt suggested Major Smith discuss Rickhey's case with Captain Albert Abels, the officer in charge of the Paperclip scientists at the Hilltop, so that Abels could clear things up. Abels told Major Smith that the stories were "petty jealousy" between scientists. Just gossip among men. Major Smith was unconvinced. He headed to Mitchel Field to interview Hermann Nehlsen and other German scientists who might have knowledge regarding the Rickhey case.

Hermann Nehlsen stood by his original story. "In 1944, twelve foreign workers were simultaneously hanged by being strung up on a cross beam and raised by a crane in the presence of the workers," Nehlsen swore in an affidavit. "Dr. Rickhey was the chief instigator for the execution," meaning it was Rickhey's idea to hang the men. A second witness emerged at Mitchel Field, a former engineer on the Nordhausen rocket assembly lines named Werner Voss. Voss also testified that Rickhey was involved in the hangings, and he provided important new details that added context to the executions. Shortly before the hangings, Voss said, British airplanes had dropped leaflets urging the Nordhausen slave laborers to revolt. A group of them did revolt, and those men were among the ones hanged. The execution was a public event, Voss said, meant to intimidate other slave workers into subservience. These were serious allegations of war crimes. Major Smith needed corroboration, and for that, he headed to Fort Bliss, Texas. Smith had been told by the army that he would be able interview Wernher von Braun as well as some of the other men who had worked closely with Georg Rickhey inside the Nordhausen tunnels.

At Fort Bliss, in the evenings, the German rocket specialists gathered in a clubhouse built in a grove of cottonwood trees. There, they played cards while drinking American cocktails and beer. They strung hammocks between the trees and enjoyed balmy desert nights. Once a month, most of the Germans went into El Paso as a group to shop and have a restaurant meal. The nightmarish environment at Nordhausen, which the majority of the rocket engineers had participated in, must have seemed like another world— until Major Smith arrived to take official depositions for the U.S. Army and ask the men to recollect what those days had been like.

Major Smith's first interviews would be with Wernher von Braun and his brother Magnus von Braun. But after Major Smith arrived at Fort Bliss, he was told that both von Braun brothers were out of town. Smith interviewed Günther Haukohl instead. Haukohl had been the original designer of the Nordhausen rocket assembly line and, like Wernher von Braun, was an officer with the SS. Günther Haukohl told Major Smith that he had no clear memories about what happened in the tunnels in the last months of the war, but, yes, prisoners were abused. Rickhey's involvement is "probably a rumor," Haukohl said. V-2 engineers Hans Palaoro and Rudolph Schlidt seconded Haukohl's position, repeating the idea that the end of the war was nothing but a blur. Engineer Erich Ball, who also worked on the rocket assembly lines at Nordhausen, told Major Smith that he had witnessed two sets of prisoner hangings in the tunnels but that Georg Rickhey had not been involved in either of them. Major Smith accepted that the men might not be able to remember details about what had gone on, but surely as engineers they could remember the layout of the facility. Smith asked Haukohl, Schlidt, Palaoro, and Ball to use their engineers' precision and help him create an accurate illustration of the work spaces inside the Nordhausen tunnels, including where the rockets were assembled and where the hangings occurred. This was an important artifact for the investigation.

Next, Major Smith interviewed the former Mittelwerk operations director Arthur Rudolph. Like Georg Rickhey, Arthur Rudolph had authority over the Mittelwerk's Prison Labor Supply office, which was the unit responsible for getting food rations to the slave laborers. In his interview, Arthur Rudolph first denied ever seeing prisoners abused. Major Smith showed

Arthur Rudolph the illustration that had been drawn by Rudolph's Nordhausen colleagues Haukohl, Schlidt, Palaoro, and Ball. Major Smith pointed out to Arthur Rudolph that Rudolph's office was directly adjacent to where the twelve so-called political prisoners had been hanged from the crane. As Arthur Rudolph continued to deny ever seeing prisoners abused, Smith found his testimony increasingly suspicious. Everyone else interviewed admitted to having seen some prisoner abuse. Rudolph was adamant: "I did not see them punished, beaten, hung or shot," he told Major Smith.

Smith approached the question in a different manner. He asked Rudolph if he could recall anything about the twelve men who had been hanged from the crane. This public execution, Smith said, had been confirmed by many of Rudolph's colleagues, and Smith was trying to put together an accurate portrait of what had happened, and when. Rudolph replied, "[O]ne [dying prisoner] lifted his knees, after I got there." In other words, Rudolph had witnessed the executions. Major Smith was now convinced that the truth about what happened at Nordhausen was being covered up, collectively, by the group, and that Arthur Rudolph knew a lot more than he let on. But as Smith noted in his report, the subject of the investigation was Georg Rickhey, not Arthur Rudolph. "Mr. Rudolph impressed the undersigned as a very clever, shrewd individual," Smith wrote. "He did not wish to become involved in any investigations that might involve him in any way with illegal actions in the underground factory and as a result, was cautious of his answers."

Smith returned to Washington and advised U.S. Army Air Forces Headquarters of his findings. In Germany, a serendipitous incident occurred at this same time. William Aalmans, the Dutch citizen who had been working on the U.S. War Crimes Investigation Team, had been hunting down SS officers from Dora-Nordhausen for an upcoming regional war crimes trial. Aalmans had managed to locate only eleven out of three thousand SS officers who had run the concentration camp. Aalmans was the man who had interviewed many of the newly liberated prisoners back in April 1945, and he was also the individual who had found the Mittelwerk employee telephone list tacked to the tunnel wall. The two names at the top of that list were Georg Rickhey and Arthur Rudolph. Aalmans had no idea that both men had been recruited into Operation Paperclip. In Germany, no one Aalmans interviewed claimed to know the whereabouts of either man. Then, one afternoon in May 1947, Aalmans was taking a break from his Nazi hunting work, reading the *Stars and Stripes* newspaper in the sunshine, when, as he told journalist Tom Bower decades later, "I just saw a tiny headline, 'German Scientist

Applies for American Citizenship.'" The name mentioned was Georg Rickhey, Aalmans explained. "I screamed for joy and rushed into the head office shouting, 'We've found him!'" Within three days the Georg Rickhey matter had been escalated to the attention of the Joint Chiefs of Staff.

On May 19, 1947, an arrest warrant was issued for Georg Rickhey on orders from the War Crimes Branch, Civil Affairs Division, in Washington. "Georg Rickhey is wanted as a principal perpetrator in the Nordhausen Concentration Camp Case," the document read. Rickhey's immediate reaction was to claim he was being falsely accused. This was a case of mistaken identity, Rickhey declared. In response, the Office of the Deputy Director of Intelligence, Headquarters, European Command, had this to say: "The Dr. Georg Rickhey now employed at Wright Field and the Georg Rickhey engineer at Nordhausen are one and the same person. Dr. Karl Rahr, chief physician at Nordhausen, has been interrogated and states that he knew Rickhey, that he is now in his late 40's [or] early 50's and that there was only one Georg Rickhey at Nordhausen. This Group is making arrangements to have Rickhey returned to Germany for trial in the Nordhausen Concentration camp case." The War Department's Intelligence Division assigned the chief of security at Wright Field, Major George P. Miller, the role of escorting Rickhey back to Germany.

Rickhey's five-year contract was terminated. He was instead going to be a defendant in the Dora-Nordhausen war crimes trial.

On June 2, 1947, JIOA deputy director Bosquet N. Wev wrote to J. Edgar Hoover, the powerful director of the FBI, alerting him of the situation and sending along Rickhey's classified dossier. If the public learned that an Operation Paperclip scientist had been returned to Germany to stand trial for war crimes, the JIOA, the Army, the War Department, and the FBI would all have a lot of explaining to do. It was in everyone's interest to keep the Georg Rickhey fiasco under wraps.

On August 7, 1947, Georg Rickhey appeared as one of nineteen defendants in the Dora-Nordhausen trial. The Nazis were charged with the deaths of at least twenty thousand laborers who were beaten, tortured, starved, hanged, or worked to death while being forced to build V-2 rockets. The trial, which took place inside a former SS barracks adjacent to the Dachau concentration camp, lasted four months and three weeks. The prosecution requested that Wernher von Braun be allowed to testify at the trial, but the army said it was too much of a security risk to allow von Braun to travel to Germany. The Russians could kidnap him, the U.S. Army said. They did not say that von Braun had recently traveled to Germany to marry his cousin and bring her back to Texas with him. At the end of the trial, fifteen of the nineteen defendants were found guilty. Four, including Georg Rickhey, were acquitted. "Then, in an unprecedented move, the Army classified the entire trial record," explains journalist Linda Hunt. The record would remain secret from the public for another forty years. There was far too much at stake for the U.S. Army to allow the information about what had really gone on in the Nordhausen tunnel complex to get out to the public. The trial record would call attention to the backgrounds of the 115 rocket scientists at Fort Bliss. The future of the United States missile program was too important. The Russians had their rocket scientists and America needed hers, too.

Shortly after Rickhey was acquitted, back in Washington, D.C., the office of the secretary of the air force received a call from the State Department. The official identified himself as Henry Cox. The air force captain who took Cox's call wrote up a memo of what was said.

"After exchanged pleasantries, [Cox] inquired if we had ever returned a specialist to Europe for trial. I thought a moment and, realizing that the Rickhey case was in the public domain and in all probability facts known to the State Department since his sister made inquiries, I did not dare deny the story. I, therefore, told Mr. Cox that one Air Force case had been returned to Europe under suspicion, that my recollection was he was cleared; that in any event we did not desire to take further chances on the case and that he had not been returned to the United States." The State Department appeared to have bought the story.

Rickhey was gone, but there was no shortage of Nazis coming to Wright Field to work. On August 22, 1947, one of the Reich's top ten pilots, Siegfried Knemeyer, arrived. This was the daring pilot and engineer whom Albert Speer had hoped would help him to escape to Greenland and whom Hermann Göring had called "my boy."

"Knemeyer is very popular with his fellow nationals since he is a good mixer and free of vanity," read a memo in his intelligence dossier. "He is a diligent worker with an inventive mind. He is absorbed by his work and especially enthusiastic about participating in aerial tests." Soon Knemeyer's wife, Doris, and their seven children arrived in America to join him and to become U.S. citizens. The family moved into a large, drafty farmhouse on Yellow Springs Road. Doris Knemeyer hated provincial life in Dayton, Ohio. In Berlin, the Knemeyers had a grand home in the Charlottenburg district, with many servants to help take care of the Knemeyer brood. Raising seven children by herself in America was not what Doris Knemeyer had in mind. The difficulties at home did not go unnoticed by Knemeyer's supervisors at Wright Field.

"Since the arrival of his family he seems harassed and neglectful of his personal appearance," read an internal security report. But Knemeyer was determined to succeed. Assigned to the Communication and Navigation Laboratory at Wright Field, Knemeyer found his stride. He began to make significant contributions to navigational instruments for his new employer, the U.S. Air Force — no longer part of the army anymore. Knemeyer is "a genius in the creation of new concepts in flight control," wrote Colonel John Martin, Knemeyer's superior at the lab.

Knemeyer's friend Werner Baumbach, Hitler's general of the bombers, had been scheduled to come to Wright Field to work alongside Knemeyer, but an entry in Baumbach's intelligence dossier noted that there had been a last-minute change: "Lt. Colonel Baumbach has since been substituted," it read. Baumbach went to Argentina instead, to train fighter pilots for Juan Perón. Whether he had been dropped from Operation Paperclip or Juan Perón had offered him a better deal remains a mystery. Werner Baumbach died a few years later after an airplane he was testing crashed into the Río de la Plata, near Uruguay.

Also arriving at Wright Field in the summer of 1947 was General Walter Dornberger, newly released from England's Special Camp XI, outside Bridgend,

South Wales (formerly Island Farm). Before turning General Dornberger over to the Americans, the British labeled him a "menace of the first order" and warned their Allied partners of his deceitful nature. While holding Dornberger for war crimes, British intelligence had eavesdropped on him and recorded what he said. When the Americans listened to these secret audio recordings, they, too, concluded that Hitler's former "chief of all rocket and research development" had "an untrustworthy attitude in seeking to turn ally against ally." Still, Dornberger signed a Paperclip contract, on July 12, 1947, just weeks after his release from prison. Dornberger's skill at manipulation was put to use by Army Ordnance, which had him write classified intelligence briefs. America needed to develop missiles regardless of what any naysayers might think, Dornberger believed. .

"Russia strives now only for time to prepare for war before the United States," Dornberger wrote in a classified budget pitch, financed by the Ordnance Department, in 1948. "The United States must decide upon a research and development program that will guarantee satisfactory results within the shortest possible time and at the least expense. Such a program must be set up even if its organization appears to violate American economic ideals and American traditions in arms development," Dornberger wrote. At least it could be said that Dornberger remained true to his totalitarian-leaning principles — his belief that democratic ideals and traditions could be ignored in the quest for military supremacy. That the U.S. Army condoned Dornberger's idea appears never to have been made public before; his pitch was presented to Ordnance Department officials at the Pentagon. A copy of the classified document was found in 2012 in Dornberger's personal papers kept in a German state archive.

Opposition to Operation Paperclip gained momentum with America's scientific elite. On February 1, 1947, the Federation of American Scientists (FAS) met in New York City to ask President Truman to put an end to it. The American scientists saw the Nazi scientist program as a "drastic step in the search for military power." When it was learned that some of the one thousand additional German scientists on the Paperclip recruiting list were being hired for short-term military work followed by longer-term positions at American universities, many were outraged. "Certainly not wishing to jeopardize the legitimate needs of the national defense, and not advocating the policy of hatred and vengeance toward our former enemies, we nevertheless believe that a large-scale importation of German scientists . . . is not in keeping with the best objectives of American domestic and foreign policy," the members of FAS wrote. One American scientist was more forthright. "Certainly any person who can transfer loyalties from one idealology [*sic*] to another upon the shifting of a meal ticket is not better than Judas!" he said.

Albert Einstein was the most esteemed figure to publicly denounce Operation Paperclip. In an impassioned letter, written on behalf of his FAS colleagues, Einstein appealed directly to President Truman. "We hold these individuals to be potentially dangerous.... Their former eminence as Nazi Party members and supporters raises the issue of their fitness to become American citizens and hold key positions in American industrial, scientific and educational institutions."

Another important figure among the opposition was the nuclear physicist Hans Bethe. Bethe had fled Nazi Germany in 1933 and had worked on the Manhattan Project during the war. In the *Bulletin of the Atomic Scientists,* Bethe and Dr. Henri Sack, a colleague from Cornell University, posed a series of simple questions about Operation Paperclip. "Was it wise, or even compatible with our moral standards to make this bargain, in light of the fact that many of the Germans, probably the majority, were die-hard Nazis?" Bethe and Sack asked. "Did the fact that the Germans might save the nation millions of dollars imply that permanent residence and citizenship could be bought? Could the United States count on [the German scientists] to work for peace when their indoctrinated hatred against the

Russians might contribute to increase the divergency between the great powers? Had the war been fought to allow Nazi ideology to creep into our educational and scientific institutions by the back door?" Their final question struck at the dark heart of the Nazi scientist program. "Do we want science at any price?"

The condemnation of Operation Paperclip by these leading American scientists and others had a ripple effect on the general public. Reporters began searching for leads about individual German scientists' wartime activities, but this proved an almost impossible effort given the program's classified nature. Frustrated by the lack of information, some Americans sent threatening letters addressed to the German scientists at Wright Field and Fort Bliss. The army tightened security and surveillance. In Washington, D.C., the War Department was rightly concerned that intense opposition and bad publicity had put the entire program in jeopardy of collapse, and in the winter of 1947 the War Department forbade the further release of information about the program.

Despite the precautions, two individuals from the program at Fort Bliss were exposed by an angry public as "real Nazis": Mr. and Mrs. Herbert Axster. Herbert Axster, a former lieutenant colonel in the Wehrmacht, had been a Nazi Party member since 1940. He was not a scientist but rather a patent lawyer and an accountant. At Peenemünde he worked as

General Dornberger's chief of staff. His wife, Ilse Axster, was a leader of NS-Frauenschaft, a Nazi Party organization for women. During a routine investigation of the Axsters conducted for their visa applications, neighbors in Germany told army intelligence officers that Mrs. Axster was a particularly sadistic Nazi. On the Axsters' estate, the neighbors said, the couple kept forty "political prisoners," Russians and Poles, whom they used as slave laborers. The neighbors said Mrs. Axster was known to use a horsewhip on her servants. This information was first passed on to Rabbi Stephen Wise of the American Jewish Congress and then to the press. "These scientists and their families are supposed to have been 'screened,' Wise wrote to the War Department. "The Axsters prove that this 'screening' is a farce and the War Department 'screeners' are entirely incapable of performing this task." The Department of Justice withdrew Herbert Axster's application for legal immigration, but the Axsters were not sent home. Eventually they left Fort Bliss and Herbert Axster opened a law firm in Milwaukee, Wisconsin.

The War Department cabled USFET demanding an explanation as to how the Axsters got through the screening process. A classified cable was sent back, stating that a few "ardent Nazis" may have slipped into the program based on "unavailability of records" immediately after the war.

CHAPTER FOURTEEN

Strange Judgment

T he doctors' trial was the first of twelve subsequent war crimes trials at Nuremberg wherein the U.S. government prosecuted individuals in specific professions—including industrialists, lawyers, and generals—who had served the Third Reich. The first trial, called the trial of the major war criminals, was prosecuted by the four Allied powers. When it ended, in October 1946, tensions between the Soviet Union and the United States had escalated to the degree that cooperation between the two nations was no longer possible. The subsequent twelve trials took place at the Palace of Justice, as the others had, but with only American judges and prosecutors. The twenty-three defendants in the doctors' trial were charged with murder, torture, conspiracy, and atrocity in the name of medical science. "Mere punishment of the defendants," said Brigadier General Telford Taylor in his opening statement, "can never redress the terrible injuries which the Nazis visited on these unfortunate peoples." But General Taylor reminded the tribunal and the world that one of the central purposes of the trial was to establish a record of proof of the crimes, "so that no one can ever doubt that they were fact and not fable; and that this court, as agent of the United States and as the voice of humanity, stamps these acts, and the ideas which engendered them, as barbarous and criminal."

Dr. Leopold Alexander, the Viennese-born American psychiatrist who had first learned of the Luftwaffe freezing experiments, helped General Taylor write his opening speech. Dr. Alexander's official title was expert consultant to the secretary of war, but in layman's terms he was the most influential person involved in the doctors' trial after Telford Taylor. Dr. Alexander worked tirelessly to supply information and answers for the prosecution team as they prepared to go to trial. He conducted hundreds of hours of pretrial interrogations with the defendants and interviewed scores of witnesses and victims. He was able to converse in both German and English and he took extensive notes in both languages. Throughout the trial Dr. Alexander delivered presentations of evidence and cross-examined witnesses. When the trial was over he would be given credit for writing the Nuremberg Code.

Of the twenty-three defendants, no one outside a small group of American medical military elite had any idea that four of the accused had already worked for the U.S. Army under Paperclip. This fact would remain secret for forty years. Neither Siegfried Ruff, nor Konrad Schäfer, Hermann Becker-Freyseng, nor Oskar Schröder mentioned this damning fact during the trial.

Siegfried Ruff was one of the few physicians who admitted he'd overseen the medical murders at Dachau but said he did not perform any of the experiments himself. In his defense, he said he didn't believe that what he did was wrong. "It was understood that concentration camp inmates who had been condemned to death would be used in the experiments, and as compensation they were to have their sentences commuted to life in prison," Ruff told the judges. "Personally, I would not consider these experiments as immoral especially in war time." This line of defense, that extraordinary times called for extraordinary measures, was also used by Schröder and Becker-Freyseng.

The strange case of Konrad Schäfer played a unique role in Operation Paperclip. Having invented the Schäfer process to separate salt from seawater, Schäfer was clearly aware that concentration camp prisoners were going to be used in the testing process. Schäfer's superior, Oskar Schröder, chief of medical services for the Luftwaffe, confirmed this under oath, "In May 1944, in order to discuss what further steps should be taken, Becker-Freyseng and Schäfer attended [a] meeting as representatives of my office. As a result of the meeting it was decided to conduct further experiments on human beings being supplied by the Reichs-führer-SS, Heinrich Himmler." It was Becker-Freyseng who drew up the letter, which was sent to Himmler, requesting more prisoners to conduct experiments on. But Schäfer said he never actually went to Dachau and no evidence placing him there survived the war. As for eyewitness testimony, all but one of the saltwater experiment victims had been killed. The sole survivor, Karl Höllenrainer, was located by Dr. Alexander and put on the witness stand. Höllenrainer did not recognize Konrad Schäfer but he recognized one of his codefendants in the dock. Höllenrainer's testimony proved to be one of the most dramatic events in the doctors' trial.

It was June 27, 1947, and Karl Höllenrainer stood trembling in the hushed courtroom. Small in stature, dark-haired, and nervous, he was a broken man.

"Now, witness," asked U.S. Prosecutor Alexander G. Hardy, "for what reason were you arrested by the Gestapo on May 29, 1944?"

"Because I am a Gypsy of mixed blood," Höllenrainer said.

Höllenrainer's crime was that he had fallen in love with and married a German girl, a violation of the infamous Nuremberg law that prohibited a non-Aryan from marrying or having sexual relations with a German citizen. After being arrested, Höllenrainer was sent to three different concentration camps, first Auschwitz, then Buchenwald, and finally Dachau, where he was selected to take part in the seawater experiments being performed by Luftwaffe.

At Dachau, Höllenrainer was deprived of food, forced to drink chemically processed seawater, and then monitored for signs of liver failure and madness. One experiment among the many stood out in his memory. Without using anesthesia, a Luftwaffe doctor had removed a piece of Karl Höllenrainer's liver in order to analyze it. Now, from the witness stand, Höllenrainer was asked to identify the Nazi doctor who'd performed this liver puncture on him.

"Do you think you would be able to recognize that doctor if you saw him today?" Prosecutor Hardy asked.

"Yes," said Höllenrainer. "I would recognize him at once."

Höllenrainer stared across the courtroom, his eyes focused sharply on one of the twenty-three defendants sitting in the dock, Dr. Wilhelm Beiglböck, 40, the Luftwaffe doctor in charge of the saltwater experiments. Beiglböck had deep crevices at each side of his mouth and five pronounced dueling scars running across his left cheek.

As Höllenrainer stared at Dr. Beiglböck, "everyone in the courtroom waited tensely," recalls Vivien Spitz, a young American court reporter whose job it was to take down the testimony. Spitz sat in front of the judges, within clear view of the witnesses and defendants. Höllenrainer "was a little man and I watched him stand up," she explains. The prosecutor asked Höllenrainer to proceed to the defendants' dock to identify the doctor who'd removed a piece of his liver without anesthesia.

"He paused for just a moment," Vivien Spitz remembers, "with his eyes seeming to be fixed on a doctor in the second row of the prisoners' dock. Then, in a [flash] he was gone from the witness stand!"

Karl Höllenrainer sprang into action. To Vivien Spitz's eye, Höllenrainer seemed to leap over the German defense counsels' tables and "appeared to be almost flying through the air toward the prisoners' dock." In his right hand, stretched up high in the air, Karl Höllenrainer clutched a dagger, Vivien Spitz recalls. "He was reaching for Dr. Beiglböck, the consulting physician to the German Air Force."

There was shock in the courtroom. Confusion and

Major General Dr. Walter Schreiber was the surgeon general of the Third Reich. "The most sinister crime in which Schreiber is involved is the introduction of intravenous lethal phenol injections," explained war crimes investigator Dr. Leopold Alexander, "as a quick and convenient means of executing troublemakers." Paperclip contracts: U.S. Army, Camp King, Germany; U.S. Air Force, Texas. (NARA)

Dr. Kurt Blome was Hitler's biological weapons maker and the deputy surgeon general of the Third Reich. He had nearly completed a bubonic plague weapon when the Red Army captured his research institute in Poland. Paperclip contract: U.S. Army, Camp King, Germany. (NARA)

Erich Traub was a virologist, microbiologist, and doctor of veterinary medicine. He weaponized rinderpest (cattle plague) at the request of Heinrich Himmler, Reichsführer-SS, traveling to Turkey to acquire a black market sample of the virus during the war. Paperclip contract: Naval Research Institute, Maryland. (NARA)

Major General Walter Dornberger was in charge of V-weapons development for the Reich. Arrested by the British and held for nearly two years on war crimes charges, Dornberger was released into U.S. custody with the warning that he was a "menace of the first order." Paperclip contract: U.S. Army Air Forces. (NARA)

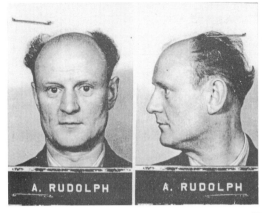

A. RUDOLPH   A. RUDOLPH

Arthur Rudolph specialized in V-weapons assembly and served as operations director at the slave labor facility in Nordhausen. In America he would become known as the Father of the Saturn Rocket. "I read *Mein Kampf* and agreed with lots of things in it," Rudolph told journalist John Huber in 1985. "Hitler's first six years, until the war started, were really marvelous." Paperclip contract: U.S. Army, Texas. (NARA)

Georg Rickhey oversaw tunnel operations for Hitler's Führerbunker headquarters in Berlin. On the V-2 program he was general manager of the slave labor facility and appeared as a defendant in the Nordhausen war crimes trial. Paperclip contract: U.S. Army Air Forces, Ohio. (NARA)

91   91

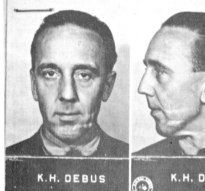

K.H. DEBUS   K.H. DEBUS

Kurt Debus was a V-weapons engineer who oversaw mobile rocket launches as well as those at Peenemünde. An ardent Nazi, he wore the SS uniform to work. In America, Debus became the first director of NASA's John F. Kennedy Space Center in Florida. Paperclip contract: U.S. Army, Texas. (NARA)

Otto Ambros was Hitler's most valuable chemist, codiscover of sarin gas (the "a" in sarin denotes his name), and chief of the Reich's Committee-C for chemical warfare. The U.S. Army coveted his knowledge. Tried at Nuremberg, Ambros was convicted of mass murder and slavery, then granted clemency by High Commissioner John J. McCloy. Paperclip contracts: U.S. Department of Energy. (NARA)

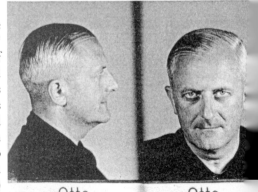

Otto   Otto
AMBROS   AMBROS
WCPL 1442   WCPL 1442

Friedrich "Fritz" Hoffmann was a chemist and philosopher. When captured by Allied forces he carried a paper signed by a U.S. diplomat stating he was anti-Nazi. In America, Hoffmann synthesized Nazi nerve gas stockpiles and worked in the CIA's assassination-by-poison program. Paperclip contract: U.S. Army Chemical Corps, Maryland. (NARA)

Jürgen von Klenck was a chemist, SS officer, and deputy chief of the Committee-C for chemical warfare. Surprised by how much information von Klenck provided, his interrogators concluded "that a lesser secret has been admitted to deflect the investigation from a more important secret." Paperclip contracts: U.S. Army, Heidelberg. (NARA)

Dr. Hubertus Strughold was in charge of the Aviation Medical Research Institute of the Reich Air Ministry in Berlin. Despite being sought for war crimes, he was hired by the U.S. Army Air Forces and became America's Father of Space Medicine. He went to great lengths to whitewash a dubious past. "Only the janitor and the man who took care of the animals," were members of the Nazi party, he told a journalist in 1961, referring to his Institute, which was filled with hardcore Nazis. Paperclip contracts: U.S. Army Air Forces, Heidelberg; U.S. Air Force, Texas. (NARA)

Dr. Theodor Benzinger directed the Experimental Station of the Air Force Research Center, Rechlin, under Hermann Göring and was an officer with the SA (Storm Troopers). While working for the U.S. Army in Heidelberg, Benzinger was arrested, imprisoned at Nuremberg, and listed as one of the defendants in the doctors' trial. Shortly thereafter he was mysteriously released. Paperclip contracts: U.S. Army Air Forces, Heidelberg; Naval Medical Research Institute, Maryland. (NARA)

Dr. Konrad Schäfer was a physiologist and chemist who developed a wartime process to separate salt from seawater in sea emergencies. Medical experiments at the Dachau concentration camp were based on the Schäfer Process. He was tried at Nuremberg and acquitted. Paperclip contracts: U.S. Army Air Forces, Heidelberg; U.S. Air Force, Texas. (NARA)

Dr. Hermann Becker-Freyseng was an aviation physiologist who worked under Dr. Strughold in Berlin and oversaw medical experiments on prisoners at Dachau. He was tried and convicted at Nuremberg, then contributed to Strughold's U.S. Army work from his prison cell. Paperclip contract: U.S. Army Air Forces, Heidelberg. (NARA)

Dr. Siegfried Ruff directed the Aero Medical Division of the German Experimental Station for Aviation Medicine in Berlin and was a close colleague and coauthor of Dr. Strughold. At Dachau, Ruff supervised medical murder experiments. Tried at Nuremberg and acquitted. Paperclip contract: U.S. Army Air Forces, Heidelberg. (NARA)

Siegfried Knemeyer was chief of German Air Force technical developments under Hermann Göring. Hailed one of the Reich's top ten pilots, Albert Speer asked Knemeyer to pilot his escape to Greenland. Paperclip contract: U.S. Army Air Forces, Ohio. (NARA)

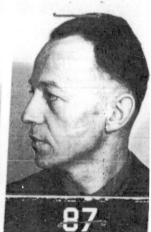

Walther Riedel was an engineer with the V-weapons design bureau and part of the von Braun rocket team. His Army interrogator classified him as an "ardent Nazi," but after Riedel threatened his handler that he would go work for the Russians he was hired and brought to America. Paperclip contract: U.S. Army, Texas. (NARA)

Emil Salmon, aircraft engineer and SS officer, was implicated in the burning down of a synagogue during the war. "This Command is cognizant of Mr. Salmon's Nazi activities and certain allegations made by some of his associates in Europe," wrote the U.S. Army Air Forces, but they found Salmon's expertise "difficult, if not impossible, to duplicate." Paperclip contract: U.S. Army Air Forces, Ohio.

Harry Armstrong set up the U.S. Army Air Forces Aero Medical Center in Germany and hired fifty-eight Nazi doctors to continue work they had been doing for the Reich. The Center violated the Potsdam Accord and was shut down after two years. Thirty-four Nazi doctors followed Armstrong to the U.S. Air Force School of Aviation Medicine in Texas. (U.S. Air Force)

Charles E. Loucks holding an incendiary bomb. Loucks oversaw the Paperclip scientists working on chemical weapons at Edgewood Arsenal. After being transferred to U.S. Occupied Germany, Loucks created an off-book working group on sarin production and invited Hitler's former chemists and Himmler's right-hand man to weekly roundtable discussions at his home. (Papers of Charles E. Loucks, U.S. Army Military History Institute)

Donald L. Putt, accomplished test pilot and engineer, was one of the first wartime officers to arrive at Hermann Göring's secret aeronautical research center at Völkenrode. Amazed by what he saw, Putt recruited dozens of Nazi scientists and engineers for Operation Paperclip and oversaw their work at Wright Field. (U.S. Air Force)

John Dolibois, a young U.S. Army officer and fluent German speaker, worked for military intelligence (G-2). He interrogated the major war criminals of the Nazi Party at the "Ashcan" internment facility in Luxembourg. (Collection of John Dolibois)

Heinrich Himmler, Reichsführer-SS, and his entourage during a tour of the Mauthausen concentration camp. Himmler's SS oversaw a vast network of state-sponsored slavery across Nazi-occupied Europe through an innocuous sounding division called the SS Business Administration Main Office. Reich slaves produced armaments, including the V-2 rocket. (USHMM)

Albert Speer (left), Adolf Hitler, and a cameraman in Paris. As minister of Armaments and War Production, Speer was responsible for all warfare-related science and technology for the Third Reich, starting in February 1942. (NARA)

As the Third Reich crumbled, Nazis stashed huge troves of scientific treasure, secret documents, and gold in salt mines across Germany. The 90th Infantry Division discovered this enormous cache of Reichsbank money and SS documents in Merkers, Germany. (NARA)

American liberators stand at the entrance of the Nordhausen underground tunnel complex where V-2 slave laborers assembled rockets. (U.S. Air Force)

In this rare photograph, IG Farben employees are seen fencing for sport at the corporation's Auschwitz facility and within sight of the three large chimneys of the death camp's crematoria. The sign behind the fencers reads, "Company Sporting Club, IG Auschwitz." Farben's plant was also called Auschwitz III or Buna-Monowitz. (Fritz Bauer Institute)

Otto Ambros laughing with his attorney during the Nuremberg trial against IG Farben executives. Ambros served as Manager of IG Auschwitz and also managed the Reich's Dyhernfurth poison gas facility in Silesia. (NARA)

In this previously unreleased photograph, Nazi doctors and scientists working for the U.S. Army Air Forces in Heidelberg gather with American officers for a group photograph circa 1946, prior to the arrest of five doctors on war crimes charges. Bottom row: at far left is Dr. Siegfried Ruff (arrested), at center is Richard Kuhn, third from right is Dr. Hubertus Strughold; Second row: third from left is Konrad Schäfer (arrested); Top row: third from left is Dr. Theodor Benzinger (arrested), far right (in tie) is Hermann Becker-Freyseng (arrested). (U.S.A.F School of Aerospace Medicine)

Dr. Kurt Blome consulting with his lawyer at the Nuremberg doctors' trial. Dr. Konrad Schäfer sits behind Blome. (NARA)

At the Nuremberg doctors' trial, defendant Dr. Kurt Blome frowns (at center). Behind him are (left to right) doctors Hermann Becker-Freyseng, Georg Weltz, Konrad Schäfer, Waldemar Hoven, Wilhelm Beiglböck. To Blome's right is Karl Gebhardt, the SS doctor who performed experiments on Janina Iwanska and was hanged. (NARA)

Dr. Wilhelm Beiglböck oversaw the salt water experiments at Dachau and removed a piece of prisoner Karl Höllenrainer's liver without anesthesia. During the trial, Höllenrainer, one of the only experiment survivors, tried to stab Beiglböck. (NARA)

Dr. Leopold Alexander, war crimes investigator and expert consultant at the Nuremberg doctors' trial, explains to the judges the nature of the medical experiment performed on Jadwiga Dzido at the Ravensbrück concentration camp. Paperclip doctor Walter Schreiber was in charge of the experiments. (NARA)

After being acquitted at Nuremberg, Dr. Konrad Schäfer took over Strughold's U.S. Army job in Germany while his Paperclip contract was sorted out. In Texas, Schäfer tried and failed to make the Mississippi River drinkable. His superiors found him "singularly unsuccessful in producing any finished work and has displayed very little real scientific acumen." Schäfer was asked to leave the country but refused because he had already received his immigrant's visa per Operation Paperclip. (NARA)

The IG Farben building in Frankfurt was taken over by the U.S. Army and became home to various American military and political organizations during the Cold War. Today it is Johann Wolfgang Goethe University. (Author's collection)

Castle Kransberg, outside Frankfurt, was Göring's Luftwaffe headquarters. The Allies captured it, code named it "Dustbin," and interrogated Nazi scientific and industrial elite here. (Author's collection)

Hitler's nerve gas–proof bunker underneath Castle Kransberg was designed by Albert Speer. In the event the Reich used chemical weapons, Hitler's high command assumed the Allies would retaliate in kind. (Author's collection)

Hitler wrote *Mein Kampf* at Landsberg Prison in 1924. After the war, it became home to 1,526 convicted Nazi war criminals until John J. McCloy granted clemency to the majority of those convicted at Nuremberg. (Author's collection)

The unmarked graves of hanged Nazi war criminals mark the church lawn at Landsberg Prison. (Author's collection)

The Cathedral inside Landsberg Prison, where convicted Nazi war criminals were allowed to pray. The benches were built at an angle so guards could observe individuals. (Author's collection)

Camp King was used as a Cold War black site. The CIA, Army, Air Force, and Naval Intelligence shared access to Soviet spies kept prisoner here using "extreme interrogation" techniques and "behavior modification programs," as part of CIA Operations Bluebird and Artichoke. Doctors Schreiber and Blome were Camp King Post physicians under Paperclip contracts from 1949 to 1952. (Author's collection)

John J. McCloy (center) with President Truman. McCloy championed the Nazi scientist program from its first days in May 1945. Between his tenure as assistant secretary of war and high commissioner of Germany, McCloy served as president of the World Bank. (NARA)

Charles E. Loucks with German scientists and others at a party at Edgewood Arsenal. Loucks befriended Nazi scientist and Nobel Prize winner Richard Kuhn, who introduced Loucks to LSD, which was later tested on hundreds of soldiers at Edgewood and used in the Army's Psychochemical Warfare program and the CIA's MKUltra mind-control program. (Papers of Charles E. Loucks, U.S. Army Military History Institute)

SS-Brigadeführer Walter Schieber served on Reichsführer-SS Himmler's personal staff and was chief of the Armaments Supply Office in the Speer Ministry where he oversaw tabun and sarin production. Shown here in 1951, Schieber worked for the U.S. Army and for the CIA as part of Operation Paperclip. (Collection of Paul-Hermann Schieber)

The "Eight Ball," at Camp Detrick. Airtight, bombproof, and weighing 131 tons, this one-million-liter chamber allowed Detrick's scientists to understand how aerosolized biological agents would work at different altitudes in the open air. Monkeys and human test subjects sat inside. (U.S. Army)

Fritz Hoffmann, the CIA's poison master, relaxes on his front lawn in suburban Maryland, circa 1948. (Collection of Gabriella Hoffmann)

Paperclip rocket scientists and specialists at Fort Bliss, Texas, circa 1946. Some were sent home when the Army learned they did not all have "rare minds." Karl Otto Fleischer, for example, claimed to have been the Wehrmacht's business manager when in reality he was in charge of food services. (NASA)

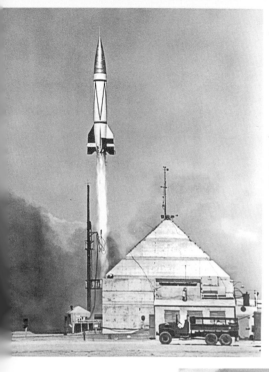

A V-2 rocket carrying the first monkey astronaut, Albert, blasts off at White Sands, New Mexico. (NASA)

Wernher von Braun and Kurt Debus confer during the countdown for a Saturn launch in 1965. (NASA)

Arthur Rudolph holds a model of the Saturn V rocket, which launched man to the moon in 1969. In 1983 the Department of Justice began its investigation of Rudolph on war crimes charges. He was told to prepare to stand trial or to renounce his U.S. citizenship and leave the country. Rudolph left. (NASA)

Von Braun and his team at the Redstone Arsenal in 1958. Pictured from left are Ernst Stuhlinger, Helmut Hoelzer, Karl Heimburg, E. D. Geissler, E. W. Neubert, Walter Haeussermann, von Braun, W. A. Mrazek, Hans Hueter, Eberhard Rees, Kurt Debus, and Hans Maus. (NASA)

Kurt Debus, director of NASA's John F. Kennedy Space Center, in an undated photo. According to his U.S. government security report, during the war Debus turned a colleague over to the Gestapo for making anti-Nazi remarks. The National Space Club in Washington, D.C., oversees the annual Dr. Kurt H. Debus Award. (NASA)

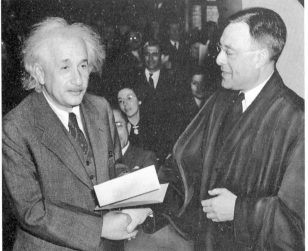

Albert Einstein accepting his certificate of American citizenship in 1940. One of Germany's most famous pre-war scientists, Einstein left Nazi Germany just months after Hitler took power, declaring that science and justice were now in the hands of "a raw and rabid mob of Nazi militia." He appealed to President Truman to cancel Paperclip, calling anyone who served Hitler unfit for U.S. citizenship. (Library Congress, World-Telegram)

mayhem. The Nazis' defense attorneys scrambled to get out of Höllenrainer's way. Three American military police rushed forward and grabbed Höllenrainer. Vivien Spitz remembers how security "subdued" Karl Höllenrainer just before he reached Beiglböck, "preventing him from delivering his own brand of justice."

It took minutes for order to be restored in the courtroom. The military police set Karl Höllenrainer before Presiding Judge Walter Beals, who was furious. Seventy years old, overworked and in failing health, Judge Beals clung to the idea that his primary role at Nuremberg was to educate the German people in the ways of American democracy and due process.

"Witness!" thundered the judge. "You were summoned before this Tribunal as a witness to give evidence."

"Yes," Karl Höllenrainer meekly replied.

"This is a court of justice," roared Beals.

"Yes," said Höllenrainer, trembling worse than before.

"And by your conduct in attempting to assault the defendant Beiglböck in the dock, you have committed a contempt of this court!"

Karl Höllenrainer pleaded with the judge. "Your Honor, please excuse my conduct. I am very upset—"

The judge interrupted and asked if the witness had anything else to say in extenuation of his conduct.

"Your Honor, please excuse me. I am so worked up. That man is a murderer," Höllenrainer begged, pointing to the expressionless Dr. Beiglböck. "He has ruined my whole life!"

The judge told Höllenrainer that his statement did not make his conduct forgivable. That he had insulted the court. That it was the judgment of the tribunal that he be confined in the Nuremberg prison for a period of ninety days, as punishment for insulting due process.

Karl Höllenrainer spoke in a tidy, pleading voice. The power and conviction that had allowed him to almost fly across the courtroom with the intent of stabbing Dr. Beiglböck with his dagger was gone. Now Höllenrainer was on the edge of tears. "Would the Tribunal please forgive me?" Höllenrainer asked. "I am married and I have a small son." He pointed at Dr. Beiglböck. "This man is a murderer. He gave me salt water and he performed a liver puncture on me. I am still under medical treatment." Karl Höllenrainer pleaded for mercy from Judge Beals. "Please, do not send me to prison."

The judge saw no room for clemency. Instead, Beals asked a guard to remove Karl Höllenrainer from the courtroom, referring to him as a "prisoner" now.

"My heart broke," Vivien Spitz recalls. All she could do was lower her head. She was a professional court reporter and it was inappropriate for anyone to see that she was crying. Sixty years later, when recollecting the incident, she still wondered why Judge Beals did what he did. "It was impossible to be dispassionate.... Why ninety days? Why not one or two days—just to make a point? After all the torture the witness had suffered it seemed to me to be an outrageous elevation of process over substance."

Karl Höllenrainer was removed from the courtroom. He was led down a long, secure corridor and into the prison complex, the same location where Dr. Beiglböck and all of the other Nazi war criminals were being held. It had been Dr. Alexander who had made the decision to put Karl Höllenrainer on the witness stand, and in his journal he wrote about how conflicted he felt about his decision. Dr. Alexander had spoken with Höllenrainer ten days before he testified against Beiglböck and was aware how upset Höllenrainer was, noting in a report how his hands shook. Karl Höllenrainer shared with Dr. Alexander that he suffered from a "tremendous feeling of inner rage" whenever he thought about what had happened to him at the Dachau concentration camp. Höllenrainer felt powerless, he said. He could close his eyes and "see the doctor in front of him who...had ruined his life and killed three of his friends." Dr. Alexander knew how powerful Höllenrainer's testimony would be. He was the only known victim of the saltwater experiments to have survived. Witness testimony was powerful, as his proved to be.

The thought of him being confined in a prison with the very doctors who had tortured him was unacceptable to Dr. Alexander, and that night he went to Judge Beals to advocate on Höllenrainer's behalf. The judge showed clemency and released him, on bail, into Alexander's custody. Four days later, on July 1, 1947, Karl Höllenrainer was allowed to continue his testimony. He rose to the occasion and provided harrowing details about what dying of thirst does to a man. He described how his friends "foamed at the mouth." How "they had fits of raving madness" before succumbing to an agonizing death.

One of the Nazi doctors' defense lawyers, Herr Steinbauer, was given an opportunity to cross-examine Karl Höllenrainer. Steinbauer accused Höllenrainer of lying.

"How can there be foam on a mouth which is completely dried out?" Steinbauer asked.

Höllenrainer said he was telling the truth.

"Listen, Herr Höllenrainer," Steinbauer said, "don't be evasive as Gypsies usually are. Give me a clear answer as a witness under oath."

Höllenrainer attempted to answer.

"You Gypsies stick together, don't you?" Steinbauer interrupted.

The exchange could have served as a metaphor for the whole trial. Through the eyes of a Nazi it was and always would be the *Übermenschen* versus the

Untermenschen. It was why General Taylor's opening remarks sent such a powerful message. It wasn't just the acts of the Nazi doctors that were "barbarous and criminal" but the very ideas that had engendered such acts.

Late at night, Dr. Alexander wrote a private letter to General Telford Taylor stating how he felt about the Nazi doctors on trial. "I feel that all of the accused today, Schaefer, Becker-Freyseng and Beiglböck, would admit that this problem [seawater drinkability] could have been solved in one afternoon with a piece of jelly [and] a salt solution." Instead, Dr. Alexander wrote, "All of these men on the dock slaughtered for gain of scientific renown [and] personal advancement. They were like Tantalus, a mythical, ambitious ruler who slew his own child for reward."

In a letter to his wife, Dr. Alexander penned a fascinating reveal: "Dr. Beiglböck from Vienna, turns out to have been in the same class with me during our final year at medical school," before the war. The way Dr. Alexander remembered it, Beiglböck was caught cheating and had to leave the school. "He does not recognize me," Dr. Alexander told his wife, "but I regularly recognize him." Two men whose professional lives had begun so similarly had taken such radically diffent paths.

There was another dramatic turning point in the doctors' trial, and that involved the case of Dr. Kurt

Blome. Blome was admittedly the director of biological weapons research for the Reich, which was not in itself a crime. But there were scores of documents in which Blome discussed the need to conduct experiments on human beings so as to further plague research at the Bacteriological Institute in Posen, where he was in charge. Blome's defense, some of which he argued himself, was that he intended to experiment on humans but that he never actually performed the experiments. Intent, said Blome, was not a crime. The prosecution had been unable to find any witnesses to testify against Blome, and Blome used this fact in his defense. There was the testimony of Major General Walter Schreiber against Blome from the first Nuremberg trial, but the Soviets had refused to let Dr. Alexander interview Dr. Schreiber himself, so the veracity of his testimony was never independently verified. The prosecution presented many documents in which Blome discussed with Himmler his plans to experiment on humans. But there were no documents showing Blome's actual guilt. There was another element of Blome's defense that the prosecution was completely unprepared for, and that was Dr. Blome's wife, Bettina, a physician and bestselling author. Frau Blome meticulously researched experiments that were conducted by the U.S. Office of Scientific Research and Development (OSRD) during the war. This included malaria experiments on Terre

Haute federal prison inmates. She also uncovered Dr. Walter Reed's famous nineteenth-century yellow fever research for the U.S. Army, in which volunteer human test subjects had died. The defense counsel, Robert Servatius, picked up on this theme of U.S. Army human experimentation where Blome's wife left off.

Servatius had located a *Life* magazine article, published in June of 1945, that described how OSRD conducted experiments on eight hundred U.S. prisoners during the war. Servatius read the entire article, word for word, in the courtroom. None of the American judges was familiar with the article, nor were most members of the prosecution, and its presentation in court clearly caught the Americans off guard. Because the article specifically discussed U.S. Army wartime experiments on prisoners, it was incredibly damaging for the prosecution. "Prison life is ideal for controlled laboratory work with humans," Servatius read, quoting American doctors who had been interviewed by *Life* reporters. The idea that extraordinary times call for extraordinary measures, and that both nations had used human test subjects during war, was unsettling. It pushed the core Nazi concept of the *Untermenschen* to the side. The Nuremberg prosecutors were left looking like hypocrites.

On August 6, 1947, the verdicts were read. Seven defendants were given death sentences, and nine were given prison sentences of between ten years and life, including Beiglböck (fifteen years), Becker-Freyseng (twenty years), and Schröder (life). Seven Nazi doctors were acquitted, including Blome, Schäfer, Ruff, and Weltz. Of Blome's acquittal, the Nuremberg judges noted, "It may well be that defendant Blome was preparing to experiment on human beings in connection with bacteriological warfare, but the record fails to disclose that fact, or that he ever actually conducted the experiments."

Konrad Schäfer returned to work at the Army Air Forces Aero Medical Center in Heidelberg where he assumed Dr. Strughold's former position as chief of staff. Strughold had already set sail for America as part of Operation Paperclip and was now in Texas, serving as "professional advisor" to Colonel Armstrong, the commandant of the School of Aviation Medicine at Randolph Field. Armstrong had set up a new aeromedical research laboratory there and was in the process of bringing dozens of Hitler's former doctors to Texas to conduct medical research. Hubertus Strughold's job description also included "overall supervision of the professional activities of the German and Austrian scientists [to be] employed by the School." Meanwhile, the FBI was having difficulty getting approval for Strughold's visa application for immigration into the United States. The Bureau had conducted a special inquiry into Dr. Strughold back in Germany. "One informant, who had known the subject," investigators learned, said that Strughold "had expressed the opinion that the Nazi party had done a great deal for Germany. He [Strughold] said that prior to Nazism, the Jews had crowded the medical schools and it had been nearly impossible for others to enroll." To counterbalance this derogatory depiction of Strughold, Dr. Konrad Schäfer was asked to write a letter of recommendation for his former boss. In a sworn statement written just a few weeks after Schäfer was acquitted at Nuremberg, Schäfer praised Dr. Strughold for his "ethical principles" while conducting research work. "I also know that he strictly refused to take part [in] or permit scientific research work which was damaging to human health," Schäfer wrote. The following year, Dr. Konrad Schäfer would be on a boat headed for America, an Operation Paperclip contract in hand.

As for Dr. Blome, he was seen as a highly desirable recruit for Operation Paperclip. Blome allegedly knew more about bubonic plague research than anyone else in the world. But, given his former position in Hitler's inner circle, coupled with the fact that Blome had worn the Golden Party Badge, bringing him to America as part of Operation Paperclip remained too difficult for the U.S. Army to justify. But as the Cold War gained momentum and intense suspicion of the Soviets increased, even someone like Kurt Blome would eventually be deemed eligible for Operation Paperclip.

PART IV

"Only the commander understands the importance of certain things, and he alone conquers and surmounts all difficulties. An army is nothing without the head."

—Napoleon

CHAPTER FIFTEEN

Chemical Menace

It had been nearly a year since State-War-Navy Coordinating Committee Paper No. 275/5 authorized JIOA to import one thousand Nazi scientists from Germany so as to deny them to the Russians. For the JIOA, things were not happening nearly fast enough. The number of Operation Paperclip scientists who had come to America in this time frame had nearly doubled, from 175 in March 1946 to 344 in February 1947, but none of these scientists had been granted visas yet. JIOA officials perceived State Department official Samuel Klaus to be the man most responsible for the holdup. On February 27, 1947, Klaus was called into a meeting in the Pentagon with his military counterparts. There, JIOA director Colonel Thomas Ford told Klaus that he had a list of German scientists already in the United States whose visas needed to be expedited. He asked Samuel Klaus to sign a waiver that would allow JIOA to begin this process.

Klaus refused to rubber-stamp anything. "I told him that of course I could do no such thing and that certification presupposed that the Department should have an opportunity to pass judgment of some kind before affixing the signature," he wrote in a State Department memo. Colonel Ford reminded Klaus of the language in a recent JIOA directive dated September 3, 1946. According to the directive, the Department of State was to "accept as final the investigation and security reports prepared by JIOA, for insuring final clearance of individuals concerned." Klaus told Ford that he would not sign a document that was essentially a blanket waiver of anything—that he would have to see the list of scientists first.

Ford told Klaus that the list was classified and that he did not have the clearance to see it. Klaus still refused to sign off on it. "Ford stated in various ways that if I did not accept the paper on behalf of the State Department he would give the information to several senators who would take care of the Department," Klaus wrote of Ford's threat. Colonel Ford accused him of standing in the way of "national interest."

Colonel Ford's request was likely influenced by information the JIOA had recently received from the Office of the Military Government in Germany. After months of negative press attention directed at Operation Paperclip, spearheaded by the Federation of American Scientists and given further momentum by the American Jewish Congress's having identified the Axsters, of Fort Bliss, as "real Nazis," the War Department had asked OMGUS to conduct an internal review of the Operation Paperclip scientists working in the United States. It was OMGUS in Germany that oversaw the reports on all the German scientists issued by Army Intelligence, G-2. The War Department told OMGUS that it did not want to have to deal with any other exposés.

In response, OMGUS sent the War Department 146 security reports that could potentially create scandals were the information contained in them to be revealed. One report pertained to rocket engineer Kurt Debus. Debus was an ardent Nazi. He had been an active SS member who, according to the testimony of colleagues, wore his Nazi uniform to work. Most troublesome was the revelation in his OMGUS security report that during the war he had turned a colleague, an engineering supervisor named Richard Craemer, over to the Gestapo for making anti-Nazi remarks and for refusing to give Debus the Nazi salute. From Nazi-era paperwork, it was clear that Debus had initiated the investigation into Craemer. No one else had heard the conversation between the two scientists; Debus had gone home after work and taken it upon himself to write up a report of what was allegedly said, which he had then submitted, in transcript form, to the SS. The specifics of the incident made it impossible for the European Intelligence Division of the U.S. Army to cast Kurt Debus as an apolitical scientist trying to survive in a fascist world. Debus had gone out of his way to have a colleague arrested by the Gestapo. Craemer was subsequently prosecuted for treason. According to the report, and as a result of Debus's actions, on November 30, 1942, "Craemer was called to Gestapo H.Q. and confronted with this damning evidence." The army intelligence officer assigned to the case, Captain F. C. Groves, explained what happened next. On April 5, 1942, "he [Craemer] was dragged before a 'Sondergericht' [a special Nazi court] and sentenced to 2 years imprisonment." Groves's lengthy summary and analysis of the incident was sent to the Joint Chiefs of Staff. Debus's "deliberate and vicious denunciation [of] Craemer" was not something army intelligence could disavow, wrote Groves. It was a matter of public record in Germany.

An OMGUS report put von Braun on equally shaky footing, and army intelligence cautioned that, given his international profile, pushing the State Department for a visa for von Braun could cause problems. The report revealed that not only had von Braun been an SS officer with the high rank of

SS-Sturmbannführer, or SS-Major, but his membership had been sponsored by Heinrich Himmler. In response, JIOA director Colonel Ford proposed contrary action to what was being advised. Ford believed that the best solution was to get these undesirables on an even faster track toward citizenship. If Klaus signed the waiver, JIOA could expedite the process.

With Klaus refusing to sign the documents, the situation escalated into a standoff. Klaus left the meeting and did some investigative work on his own. In the months that followed, he learned what had transpired with Georg Rickhey at Wright Field—that Rickhey had been returned to Germany to stand trial for alleged war crimes. The information had purposely been withheld from the State Department. In a memo to the undersecretary of state, Klaus defended his opposition to the JIOA's bullish requests to have him sign a blind waiver. "That the department's security fears were not baseless was recently demonstrated when a war criminal, wanted for war crimes of a bestial kind, was found here among these scientists, to be returned to Germany." Klaus was determined to hold his ground, but so was the JIOA. The difficulty for Klaus was that he was outnumbered.

The same month that the standoff between the State Department and the JIOA began, the Chemical Corps finally imported its first German scientist, an expert in tabun nerve agent synthesis named Dr. Friedrich "Fritz" Hoffmann. The U.S. Army had been interested in stockpiling tabun ever since it obtained its first sample from the Robbers' Lair, in the British zone in Germany, in May of 1945. The man in charge of the tabun nerve agent program for the Chemical Corps was Colonel Charles E. Loucks, commander of the Army Chemical Center at Edgewood Arsenal, in Maryland.

The fifty-one-year-old Colonel Loucks had dedicated his life to chemical warfare. Born and raised in California, Charles E. Loucks received a Bachelor of Arts degree from Stanford University and first became fascinated by war gas while serving in World War I. During his assignment with the first Gas Regiment at Edgewood Arsenal, in 1922, his life's path seemed to have been set. While Loucks distinguished himself as an expert rifleman, competing in National Rifle and Pistol matches and winning awards, chemical weapons fascinated him the most. He went back to school, to the Massachusetts Institute of Technology, to study chemical engineering, and in 1929 he received a Master of Science degree. By 1935 Loucks was the technical director of Chemical Warfare Service, Research and Development, at Edgewood. He had served as an officer with chemical weapons ever since.

Major Loucks spent the first year of World War II refining the U.S. military's standard-issue gas mask.

He even worked with Walt Disney and the Sun Rubber Company to transform the spooky, apocalyptic-looking face protector into a more kid-friendly version with a Mickey Mouse face. In August 1942, Loucks was made commander of the Rocky Mountain Arsenal, outside Denver, Colorado, where he was in charge of planning, building, and operating the largest-capacity toxic manufacturing plant in the United States. Rocky Mountain produced mustard gas and incendiary bombs on an industrial scale—incendiary bombs that were dropped on Germany and Japan. Loucks was awarded the Legion of Merit for his war service.

During the war the U.S. Army had more than four hundred battalions ready for chemical combat. If chemical warfare broke out, the army expected it would involve mustard gas. Sixty thousand soldiers trained in chemical warfare were sent into battle with gas masks and protective suits, carrying a small card that explained what to do in the event of a chemical attack. Fortunately, World War II ended without the use of chemical weapons, but experts like Charles Loucks were caught off guard when they learned just how outperformed America's chemists had been by Hitler's. With the discovery that the Nazis had mass-produced previously unknown agents like tabun and sarin gas came the realization that if Germany had initiated chemical warfare it would have been a grossly uneven match. Since war's end, the army's Chemical Corps had received 530 tons of tabun, courtesy of the British, who had seized the Reich's colossal cache from the Robbers' Lair. But the Chemical Corps had yet to produce much tabun on its own, which is why Dr. Fritz Hoffmann had been brought to America under Operation Paperclip. Hoffmann arrived in February 1947 and immediately got to work synthesizing tabun. Nerve agents were Hoffmann's area of expertise; he had synthesized poison gas at the Luftwaffe's Technical Academy, Berlin-Gatow, and also at the chemical warfare laboratories at the University of Würzburg during the war.

Fritz Hoffmann was a gigantic man, six foot four, with dark eyes, sharp, angular features, and high cheekbones. He kept his hair combed back behind the ears, perhaps with the help of hair oil. Hoffmann was a gifted organic chemist and a contemplative man who also had a PhD in philosophy. He had suffered from polio as a child. According to the State Department, he was one of the minority of individuals recruited into Operation Paperclip who had been anti-Nazi during the war. When Hoffmann was captured at war's end he carried with him an unusual document, an affidavit from the U.S. Embassy in Zurich, Switzerland, signed by American Consul General Sam E. Woods. "The bearer of this document...Dr. Frederick Wilhelm Hoffmann [was] anti-Nazi during the whole war," the document stated. It further explained that Hoffmann's father-in-law, the German economist Dr. Erwin Respondek, had risked life and limb to work as a spy for the Americans during the war. Respondek "rendered extremely valuable services to the Allied war effort," wrote Consul General Woods. "These services, which are known to the former Secretary of State, the Honorable Cordell Hull, were rendered at great personal risk to himself and family, were performed without compensation and were so valuable to our cause that...every courtesy and help should be accorded" to all members of the family, including Respondek's son-in-law, Fritz Hoffmann, the affidavit read.

After being interrogated by army intelligence in Germany, Hoffmann was found "fit for exploitation for the Office of the Military Government, United States." He was sent to the Army Chemical Center in Berlin, where he started working for the Chemical Warfare Service until his Operation Paperclip contract was finalized.

When Friedrich Hoffmann arrived at Edgewood Arsenal in February 1947, America's premier chemical weapons research and development facility had been producing war gases for thirty years. The three-thousand-acre peninsula, thick with field and forest, was located twenty miles northeast of Baltimore, Maryland, on the Chesapeake Bay. Because Fritz

Hoffmann was the first German at Edgewood thus far, he was quartered inside a barracks with American soldiers. The Top Secret research and development division he had been assigned to was called the Technical Command, and it was there that he began work "synthesizing new insecticides, rodenticides and miticides," which meant tabun.

Colonel Loucks had been ordered by the Chemical Corps to determine how to produce tabun on an industrial scale under a classified program codenamed AI.13. Army intelligence believed the Soviets' chemical weapons program to be considerably ahead of its own, and the Chemical Corps was under pressure to catch up. The Soviets had captured the entire IG Farben laboratory at Dyhernfurth and had since reassembled it outside Stalingrad, in the town of Beketovka, under the codename Chemical Works No. 91. In addition to having the laboratory and the science, the Soviets had also captured some of the Farben chemists who had worked at Dyhernfurth. Through their own version of Operation Paperclip, a parallel exploitation program called Operation Osoaviakhim, the Soviets captured German chemist Dr. von Bock and members of his team. The group was taken to Chemical Works No. 91, where von Bock, an expert in air filtration systems, decontamination systems, and hermetically sealed production compartments, was put to work. Von Bock's esoteric knowledge gave the Soviets access to critical technical aspects of producing tabun. The hope was that Fritz Hoffmann could bring the Americans up to speed at Edgewood.

Indeed, Fritz Hoffmann proved to be a great asset for the classified program. Within months of his arrival at Edgewood he was delivering "work of a high order" and had "shown considerable ingenuity and excellent knowledge," according to a declassified internal review of his performance. Tabun work progressed. When military intelligence learned that the Soviets' Chemical Works No. 91 was also producing sarin, Edgewood was told to continue working on tabun but also to redouble efforts in sarin production, with a plan to start producing it on an industrial scale as soon as possible. Project AI.13 was given even higher priority and a new code name, Project AI.13-2.1. In Germany, Nazi scientists with knowledge of tabun and sarin were now being even more aggressively sought for recruitment into Operation Paperclip.

Declassified documents reveal that the Chemical Corps wanted to employ Otto Ambros, but he was not available. Ambros was incarcerated inside the prison complex at Nuremberg, awaiting a war crimes trial. Imprisoned alongside Ambros were fellow Farben board members Fritz Ter Meer and Karl Krauch as well as Hermann Schmitz, Farben's powerful CEO and the man who kept the Auschwitz scrapbook hidden in a secret wall safe. SS-Brigadeführer Dr. Walter Schieber, Ambros's liaison to the Speer ministry and the man who oversaw the work at Dyhernfurth, was also high on the Paperclip target list, but recruitment of Schieber would have to wait. Schieber was also being held at Nuremberg. The IG Farben trial, officially called *United States of America v. Carl Krauch et al.*, was scheduled to begin in a few months, in the summer of 1947. Time would tell who, if any, of Hitler's chemists would become available.

It had been a little over four months since the *New York Times* had broken the story of Operation Paperclip, and the negative attention to the program had not subsided. Then, on March 9, 1947, journalist Drew Pearson reported the most outrageous news story about Operation Paperclip to date. According to Pearson, the U.S. Army had offered Farben executive Karl Krauch a Paperclip contract while he was incarcerated at Nuremberg. Krauch was the lead defendant in the upcoming war crimes trial. He had served as Göring's plenipotentiary for chemical production and had advocated for the use of nerve agents against the Allies. Krauch was the man who galvanized his fellow German industrialists to mobilize resources to help the Nazis go to war.

Drew Pearson's column, "Washington Merry- Go- Round," was widely read and highly influential, and the report about Krauch caused such a scandal that General Eisenhower, the U.S. Army chief of staff, demanded to know what was going on with Operation Paperclip. The man who briefed Eisenhower was chief of military intelligence Stephen J. Chamberlin. After what was reported to be a twenty-minute meeting, General Eisenhower remained in support of the Paperclip program. For the military chiefs inside the JIOA, the problems with the public's perception of Paperclip were almost universally blamed on the State Department. Dean Rusk, who worked in the office of the assistant secretary of war, summed up JIOA's attitude toward the State Department in a memorandum the day after the Eisenhower briefing. "The public relations people are feeling mounting pressure on the German scientist business.... Our position is inherently weak because the State Department finds this whole program difficult to support."

The JIOA decided it was time to get rid of Samuel Klaus. Director Thomas Ford wrote General Stephen Chamberlin saying that Klaus was "obnoxiously difficult" and needed to be removed. Within the week, Klaus was transferred from the visa section of the State Department to the Office of the Legal Adviser. He no longer had any say over the policies and procedures regarding Operation Paperclip. It remains a mystery if Klaus had anything to do with the public's perception of the program, but in years to come, and as

McCarthyism expanded, Klaus would be accused of having leaked negative stories about the German scientist program to the press. But even with Klaus banished from the program, for the first time since Paperclip's inception in the spring of 1945, the entire operation was now in danger of collapse. Assistant Secretary of War Howard Petersen worried with colleagues that the War Department would be buried in scandals involving Nazi scientists. Petersen predicted that the whole program would be shut down in a matter of months.

Instead the opposite happened. With the signing of the National Security Act by President Truman, on July 26, 1947, America's armed services and intelligence agencies were restructured. The War Department was reconstituted into the Department of Defense, the State-War-Navy Coordinating Committee became the National Security Council, and the Central Intelligence Agency was born.

A new age was dawning for the controversial Paperclip program. On the one hand, it struggled to hold up a false face of scientific prowess behind which lay a tawdry group of amoral war opportunists, many of whom were linked to war crimes. But just as von Braun admitted to *New Yorker* writer Daniel Lang that what he really cared most about was seeing "how the golden cow could be milked most successfully," the newly created CIA saw the Paperclip scientists in similar quid pro quo terms. There was advantage to be had in using men who had everything to lose and were, at the same time, uniquely focused on personal gain.

In Operation Paperclip the CIA found a perfect partner in its quest for scientific intelligence. And it was in the CIA that Operation Paperclip found its strongest supporting partner yet.

Sometime in the spring of 1947, scientists at Edgewood Arsenal began conducting human experiments with tabun nerve agent. All soldiers used in these experiments were so-called volunteers, but the men were not made privy to the fact that they were being subjected to low-level concentrations of tabun. Some of the tests took place in Utah, at the Dugway Proving Ground. Other tests took place inside Edgewood's "gassing chamber for human tests," a 9 x 9-foot tile-and-brick cube with an airtight metal door. One of the people observing the tabun tests was Dr. L. Wilson Greene, technical director of the Chemical and Radiological Laboratories at the Army Chemical Center at Edgewood and a close collaborator of Fritz Hoffmann's. Greene was a short man with a square jaw and a barrel chest. What he lacked in height he made up for in vision. While observing the behavior of the soldiers in the tabun gas experiments, L. Wilson Greene had a revelation.

Greene noticed that, after soldiers were put in the "gassing chamber," they became "partially disabled for from one to three weeks with fatigue, lassitude, complete loss of initiative and interest, and apathy." What struck Greene most was that the men were wholly incapacitated for a period of time but not permanently injured. The men recovered entirely on their own; the antidote was time. Dr. L. Wilson Greene saw in this a new kind of warfare. He sat down and began outlining his idea for America's war-fighting future in an opus that would become known as "Psychochemical Warfare: A New Concept of War." In the monograph, Greene wrote, "The trend of each major conflict, being characterized by increased death, human misery, and property destruction, could be reversed." His seminal vision for psychochemical warfare—a term he coined—was to incapacitate a man with drugs on the battlefield but not to kill him. Greene believed that in this way the face of warfare could change from barbaric to human. Incapacitating agents were "gentle" weapons; they knocked a man out without permanent injury. With psychochemical warfare, Greene explained, America could conquer its enemies "without the wholesale killing of people or the mass destruction of property."

Greene was not proposing to use low levels of tabun gas on the battlefield. He was talking about using other kinds of incapacitating agents, drugs that could immobilize or temporarily paralyze a person, "hallucinogenic or psychotomimetic drugs...whose effects mimic insanity or psychosis." "There can be no doubt that their will to resist would be weakened greatly, if not entirely destroyed, by the mass hysteria and panic which would ensue," Greene explained.

Greene proposed that an immediate "search be made for a stable chemical compound which would cause mental abnormalities of military significance." He sought drugs that made people irrational. In his monograph, Greene provided the army with a list of "61 materials known to cause mental disorders." These sixty-one compounds, he said, should be studied and refined to determine which single compound would be the best possible incapacitating agents for U.S. military use. Greene requested a budget of fifty thousand dollars, roughly half a million dollars in 2013, which was granted. Research began. Greene assigned his colleague and friend Fritz Hoffmann the job of researching a multitude of toxins for potential military use.

Fritz Hoffmann was by now recognized at Edgewood as one of the most gifted organic chemists in the Chemical Corps. If anyone could find and prepare the ideal incapacitating agent for the battlefield, Hoffmann could. He began a broad spectrum of research on everything from well-known street drugs to highly obscure toxins from the third world. There was mescaline, obtained from the peyote cactus and used by

Native American Indians, with side effects ranging from divination to boredom. He studied fly agaric, a hallucinogenic mushroom found on the barren slopes of Mongolia and rumored to facilitate contact with the spirit world, and piruri, a toxic vegetable leaf from Australia, used by Aborigines, that was found to suppress thirst. Yaxee and epena, from Venezuela, Colombia, and Brazil, caused people to see things that weren't really there. It was a drug that for centuries had been "used by primitive tribes to escape the realities of their plight by using hallucinogenic properties." Soon Hoffmann would travel the world in search of these incapacitating agents on behalf of the Chemical Corps.

Dr. Greene's idea of psychochemical warfare would have a profound effect on the future of the U.S. Army's Chemical Corps, but it would also greatly affect the direction of the newest civilian intelligence organization in Washington, the CIA. The Agency had deep pockets and big ideas. For the CIA, using drugs to incapacitate individuals had many more applications than just on the battlefield, and the Agency began developing programs of its own. Fritz Hoffmann and L. Wilson Greene were at the locus of a growing partnership being forged between the Chemical Corps and the CIA. Soon, biological warfare experts from Camp Detrick would also be brought into the fold. This particular biological weapons program, which would be run by a group called Special Operations

Division, or SO Division, was fueled by Operation Paperclip and would develop into one of the most controversial and collaborative efforts in the history of the CIA.

It had been a year and a half since the Merck Report on the biological weapons threat had been released, and an influx of congressional funding had transformed Camp Detrick into a state-of-the-art bioweapons research and development facility. The army purchased 545 acres of land adjacent to what had been called "Area A" and created a new area, designated "Area B," where some of Detrick's first postwar field tests with crop dusters and spray hoses would occur. During the war, dangerous pathogens like anthrax and "X" had been tested and cultured inside Detrick's germ lab, a rudimentary wooden building covered in black tarpaper and nicknamed the Black Maria by scientists. During the war, an industrial-size boiler, used for fermenting, sat on the lawn outside the germ lab. Now, given the scope of work planned for the immediate future, Detrick needed an aerosol chamber that was bigger and better than anything else like it in the world. The job of designing such a structure was assigned to a bacteriologist named Dr. Harold Batchelor.

Detrick's British counterparts, at Porton Down, had an excellent chamber of their own, but it fit only two or three mice. What Batchelor came up with was a monstrous spherical one-million-liter chamber called the Eight Ball, shaped like a giant's golf ball and held upright by iron "legs." The Chicago Bridge and Iron Works was commissioned to build the Eight Ball to specifications that made it airtight and bombproof. The Eight Ball was to have portholes, doors, and hatchways and steel walls of one and a half inches.

Inside the Eight Ball, airflow would simulate weather systems, with scientists on the outside controlling temperatures on the inside within a range of 55 to 90 degrees Fahrenheit. Humidity could be controlled inside the Eight Ball to fluctuate between 30 and 100 percent. This state-of-the-art environmental control would allow Detrick's scientists to understand how aerosolized biological agents would work at different altitudes in the open air. The sphere would weigh more than 131 tons and would stand four stories tall. A catwalk around its center would allow scientists to observe, through portholes, the test subjects sitting inside as they were exposed to the world's deadliest germs. The Chicago Bridge and Iron Works agreed to a delivery date of 1949.

With the chamber's design complete, Dr. Batchelor prepared to travel to Germany. There was an important German scientist who was just now becoming available for an interview. This was a man who knew more than almost anyone else in the world about biological weapons. He was particularly knowledgeable about weaponized bubonic plague.

The physician was Dr. Kurt Blome, the former deputy surgeon general of the Third Reich. He had just been acquitted of war crimes at the Nuremberg doctors' trial. Now he was back on the Paperclip list.

The doctors' trial had been over for forty-two days. It was October 2, 1947, and a message from Heidelberg, marked "Secret-Confidential," arrived on the desk of the chief of the Chemical Corps. It read: "Available now for interrogation on biological warfare matters is Doctor Kurt Blome."

A meeting was arranged for November 10, 1947, between Blome and Batchelor. Present alongside Dr. Batchelor were Detrick doctors Dr. Charles R. Phillips, a specialist in desterilization, Dr. Donald W. Falconer, an explosives expert, and Dr. A. W. Gorelick, a dosage expert. Lieutenant R. W. Swanson represented the U.S. Navy and Lieutenant Colonel Warren S. LeRoy represented the army's European Command Headquarters. An interpreter and a stenographer were also present. Dr. Blome was told in advance that everything discussed would be classified.

Dr. Batchelor spoke first, setting the tone for the all-day affair. "We have come to interview Dr. Blome personally as well as professionally," Batchelor said. "We have friends in Germany, scientific friends, and this is an opportunity for us to enjoy meeting [Dr. Blome] and to discuss our various problems with him." To begin, Batchelor asked, "Would it be possible for Dr. Blome to give us an overall picture of the information that he has? The nature of the world under discussion?"

Blome spoke in English, pausing on occasion for the interpreter to help him with a word. "In 1943 I received orders from Goering for all the research of Biological Warfare," Blome explained, "all the research for BW [would fall] under the name Kanserreseach.... Cancer Research had already started long before that, and I was already working all the time but in order to keep this development secret [the Reich] disguised it."

Dr. Blome laid out the command structure of those involved in biological weapons work under Himmler, and where the men were now. It was a surprisingly small group of "around twenty" men. As head of the Reich Research Council, Blome explained, Göring was at the top, as the Reich's dictator of science. There were three men in equal positions directly under Göring, Dr. Blome explained. Blome was in charge of all research and development of pathogens. Doctors and scientists working in this area reported to him. Major General Dr. Walter Schreiber — the Russians' surprise witness at Nuremberg and the man who had pointed the finger at Blome — was in charge of vaccines, antidotes, and serums for biological weapons. All doctors and scientists working in these areas ultimately reported to Schreiber. Finally, Blome explained, Field Marshal Wilhelm Keitel oversaw the Lightning Rod Committee, *Blitzableiter*, the code name for the ordnance experts who worked on delivery systems for biological bombs. Anyone conducting tests with these kinds of weapons had to go through Keitel.

Göring had committed suicide and Keitel had been hanged at Nuremberg after the trial of the major war criminals. Major General Walter Schreiber was working for the Russians now. It appeared that Dr. Blome was the last available man standing with extensive inside knowledge of the Third Reich's bioweapons program.

Dr. Gorelick asked, "Can Dr. Blome give us actual locations of various laboratories?"

Blome spoke of the Reich's outpost on the island of Riems, a facility that specialized in "Sickness of Cattle" research, including rinderpest and foot-and-mouth disease. Blome said that because "the isle was completely isolated except by wire," it was a perfect place to conduct this kind of dangerous research. The scientist in charge of the laboratory was Professor Otto Waldmann and his assistant was Erich Traub "of international fame." Blome was referring to the fact that before the war, Traub spent several years in America doing research at the Rockefeller Institute, in New Jersey.

Rinderpest was a terrible disease, Blome said. In many ways it was the biological weapon he feared most. "Germany depended on milk and butter for

60% of her fat resources," said Blome. "In 1944 it would have resulted in a great catastrophe if foot-and-mouth disease had been used against Germany. It would have been the greatest catastrophe ever faced," according to Blome, "[i]f a country relies on all its fat resources to get milk and butter. Once the disease starts there is no stopping it." The Detrick scientists were very interested to learn more.

How did the Reich acquire and develop the pathogen, Batchelor asked? Blome had already explained this to the Operation Alsos interrogators, but that was two years ago, before the doctors' trial, and apparently these Detrick scientists were not familiar with what Blome had said when he was at Dustbin. "By international law it was prohibited to have the virus for this sickness in Europe," Blome said. "The virus was in Turkey and Himmler ordered that for the Isle." Blome confirmed that Dr. Erich Traub went to Turkey on direct orders from Himmler and acquired a strain of the dangerous virus there. At Riems, Traub then succeeded in producing a dry form of the virus. Dry forms were the deadliest of all, said Blome. "After a period of seven months based on experimentation on the Isle, this virus was still effective. After seven months they still spread and the cattle were all infected."

Blome then spoke of experimental tests conducted by Luftwaffe pilots in Russia, where the disease was sprayed from low-flying aircraft over fields of grazing cattle.

"Positive results," said Blome.

Where were Dr. Traub and Dr. Waldmann now?

"I believe that they have been taken prisoner of the Russians and they are still active in their research for the Russians," Blome said.

The conversation shifted to plague research at Posen, where Blome had set up an institute during the war. "Perhaps we would like to talk about the human angle," Dr. Phillips asked, trying to veil the uncomfortable subject by using the royal "we."

Dr. Blome had been acquitted of war crimes charges five weeks earlier. Seven of his codefendants were to be hanged. Clearly, the subject of experiments on humans was not something he was going to discuss. The question was rephrased: What did Blome believe was the most groundbreaking biological weapons work conducted by the Reich? Blome said that at Posen, he had been working at dispersing biological agents in a "combination with gas that [a]ffects the throat. When membranes are hurt [that is, damaged]...bacteria have a better chance to infect," Blome said.

Dr. Phillips rephrased his question again. Everyone wanted to know about the human experiments; it was the ultimate forbidden knowledge the Nazis possessed. "Perhaps we would like to talk about the human angle, which was, I think, Kliewe's work," he said.

Blome was no fool. "Professor Kliewe has not worked himself with experiments," Blome said. Heinrich Kliewe performed intelligence work. He was in charge of keeping tabs on the biological weapons programs being pursued by nations at war with the Reich. "Kliewe investigated Polish and Russian sabotage," Blome said. "The activity here was not to cause an epidemic amongst the population, merely to kill certain people."

The Detrick doctors were intrigued by this line of conversation and asked Dr. Blome to explain further.

During the war, Polish resistance fighters in Posen had succeeded in assassinating "about twenty people," Blome said, most of whom were SS officers. "A lot of cases of Typhoid," said Blome. "The waiters in restaurants took fountain pens filled with the inoculum and injected the inoculum into the soup or the food on their way to and from the dining room. This has been proved. Polish resistance movement was heading all these activities. A German-Polish woman doctor was working in the hospital and got hold of the bacteria and forwarded it to the other people."

"Were any preventative measures taken?" Dr. Phillips asked.

Blome explained that there was an extensive and

well-funded Reich research program going on in the field of vaccines, antidotes, and serums—against many pathogens and diseases, from cholera to parrot virus to plague. But Blome explained that the holy grail of biological weapons research was the plague. "I believe the only thing of danger to Germany would have been the plague. Because for propaganda reasons the plague got more attention tha[n] any others," Blome added. "People all over the world believe it is the worst sickness there is."

Dr. Phillips wanted to learn more about the use of "live vaccines" in Blome's plague research.

"Schreiber, as the head of the department [epidemics] had very good vaccine material on hand for typhoid, paratyphoid, cholera, diphtheria and Ruhr," Blome explained. "For the plague, they had serum on hand but the serum was not very powerful, it was very weak. They did not have any vaccine so I started to build an institute which has never been completed." On occasion, Blome said, he worked in concert with Schreiber's vaccine research laboratory to provide them with "good vaccine material," meaning the germs.

"Which laboratories?" asked Dr. Falconer.

"In the Medical Laboratory in Berlin," said Blome. "There was an official appointed only for research of epidemics. Prof. Schreiber."

"Was he under Kliewe?" Dr. Phillips asked, apparently still unaware that Professor Schreiber was the same Dr. Schreiber who had testified at Nuremberg. If, at Nuremberg, the Russians meant to send a message to American biological weapons makers with Schreiber's testimony, this had gone over the head of these bacteriologists from Camp Detrick.

Blome repeated that Schreiber was "in the same position as I would be under Göring for my cancer researches, so this man was [directly] under Göring." Meaning Blome and Schreiber were equals in the Reich's chain of command. They were also archenemies. "Schreiber is a Russian PW [prisoner of war] and everybody who knew Schreiber is convinced he is working for them," said Blome. In this world of intense suspicion, of deviance and trickery, it was impossible to know who was lying and who was telling the truth.

The time came to have a meal. "In closing this particular angle we express our gratitude to Dr. Blome for his wholehearted cooperation," Dr. Batchelor said. He suggested that they all have dinner together. Pleasant talk. "During that time we will not discuss this matter," Batchelor said. As if the Americans had not just tried to hang Dr. Blome for war crimes. As if Blome had not been a hard-core Nazi ideologue and member of the inner circle, or had not worn the Golden Party Badge. Pleasantries were exchanged and the meeting was adjourned.

* * *

After the meeting, the Detrick doctors requested from the Army Chemical Corps everything they knew about Dr. Erich Traub. That Dr. Blome said he was most afraid of an outbreak of cattle plague was serious news. Traub was the world's leading expert in the disease. Now the U.S. Army wanted him as their own.

Dr. Traub was a virologist, microbiologist, and professor and a doctor of veterinary medicine. He had been the second in command at the Reich's State Research Institute at Riems since 1942. He was also an expert in Newcastle disease, a contagious bird flu, that he was rumored to have weaponized. The Chemical Corps knew Traub spoke fluent English, that he had dark brown hair, gray-brown eyes, and two pronounced saber scars on his face—on the forehead and upper lip. And they knew that from 1932 until 1938, Traub had been a staff member at the Rockefeller Institute for Medical Research, in New Jersey. But after meeting with Dr. Blome, the Chemical Corps wanted more information on Traub, and they tracked down two of his former American colleagues for interviews. One of them, Dr. Little, described Traub as a "domineering German and a surly type individual with a violent temper." Another colleague, Dr. John Nelson, found that despite "long training in the care of animals, [Traub] went out of his way to be cruel to animals." This troubled Nelson, who felt that "any person who is cruel to animals shows little distinction and difference to his treatment of his fellow human beings."

Before the war, Traub was given the opportunity to stay in America and continue his research full-time. He chose to return to Germany, citing loyalty to the Reich. In 1939, Traub was drafted into the Wehrmacht, Veterinary Corps, and in 1940, he was elevated to captain and fought in the campaign against France. He was a member of several Nazi organizations, including the Nationalsozialistisches Kraftfahrerkorps (NSKK), or National Socialist Motor Corps; the National Socialist People's Welfare (NSV); and the Reichsluftschutzbund (RLB), or State Air Protection Corps. Dr. Traub's talents as a virologist had been identified, and he had been pulled back from the front lines and assigned to biological weapons work. According to Blome, Traub was the most talented scientist working on anti-animal research — biological weapons designed to kill the animals that a nation relies upon most for food. At the end of the war, when the Riems lab fell to the Russians, with it went Dr. Traub, his wife, Blanka, and their three children. The laboratory was renamed the Land Office II for Animal Epidemic Diseases and the Soviets put Traub back to work on bioweapons research. Now, in 1947, the Army began a plan to lure Traub away.

As for Dr. Blome, the scientists from Detrick knew

it was far too risky to offer him a Paperclip contract just weeks after he'd been acquitted of capital war crimes at Nuremberg. Drew Pearson's reporting on the Paperclip contract offered to Karl Krauch in prison had upset General Eisenhower and nearly brought the program to its knees. While Blome made clear his willingness to work for the Army, the Detrick scientists knew he would have to remain an under-the-radar consultant, relegated to the JIOA target list for potential hire at a future date.

The army bacteriologist and biological weapons expert Dr. Harold Batchelor returned to Camp Detrick with many new ideas to explore, including assassination-by-poison techniques that had been shared with him by Dr. Blome. The Reich had been researching biological weapons that would initiate epidemics but could also "kill certain people," according to Blome. In 1947, however, with scientific frontiers opening up wide, there would be hundreds of new ways to assassinate individuals using a single device carrying a biological or chemical agent. This was an area in which the CIA was interested. The Chemical Corps had the perfect scientist for the job of exploring poisons that could be used for individual assassinations, Fritz Hoffmann. But Hoffmann had a backlog of work. He still had to figure out how to synthesize tabun gas so the Chemical Corps could hurry up and begin producing it on an industrial scale.

CHAPTER SIXTEEN

Headless Monster

In June 1948, Colonel Charles E. Loucks, the man who oversaw the Paperclip scientists at Edgewood, was made brigadier general and transferred to Heidelberg. Loucks now served as chief of intelligence collection for Chemical Warfare Plans, European Command. In Heidelberg, he had access to a whole new group of Hitler's former chemists, from those who had been at the top of the chain of command on down. Within weeks of his arrival, Loucks formed a working relationship with Richard Kuhn, the former director of the Kaiser Wilhelm Institute for Medical Research in Heidelberg. Because Kuhn had an international profile before and during the war, he was a problematic candidate for Operation Paperclip. Though Kuhn had once been revered among scientists, Samuel Goudsmit of Operation Alsos was never afraid to remind colleagues that Kuhn had become an active Nazi during the war, and that he began his lectures with "Sieg Heil" and the Nazi salute. Kuhn had lied to Goudsmit in a postwar interview, swearing that he had never worked on Reich projects during the war. In fact Kuhn was a chemical weapons expert for the Reich and developed soman nerve agent. Soman was even deadlier than sarin and tabun but considered too delicate and therefore too costly to industrialize.

Richard Kuhn began working with General Loucks in Heidelberg on chemical weapons projects for the Chemical Corps. General Loucks's friendly relationship with Richard Kuhn drew ire from the British. When formally queried about his professional partnership with Kuhn, Loucks replied, "I was under the impression that Professor Kuhn had been cleared of his Nazi complicity or had suffered the penalty and is now in the good graces of both the British and the Americans." Further, wrote Loucks, "I am sure our people are certainly familiar with his background." To General Loucks, moving forward on military programs considered vital to U.S. national security was more important than dredging up an individual's Nazi past. Through the lens of history, this remains one of the most complicated issues regarding Operation Paperclip. When working with ardent Nazis, some American handlers appear to have developed an ability to look the other way. Others, like General

Loucks, looked straight at the man and saw only the scientist, not the Nazi.

Richard Kuhn had a connection with a scientist in Switzerland with whom General Loucks was particularly interested in working. The scientist had been investigating a little-known incapacitating agent that was far more potent than anything the Chemical Corps was working on at the time. This Swiss chemist had recently given a lecture, "New Hallucinatory Agent," to a gathering of the Swiss Society of Psychiatry and the Association of Physicians in Zurich. On December 16, 1948, General Loucks took a trip to Switzerland. So as not to draw attention to himself or the U.S. military, he took the unusual step of taking off his military uniform. This story does not appear in any known declassified army record but is told in Loucks's own words, in his personal diaries that were left to the U.S. Army Heritage Center in Pennsylvania. "Went back to the house and put on civilian clothes," Loucks wrote in his diary. His wife, Pearl, took the plainclothed general to the train station. En route to his destination in Switzerland, Loucks shared a compartment with a "foreigner, dark, nationality unknown," and a "Dutchman who acted Jewish in quizzing me all about myself." The train compartment, Loucks noted, provided a clean, comfortable, well-lit ride, and General Loucks arrived in Bern at 8:55 that night. Before heading to bed, Loucks enjoyed a fire in the fireplace in his hotel room. General Loucks did not note in his journal any of the details of the meeting with the mysterious Swiss scientist. The mission was classified.

Loucks returned to Germany the following day, and he did note in his journal that Richard Kuhn came over to his house for lunch with a special guest, Dr. Gerhard Schrader, the inventor of Preparation 9/91, or tabun gas. With snow coming down outside the Louckses' home in flurries, Richard Kuhn, Dr. Gerhard Schrader, and General Loucks had a pleasant chat and "lunch of pork chops."

Decades later, in a speech prepared for the Amherst chapter of the Daughters of the American Revolution and the Sons of the American Revolution, General Loucks revealed that this Swiss chemist referred to him by Richard Kuhn had been Professor Werner Stoll, a psychiatric researcher at the University of Zurich. The hallucinatory agent that Loucks was after in Switzerland would be the ultimate "incapacitation chemical" also sought by L. Wilson Greene at Edgewood "to knock out not kill." The chemical, said Loucks, was called "Lysergic Acid Diethylamide," or LSD. Stoll did not discover LSD. That distinction went to Albert Hofmann, a chemist for Sandoz pharmaceutical company in Basel. Werner Stoll, a colleague of Hoffmann's (and the son of Sandoz chief chemist, Arthur Stoll), repeated Albert Hofmann's original LSD experiment and concluded, "modified LSD-25 was a psychotropic compound that was non-toxic and could have enormous use as a psychiatric aid." In 1947, Werner Stoll had published the first article on LSD, in the *Swiss Archives of Neurology*. Stoll's second paper, entitled "A New Hallucinatory Agent, Active in Very Small Amounts," was published two years later, in 1949. But General Loucks did not see LSD as a psychiatric aid but rather as a weapon, an incapacitating agent with enormous potential on the battlefield. Soon the army and the navy would all be experimenting with LSD as a weapon, and the CIA would be experimenting with LSD as a means of controlling human behavior, an endeavor that soon came to be known as mind control.

Eventually, physicians and chemists from Operation Paperclip would work on jointly operated classified programs code-named Chatter, Bluebird, Artichoke, MKUltra, and others. LSD, the drug that induces paranoia and unpredictability and makes people see things that are not really there, would become a strange allegory for the Cold War.

One day in the late summer of 1948 a call came in to Brigadier General Charles Loucks's new office in Heidelberg. A lieutenant answered the phone. The caller, a German, left a cryptic message to be relayed to General Loucks. It was short and to the point.

"I can help," the caller said.

He left a return telephone number and his name, Schieber. General Loucks had been in Germany since June. As chief of intelligence collection for chemical warfare plans in Europe, it was Loucks's job to determine which Western European countries were developing chemical weapons and to monitor their progress. Loucks was also working on unfinished business back at Edgewood, namely, the continuing failure by the Chemical Corps to develop industrial-scale production of nerve gas, an effort that by now had officially switched from the pursuit of tabun to the pursuit of sarin. The American university professors Loucks had hired back in the U.S. were making very little progress. "We put samples [of sarin] in front of them and everything, but yet they...could not come up with any process to make some Sarin," General Loucks explained decades later in an oral history for the U.S. Army. A fire in the sarin plant at Edgewood had set the program back even further.

Along came the Schieber call.

In 1947, a thick OMGUS security report had been compiled on SS-Brigadeführer Dr. Walter Schieber. He had been involved with the U.S. Army since the fall of the Reich.

Schieber was a Nazi Bonzen, a big wheel. He was unattractive, fat, and wore a Hitler mustache and false teeth. Since the 1920s he had been regarded as one of

Hitler's Alte Kämpfer, the Old Fighters, trusted members of Hitler's inner circle who wore the Golden Party Badge. Dr. Schieber was also a loyal SS man and served on the personal staff of Heinrich Himmler.

SS-Brigadeführer Dr. Walter Schieber had been a dedicated and loyal member of the Nazi Party since 1931. He had also been frequently photographed alongside Hitler, Himmler, Bormann, and Speer as part of the inner circle. A number of these photographs survived the war, which made public dealings with Schieber impossible. U.S. Army transactions with him were classified Top Secret. As head of the Reich Ministry of Armaments Supply Office (Rüstungslieferungsamt), first under Fritz Todt when the ministry was called the Munitions Ministry, or Organisation Todt, and then under Albert Speer, Schieber was an engineer and a chemist and handled business in both areas for the Third Reich. As an engineer, he oversaw many of the Reich's underground engineering projects. "Designs for concentration camp armaments factories remained almost the exclusive work of Schieber," writes Michael Thad Allen, an expert on the SS and slave labor.

Schieber also wore his chemist's hat for at least one concentration camp experiment. In an attempt to save the SS money and address the growing problem of food shortages among slave laborers, Schieber designed a "nourishment" program called Eastern Nutrition

(Östliche Kostform). It was tested at the Mauthausen concentration camp. For a period of six months, starting in December 1943, a group of one hundred and fifty slave laborers were denied the watery broth they usually received and instead were fed an artificial paste designed by Schieber and made up of cellulose remnants, or pieces of used clothing. One hundred and sixteen of the one hundred and fifty test subjects died. After the war, there was a judicial inquiry into Schieber's Eastern Nutrition program. The West German courts determined that the resulting deaths could not necessarily be "attributed to nutrition" because there were so many other causes of death in the concentration camp.

Walter Schieber was further linked to the deaths of hundreds, if not thousands, of slave laborers through the various chemical weapons programs that were carried out at Farben's multiple production plants. Schieber was not tried at Nuremberg but was used as a witness for the prosecution instead. During the war, with his expertise as a chemist, SS-Brigadeführer Dr. Walter Schieber was the Speer ministry's liaison to IG Farben and he oversaw the industrial production of tabun and sarin gas. According to his intelligence file, one of his titles was "confidential clerk of IG Farben AG." Dr. Walter Schieber and Dr. Otto Ambros worked together at the Dyhernfurth nerve agent production facility.

By the late summer of 1948, Otto Ambros had been tried and convicted at Nuremberg and was serving an eight-year sentence at Landsberg Prison for mass murder and slavery. Schieber had been released from his obligations as a witness for the tribunal and was a free man. Now, here he was on the telephone, requesting to speak with Brigadier General Loucks. Schieber was looking for work with the U.S. Army Chemical Corps.

"I'm free now," Schieber told Loucks's lieutenant during the call. "They have nothing against me."

It was, of course, more complicated than that.

Four months prior, in February 1948, former SS-Brigadeführer Walter Schieber had signed a Top Secret Operation Paperclip contract with the JIOA. He had been recruited by Colonel Putt at Wright Field, now Wright-Patterson Air Force Base. Putt wanted to make use of Schieber's underground engineering skills — just as Georg Rickhey's skills had been used before Rickhey was returned to Germany to stand trial for war crimes. Hiring a top Nazi like Walter Schieber was risky, and the potential problems facing the U.S. Air Force for doing so were candidly discussed in an exchange of memorandums. A JIOA case officer wrote to military headquarters in Frankfurt summarizing how things were progressing. "Subject is Dr. Walter Schieber . . . requested priority shipment to Wright Field. Schieber's ability is outstanding and his potentialities believed invaluable to the United States." For the past three months, the case officer explained, Schieber had been working on "the underground factory project," a massive undertaking, at U.S. military headquarters in Germany. The project was spearheaded by Franz Dorsch, and Schieber had been working as Dorsch's first assistant. In Germany, Schieber and Dorsch had supervised "150 scientists and technicians" who had built underground factories for the Reich. The result was a thousand-page monograph for U.S. Air Force. Schieber, the report explained, "has been especially cooperative and is agreeable to going to [the] United States. He is well known to the Soviets and is desired by them. His exploitation in the United States is therefore believed highly desirable."

There was, however, a problem that needed to be addressed, wrote the case officer. "His party record is as follows: Entered party in June 31, Brigade Fuhrer in the SS, holder of the Golden Party Badge, member of SS, DAF [Deutsche Arbeitsfront, or German Labor Front], NSV [Nationalsozialistische Volkswohlfahrt, or National Socialist People's Welfare], and VDCH [VDCh, or Association of German Chemists]." There was no way to disguise that Schieber had been in Hitler's inner circle, but the JIOA also proposed a solution. "In order to minimize the possibility of unfavorable publicity in the United States your views requested on advisability of shipping subject via air under escort and or under an alias." Schieber could be part of Operation Paperclip as long as no one knew who Schieber really was. Unlike standard operating procedure with Operation Paperclip applicants, there was to be no photograph of SS-Brigadeführer Dr. Walter Schieber attached to his intelligence file.

The air force agreed with JIOA's suggestions. "Ship Dr. Walter Schieber to Wright Field Dayton Ohio. Air Force requests case be expedited and given top priority." There was one caveat that JIOA and the air force agreed upon. Dr. Schieber had to undergo a denazification trial first. "Trusting he would be placed in Class 3"—the category for individuals who were "less incriminated"—after his denazification trial, Dr. Walter Schieber packed his bags and awaited transport to the United States. Instead, on March 11, 1948, a different verdict came in: "the civilian Internee Dr. Walter Schieber, born 13-09-1897 at Beimerstetten, former Chief of Armaments Deliveries on the Reich Ministry for Armaments and War Production, Head of the Central Office for Generators, Deputy Head of the Reich Organization Industry, SS-Brigadefueher [*sic*], PSt [presently] held at this enclosure," read the decree, "was tried by the Sonder-Spruchkammer for Hesse, Neustadt-Lager, and sentenced to Group II, 2 years Labor Camp and restriction to 5 years ordinary labor." The Group 2 designation meant that Schieber was in a category with "party activists, militarists, and profiteers." His Paperclip contract would have to wait.

Five days after his trial, Schieber contacted the U.S. Army Counter Intelligence Corps to relay to his Paperclip handler his version of events. "Herr Berbeth [the judge] had subject brought into the chamber and asked him about his trip to the United States, to which Schieber replied that he did not think it would be wise for him to go to the U.S. as an offender. Herr Berbeth suggested that he should immediately apply for a re-trial or ask the Minister for Political Liberation for a pardon. Herr Berbeth also told Schieber that he would recommend such a pardon." Schieber told his Paperclip handler that, after leaving the courtroom, he had returned to the internee camp where he was being held. Then, according to Schieber, something rather shocking happened: "Berbeth of the court joined Schieber and told him, contrary to his previous statement, not to apply for another trial or a pardon, but to go to Frankfurt immediately, contact the Russian Liaison Officer and ask for an Interzonal Pass. With this pass, subject [Schieber] was to enter the Russian Zone and proceed to one of his factories in the Schwarza/Saale [where] he could be assured of every possible help, should he have the desire to work in his former position." In other words, according to Schieber, the judge at this trial worked for the Russians and was offering him a job. Schieber claimed to have told the judge that his offer sounded impossible. Was he really being advised to ignore the court's judgment, to flee the camp, get an interzonal pass, and begin working for the Russians "at a high pay grade" comparable to what he was paid during the Hitler regime? "The President of the Spruchkammer assured [me] that arrangements would be made," Schieber said. Did this really happen? Was the judge a Russian mole? Or was Schieber playing the Americans with the proverbial Russian card?

Schieber's Counter Intelligence Corps handler asked him what he thought of the judge's black market offer. "Schieber believes that the [Spruchkammern] sentence was imposed to prevent his contract with the United States and to make his residence in Germany so difficult as to force him to accept the Russian Offers." The handler had his own thoughts. "Cancellation of Schiebers [*sic*] contract after he has possibly jeopardized his safety and after he has cooperated so whole-heartedly with intelligence agencies here is certain to have an adverse effect on the future contracting or exploitation of specialists and will only serve as another example of broken faith on the part of the United States." The CIC officer felt that when "considering the magnitude of Schiebers [*sic*] potential value to the United States either for military or civil exploitation," it was obvious that his contract needed to be honored one way or other. The officer recommended that Schieber be paid as a consultant to the U.S. Army in Heidelberg. In the meantime, he could appeal the judgment of the Spruchkammern trial and become part of Operation Paperclip after the attention died down.

The air force had already soured on Schieber. At the Pentagon, his case had been reviewed again. The air force now saw Schieber's Nazi Party history in a wholly different light. In their eyes, Schieber had transformed from a brilliant engineer to a ruthless, greedy war profiteer. "Walther Schieber started his business career at the Gustloff Werke in Weimar, a combine owned by the Nazi Party and comprising five industrial corporations," read the air force response. "He soon became the leader in the German Cellulose and Rayon Ring, which was the second largest fiber combine in Germany. He founded Thuringische Zellwolle A.G. in Schwarza and subsequently gained control of the French Synthetic Fiber combine," meaning that with the help of SS officers, Schieber had confiscated Jewish-owned businesses in France and made them the property of the Third Reich. Synthetic fibers were imperative to the Reich, used for everything from soldiers' uniforms to blankets to parachutes. Photographs of Schieber displaying bolts of synthetic cloth to Hitler, Himmler, and Bormann show all parties smiling with pride.

"Schieber was appointed by Speer to the position of

Chief of the Office for Delivery of Armament Goods in the Ministry for Armaments and Munitions and was named deputy leader of the National Group Industry by [Walter] Funk," read the air force report. How he went from dealing fabrics to weapons production remained unclear, but his ambition had to have come into play. "Described as a top-ranking figure in the Nazi war economy, he is said to have constantly profited from being a party man," noted the air force. There was only one conclusion: "Cancel Air Force request for Walter Schieber. Subject considered ineligible Paperclip."

United States European Command (EUCOM) headquarters in Frankfurt—various commands had different headquarters—cabled the JIOA office at the Pentagon to let them know that cancellation of the request was an unsound idea: "It is believed here that these considerations outweigh the risk of possible criticism of his being sent to the United States, and if necessary, warrant an exception to present policy. Further, it is considered advisable that for Schieber's future personal safety and for intelligence reasons, he not be present…when and if official action against him occurs. Therefore it is strongly recommended that reconsideration be given the decision to cancel Schieber's contract and that he be sent to the United States regardless of the outcome of his appeal." EUCOM wanted Schieber moved out of Germany and sent to the United States—now. Nothing happened for three months—until Brigadier General Charles Loucks arrived in Heidelberg.

After the second Schieber call came in to General Loucks's office, Loucks told his lieutenant to set up a meeting with the man. As of 2013, their official meeting remains classified. But General Loucks wrote regularly in his desk diary, which fills in what the declassified Operation Paperclip case file on Schieber leaves out. On October 14, 1948, Dr. Schieber was invited to attend a roundtable conference at Army Chemical Corps headquarters in Heidelberg. "Classified matters" were discussed, Loucks penned in his journal that night. Schieber told Loucks he had been in on the Reich's manufacturing of nerve gas "from the beginning." He appealed to General Loucks's patriotic side. "I want you to know that if there is anything I can do to help the West I shall do it," Schieber said. Loucks liked Schieber's willingness to help the cause so much that he invited Schieber to join him at his house for a drink.

That evening Loucks recorded his thoughts: "Attended conf. with…Dr. Walter Schieber—classified matters. No particular info but hope for more later, possibly when better acquainted. I'll try to see him next time he reports in to Div. of Intell. He directed the production of war gases on a rather high echelon so doesn't have the detailed knowledge that I want, but possibly I can get the names of useful people from him. Took him to the house for a drink."

Chemical weapons, like biological weapons, were perceived by some to be "dirty business," as President Roosevelt once said. But to men like General Loucks and Walter Schieber, to advance a nation's arsenal of chemical weapons was a challenging and necessary job. It "was more interesting than going down to Paris on weekends," Loucks wrote in his journal.

General Loucks asked to meet with Dr. Schieber again, this time to ask Schieber if he could assist with a "problem" the U.S. Army was having producing sarin gas. Schieber was happy to help. He told Loucks that during the war it was the Farben chemists who had produced sarin gas and that he knew all these chemists very well. They were his friends. "They worked with me during the war," Schieber explained.

"We wouldn't expect you to do this for free," Loucks told Schieber, meaning provide the U.S. Chemical Corps with secrets. The two men arranged to meet again in the following weeks.

On October 28, 1948, Loucks and his wife, Pearl, hosted a dinner party in their home. Again, Schieber was a guest. Loucks had by now taken an extraordinary liking to Dr. Schieber and wrote his impressions of the man in his diary that night. "Schieber is interesting—an independent thinking, intelligent and very competent man. He related much of his experience with the Russians. A prisoner of war after the 1st World War for a year. He was an honorary (?) Brigade Fuehrer of SS this last war. In confinement in Nuremberg for seven months. Quartered next to Goering until the latter killed himself. Was an admirer of Todt, later worked for Speer, was directed to report to Hitler frequently. He has many anecdotes and is a loyal German. Is willing to do anything for the future of the world and Germany." Why was General Loucks so willing to overlook SS-Brigadeführer Dr. Walter Schieber's criminal past and his central role inside the Third Reich? A story that Charles Loucks told an army historian decades later sheds light on this question.

At the end of World War II, after the Japanese surrendered, Colonel Loucks went to Tokyo, where he served as the chief chemical officer for the U.S. Army. Sometimes Loucks took day trips into the countryside. In the last five months of war in Japan, American bombers conducted a massive incendiary bombing campaign against sixty-seven Japanese cities that killed nearly a million citizens, most of whom burned to death. Still, the Japanese refused to surrender, and it took two atomic bombs to end the war. The incendiary bombs dropped on those sixty-seven cities were produced at the Rocky Mountain Arsenal. Colonel Loucks oversaw the production of tens of thousands of them. In Japan, after the fighting was over, and when Loucks took day trips, he often brought his camera along and took photographs of the landscape, the damage, and the dead. When Loucks returned home to America, he compiled these photographs into an album of more than one hundred black-and-white snapshots. One photograph in the album, which is archived at the U.S. Army Heritage Center in Pennsylvania, shows Colonel Loucks standing next to an enormous pile of dead bodies.

Years later Colonel Loucks explained to the army historian what the photograph meant to him. "Driving one day in a Jeep from Yokohama to Tokyo, I stopped along the side of a road. The incendiary attacks had done their work," Loucks explained. The area "was all burned out; a wasteland all the way through. We dropped tens of thousands of them [incendiary bombs] on the whole area between Yokohama and Tokyo."

Out there in the Japanese countryside, said Loucks, "I noticed a great stack of incendiary bombs—small ones. I went over to take a look at them. They looked like something that we had made at Rocky Mountain. Sure enough, they were. Here in one place they had a great stack of them. They were burned out but the bodies were still there because they didn't burn. They stacked them up in this big high pile. I had a picture of me standing beside them, because I had been responsible for the manufacturing of them. That was just one of those incidents that didn't mean anything, but I just happened to see what had happened to some of our incendiary bombs that were over there."

In describing the photograph—an enormous pile of dead bodies next to a stack of incendiary bombs—Colonel Loucks expressed a peculiar kind of detachment. To the army historian interviewing him, Loucks made clear that what interested him in the photograph was noting the effectiveness—or in this case ineffectiveness—of the bombs he had been responsible for manufacturing. Similarly, Loucks expressed detachment as far as Dr. Schieber was concerned, as evidenced in his journal entries. It was as if Loucks could not, or would not, see Schieber in the context of the millions of Jews murdered on the direct orders of Schieber's closest wartime colleagues. What interested Loucks about Schieber was what an effective chemical weapons maker he was.

During the next meeting between General Loucks and Dr. Schieber, Loucks got very specific with Schieber in terms of what he was after. "Could you develop the process and put it on paper with drawings, specifications and tables and safety regulations to make Sarin?" Loucks asked, as noted in his desk diary.

"Yes, I could do that," Schieber said.

Loucks recalled the next conversation the two men had. "He said they had a big works outside of Berlin that was just completed when the war was over and made little token amounts but no production. When [the Reich] announced that the Russians were taking over, those engineers and chemists both came West into the American and British occupied zones." Schieber was lying to General Loucks. The nerve gas production plant outside Berlin to which Schieber referred was Falkenhagen, and it was run by Otto Ambros's deputy and the man who had stashed the steel drum outside Gendorf, Jürgen von Klenck. By war's end, the factory at Falkenhagen had produced more than five hundred tons of sarin gas, hardly "little token amounts," as Schieber claimed. If Loucks had read von Klenck's OMGUS security report, or any of the CIOS reports written by Major Edmund Tilley, he would have learned that Schieber was lying to him. Instead, General Loucks asked Schieber if he could locate these chemists who knew so much about sarin production and bring them to Heidelberg. He held out to Schieber the promise of a U.S. Army contract. Further, "We will pay all their [the chemists'] expenses and give them something for their work," Loucks said.

"Yes, I can do that." Schieber replied. He said that he knew all of the Farben chemists and could easily get them to tell the Americans everything. He listed their names for Loucks. One of the six Farben chemists on the list was Ambros's deputy and the man who ran Falkenhagen, Jürgen von Klenck.

On October 29, 1948, Colonel Loucks wrote a memorandum to the chief of the Army Chemical Corps. The best, fastest way to get German technical information on tabun and sarin gas was to hire Dr. Schieber, Colonel Loucks advised. In his diary, Loucks wrote, "Hope the chief will support us. If he does, we'll be able to get all of the German CW technical ability on our side and promptly. They know on what side they belong. All we need to do is treat them as human beings. They recognize the military defeat and the political and ideological defeat as well and accept it."

One week later, General Loucks told Schieber that he had been authorized to pay him 1,000 marks a month for consulting work. Schieber gave Loucks the contact information for the six chemists and technicians who would join him in his efforts to explain precisely how to produce sarin gas. On December 11, 1948, Loucks hosted the first roundtable meeting of Hitler's chemists in his Heidelberg home, secrecy assured. For the next three months, the chemists met every other Saturday at Loucks's home. There, they created detailed, step-by-step reports on how to produce industrial amounts of sarin gas. They drew charts and graphs and made lists of materials and equipment required. Years later, Loucks reflected, "One of the team [members] was a young engineer who had an excellent command of English which helped greatly and was his major contribution [and that was] Jurgend [*sic*] von Klenck."

When the work was finally compiled and sent to Edgewood, the results were the perfect recipe for the deadly nerve agent. According to General Loucks, without Hitler's chemists, the American program had been a failure. With them, it was a success. "That's when we built the plant out in Rocky Mountain Arsenal," Loucks explained. The incendiary bombs that Colonel Loucks oversaw at Rocky Mountain Arsenal during World War II would now be replaced by M34 cluster bombs filled with sarin gas. The Top Secret program was code-named Gibbett-Delivery.

A friendship between two brigadier generals, Loucks and Schieber, had been solidified. The following summer, Schieber sent Loucks a thank-you note and a gift, not identified in the records but described by Schieber as a piece of "equipment...that once stood at the beginning of the same work group." The unknown item had been used by Schieber during the Nazi era, when sarin gas was first developed for Hitler. Over the next eight years the two brigadier generals exchanged Christmas cards.

In January 1950, General Loucks was called to Washington, D.C., for several meetings at the Pentagon. According to Loucks's desk diary, during his first meeting there he was reprimanded by a Pentagon official for cultivating friendly relationships with Hitler's chemists.

"'I don't like this,'" Loucks wrote that his superior had informed him. "'I don't want to be made a fool of over this. Everyone seems to have cut them [the Nazis] off their list. To be friendly with them seems bad form.'"

But General Loucks noted in his diary that he had every intention of defying this superior's request. He had become good friends with the German chemists. He regularly had meals with Walter Schieber and Richard Kuhn, and, on at least one occasion, Schieber had spent the night at Loucks's house. "I'll see them anyhow," he wrote, in a diary entry dated February 1, 1950. The following day, Loucks was called back in to the Pentagon. "Went to Pentagon," he wrote, "long session with H.Q. Int. [headquarters, intelligence] people...seemed interested in what we are doing [in Heidelberg]—would give me money necessary to exploit the Germans for scientific and technical intelligence." In other words, what some at the Pentagon refused to condone, others were willing to support through covert means.

General Loucks's secret Saturday roundtable at his house in Heidelberg with the Nazi chemists remained hidden from the public for six decades. Here was a brigadier general with the U.S. Army doing business with a former brigadier general of the Third Reich allegedly in the interests of the United States. It was a Cold War black program that was paid for by the U.S. Army but did not officially exist. There were no checks and no balances. Operation Paperclip was becoming a headless monster.

The CIA's working relationship with the JIOA and Operation Paperclip had begun within a few months of the Agency's creation. Within the CIA, Paperclip was managed inside the Office of Collection and Dissemination, and one of the first things requested by its administrator, L. T. Shannon, was "a photostatted copy of a set of files compiled by Dr. Werner Osenberg and consisting of biographical records of approximately 18,000 German scientists." By the winter of 1948, hundreds of memos were going back and forth between the JIOA and the CIA. Sometimes the CIA would request information from the JIOA on certain scientists, and sometimes the JIOA would ask the CIA to provide it with intelligence on a specific scientist or group of scientists.

Also in the first three months of the CIA's existence, the National Security Council issued Directive No. 3, dealing specifically with the "production of intelligence and the coordination of intelligence production activities within the intelligence community." The National Security Council wanted to know who was producing what intelligence and how that information was being coordinated among agencies. In the opinion of the CIA, "the link between scientific planning and military research on a national scale did not hitherto exist." The result was the creation of the Scientific Intelligence Committee (SIC), chaired by the CIA and with members from the army, the navy, the air force, the State Department, and the Atomic Energy Commission. "Very early in its existence the SIC undertook to define scientific intelligence, delineate areas of particular interest and establish committees to handle these areas," wrote SIC chairman Dr. Karl Weber, in a CIA monograph that remained classified until September 2008. "Priority was accorded to atomic energy, biological warfare, chemical warfare, electronic warfare, guided missiles, aircraft, undersea warfare and medicine" — every area involving Operation Paperclip scientists. Eight scientific intelligence subcommittees were created, one for each area of warfare.

Despite the urgency, the JIOA's plan to make Operation Paperclip over into a long-term program was still at a standstill. By the spring of 1948, half of the one thousand German scientists bound for America had arrived, but not a single one of them had a visa. Troublemaker Samuel Klaus was gone from the State Department, but the JIOA could still not get the visa division to make things happen fast enough. On May

11, 1948, military intelligence chief General Stephen J. Chamberlin, the man who had briefed Eisenhower in 1947, took matters into his own hands. Chamberlin went to meet FBI director J. Edgar Hoover to enlist his help with visas. Cold War paranoia was on the rise, and both men were staunch anti-Communists. The success of Operation Paperclip, said Chamberlin, was essential to national security. The FBI had the Communists to fear, not the Nazis. Hoover agreed. Paperclip recruits needed the promise of American citizenship now more than ever, Chamberlin said, before any more of them were stolen away by the Russians. Chamberlin asked Hoover to put pressure on the State Department. J. Edgar Hoover said he would see what he could do. What, if anything, Hoover did remains a mystery. Three months later, the first seven scientists had U.S. immigrant visas. Now it was time to put the transition process to the test.

For Operation Paperclip, moving a scientist from military custody to immigrant status required elaborate and devious preparation, but in the end the procedure proved to be infallible. Scientists in the southwestern or western United States, accompanied by military escort, were driven in an unmarked army jeep out of the country into Mexico either at Nuevo Laredo, Ciudad Juárez, or Tijuana. With him, each scientist carried two forms from the State Department, I-55 and I-255, each bearing a signature from the chief of the visa division and a proviso from the Joint Chiefs of Staff, Section 42.323 of Title 22, signifying that the visa holder was "a person whose admission is highly desirable in the national interest." The scientist also had with him a photograph of himself and a blood test warranting that he did not have any infectious diseases. After consulate approval, the scientist was then let back into the United States, no longer under military guard but as a legal U.S. immigrant in possession of a legal visa. The pathway toward citizenship had begun. If the scientist lived closer to the East Coast than the West Coast, he went through the same protocols, except that he would exit the United States into Canada instead of Mexico and reenter through the consulate at Niagara Falls.

It was an international crisis in June of 1948 that finally gave Operation Paperclip its long-term momentum. Early on the morning of June 24, the Soviets cut off all land and rail access to the American zone in Berlin. This action would become known as the Berlin Blockade, and it was seen as one of the first major international crises of the Cold War. "The Soviet blockade of Berlin in 1948 clearly indicated that the wartime alliance [between the Soviets and the United States] had dissolved," explained CIA deputy director for operations Jack Downing. "Germany then became a new battlefield between east and west." The CIA presence in Germany was redoubled as its plans for covert action against the Soviets shifted into high gear. The CIA needed to hire thousands of foreign nationals living in Germany to help in this effort—spies, saboteurs, and scientists—many of whom had spent time in displaced-persons camps and interrogation facilities operated by the U.S. Army in the American zone of occupied Germany. Initially, the CIA and the JIOA worked hand in glove inside Germany to thwart Soviet threats, but soon the two agencies would start competing for German scientists and spies.

The two agencies worked together inside a clandestine intelligence facility in the American zone informally called Camp King. The activities there between 1946 and the late 1950s have never been fully accounted for by either the Department of Defense or the CIA. Camp King was strategically located in the village of Oberursel, just eleven miles northwest of the United States European Command (EUCOM) headquarters in Frankfurt. Officially the facility had two other names: the U.S. Military Intelligence Service Center at Oberursel and the 7707th European Command Intelligence Center. A small plaque in a park outside the officers' club explained to visitors the significance of the informal name. Colonel Charles B. King, an intelligence officer, had been in the process of accepting the surrender of a group of Nazis on Utah Beach in June 1944 when he was double-crossed and slain by a "strong and concentrated barrage of enemy artillery fire." There was tragic irony in the name. Camp King had become home to many well-intentioned American officers trying to make deals with untrustworthy enemies. Many of these American officers would be double-crossed and at least one of them would be killed.

A lot had changed at Camp King since John Dolibois had personally delivered six Nazi Bonzen here in August 1945. The interrogation facility had become one of the most clandestine U.S. intelligence centers in Western Europe, and for more than a decade it would function as a Cold War black site long before black sites were known as such. Camp King was a joint interrogation center and the intelligence agencies that shared access to prisoners here included Army Intelligence, Air Force Intelligence, Naval Intelligence, and the CIA. By 1948, most of its prisoners were Soviet-bloc spies.

How the CIA used Camp King remains one of the Agency's most closely guarded secrets. It was here in Oberursel that the CIA first began developing "extreme interrogation" techniques and "behavior modification programs" under the code names Operation Bluebird and Operation Artichoke. The unorthodox methods the CIA and its partner agencies explored included hypnosis, electric shock, chemicals, and illegal street drugs. Camp King was chosen as an ideal place to do this work in part because it was "off-site" but mainly because of its access to prisoners believed to be Soviet spies.

When the Americans captured the facility in the spring of 1945, the Nazis had been using it as an interrogation facility for Allied fliers. Camp King's first commanding officer was Colonel William Russell Philp, and through the fall of 1945, Philp shared the Military Intelligence Service Center at Oberursel with General William J. Donovan, founding director of the Office of Strategic Services. General Donovan oversaw an operation here whereby high-ranking Nazi generals, including those dropped off by John Dolibois, were paid to write intelligence reports on subjects like German order of battle and Nazi Party chain of command. Dolibois, fluent in German, acted as Donovan's liaison to the Nazi prisoners during this time. General Donovan kept an office at Oberursel until the OSS was disbanded in September 1945, after which he returned to Washington, D.C., and civilian life.

Colonel Philp's job was to handle the rest of the prisoners. In the months that followed war's end, the Camp King prisoner population grew to include Russian defectors and captured Soviet spies. There was valuable intelligence to be gained from these individuals, Philp learned, willingly or through coercion. But Philp also found that his officers lacked a greater context within which to interpret the raw intelligence being gathered from the Soviets. Russia had been America's ally during the war. Now it was the enemy. The Soviets were masters of disinformation. Who was telling the truth? The Nazi prisoners claimed to know, and Colonel Philp began using several of them to interpret and analyze information from Soviet defectors. These Nazis were "experts in espionage against the Russians," Philp later said. Two of them seemed particularly knowledgeable: Gerhard Wessel, who had been an officer in the German intelligence organization Abwehr, and Wessel's deputy, Hermann Baun. Philp put the men to work. What started out as a "research project using POWs" became a "gradual drift into operations," said Philp. He moved the Nazis into a safe house on the outskirts of Camp King, code-named Haus Blue, where they oversaw counterintelligence operations against the Soviets under the code name Project Keystone. Philp found that working with Nazis was a slippery slope, and in a matter of months the Germans had transformed from prisoners to paid intelligence assets of the U.S. Army.

In the summer of 1946 a major event occurred that influenced the CIA's future role in Operation Paperclip and Camp King. Major General Reinhard Gehlen, former head of the Nazis' intelligence operation against the Soviets, arrived at Camp King. Gehlen had been in the United States under interrogation since 1945. Here at Oberursel, Army Intelligence decided to make

Gehlen head of its entire "anti-Communist intelligence organization," under the code name Operation Rusty. Eventually the organization would become known simply as the Gehlen Organization. A network of former Nazi intelligence agents, the majority of whom were members of the SS, began working out of offices at Camp King side by side with army intelligence officers. Colonel Philp was in charge of overall supervision. By late 1947, the Gehlen Organization had gotten so large it required its own headquarters. Army intelligence moved the organization to a self-contained facility outside Munich, in a village called Pullach. This compound was the former estate of Martin Bormann and had large grounds, sculpture gardens, and a pool. The two facilities, at Oberursel and Pullach, worked together. Gehlen and Baun claimed to have six hundred intelligence agents, all former Nazis, in the Soviet zone of occupied Germany alone. According to documents kept classified for fifty-one years, relations between Gehlen and Philp declined and became hostile as Philp finally realized the true nature of who he was dealing with. The Gehlen Organization was a murderous bunch, "free-wheeling" and out of control. As one CIA affiliate observed, "American intelligence is a rich blind man using the Abwehr as a seeing-eye dog. The only trouble is — the leash is much too long."

The army became fed up with the Gehlen Organization, but there was no way out. Its operatives were professional double-crossers and liars—many were also alleged war criminals—and now they had the army over a barrel. Decades later it would emerge that General Gehlen was reportedly earning a million dollars a year. In late 1948, CIA director Roscoe Hillenkoetter met with army intelligence to discuss the CIA's taking charge of the Gehlen Organization. The two parties agreed, and on July 1, 1949, the CIA officially assumed control of Gehlen and his men.

That same summer, the CIA created the Office of Scientific Intelligence (OSI), and its first director, Dr. Willard Machle, traveled to Germany to set up a program for "special interrogation methods" against Soviet spies. The CIA had intelligence indicating that the Soviets had developed mind control programs. The Agency wanted to know what it would be up against if the Russians got hold of its American spies. In an attempt to determine what kinds of techniques the Soviets might be using, the CIA set up a Top Secret interrogation program at Camp King. The facility offered unique access to Soviet spies who had been caught in the Gehlen Organization's web. Revolutionary new interrogation techniques could be practiced on these men under the operational code name Bluebird.

A limited number of official CIA documents remain on record from this program. Most were destroyed by CIA director Richard Helms. Initially the CIA envisioned Operation Bluebird as a "defensive" program. Officers from Scientific Intelligence were "to apply special methods of interrogation for the purpose of evaluation of Russian practices." But very quickly the Agency decided that in order to master the best defensive methods it needed to first develop the most cutting-edge offensive techniques. This sounded like doublespeak and was indicative of the Cold War mind-set that was taking hold in intelligence circles and also in the military. The CIA believed it needed to develop the sharpest sword to create the strongest, most impenetrable shield. Operation Bluebird was just the beginning. Soon the program would expand to include mind control techniques and Nazi doctors recruited under Operation Paperclip.

CHAPTER SEVENTEEN

Hall of Mirrors

In the fall of 1948, in Germany, one of the most unusual press conferences of the Cold War took place. Major General Walter Schreiber, the former surgeon general of the Third Reich, had last been seen on the stand at Nuremberg testifying against fellow members of the Nazi high command. Then, on November 2, 1948, he reappeared at a press conference. After three years, five months, and three days in Soviet custody, Schreiber had allegedly "escaped" from his Soviet captors. Now the vaccine specialist said he had important news to share with the free world. The press conference opened with Schreiber delivering a brief statement about what had happened to him—he'd been a prisoner of the Soviets since the fall of Berlin and had recently escaped—followed by a lengthy question-and-answer period with an

American official acting as translator. The first question asked by a reporter was, How did Schreiber manage to escape?

Schreiber said he had "broken free" of his Communist guards in "a life or death situation" but hesitated to say more. With him now, in the safety of U.S. protective custody, he said, were his wife, Olga, his fourteen-year-old son, Paul-Gerhard, and one of his two grown daughters.

"How was it possible the Russians let him get out?" asked another member of the press, a question on everyone's mind. In the two years since General Schreiber's stunning testimony at Nuremberg, he'd been made *starshina,* or elder, in the Soviet military. It was almost inconceivable that a major player like Schreiber simply slipped away from his Soviet guards. Yet here he was.

"I'm not asking the details," the reporter clarified, "but how was it possible he was able to escape?"

In November 1948, Berlin was a city under psychological and physical siege. For more than four months now, the Soviets had blocked all rail, canal, and road access between East and West Berlin. To feed the civilians in the western zone, the Americans were flying in airplanes full of food. Schreiber's "escape" happened during the height of the Berlin Blockade.

"For reasons of security, [I] would not like to answer this question," Schreiber said.

"I don't want to ask any details," repeated the newsman. "But is it possible for others who are in the same position to get out?"

"The question was answered," said Schreiber.

Pressed further, Major General Dr. Schreiber reconstructed some of the events in his tale of escape. He'd been in Soviet Russia until the summer, he said. There, he and a group of other former Nazi generals lived together in a villa outside Berlin. In July or August, six of the generals, including Schreiber, were unexpectedly transported to a country house on the German-Polish border, near Frankfurt on the Oder, east of Berlin (not to be confused with Frankfurt on the Main, located in the American zone, southwest of Berlin). With regard to this mysterious journey and its greater purpose, Major General Schreiber said, "We were not asked, but we were told that we were going to join the police." Only then, Schreiber explained, did he learn he had been "appointed Chief Medical officer for the newly formed [East] German police." Schreiber said he was offered "food, clothing, housing, furniture...for advantages."

Four of the Nazi generals agreed to take the job. Schreiber said he objected. He was a scientist, not a policeman, he claimed to have told his Soviet handlers. The group of generals was transported to a home in Saxony, close to the Czechoslovakian border. Finally, "The last day of September, the four [generals] who had agreed were sent to Berlin in order to start their jobs," Schreiber said. He and another general, who had by this time also voiced objection, remained "guarded by police." Two days later Schreiber was sent to Dresden, in the Soviet zone. "There we were very well received, and I was offered the chance to become professor at the University of Leipzig," Schreiber told the press corps. "I demanded the University of Berlin. I had special reason for this demand. That was denied of me. For this reason, I made myself free."

So that was that. A chorus of West German reporters wanted more details. How does one simply make oneself free of Soviet military police, especially if one is the former surgeon general of the Third Reich? For the Soviets, turning high-ranking Nazi generals into Communist officials was an immense propaganda coup in the early days of the Cold War. One had to assume that Soviet military intelligence (Glavnoye Razvedyvatel'noye Upravleniye, or GRU) was keeping a watchful eye on each of the generals through their transition from Soviet Russia into the East German zone. The GRU's notorious official emblem featured an omnipotent bat hovering above the globe. The GRU kept radarlike track of people. They had eyes in the night. To allow Dr. Schreiber to get away sounded implausible.

"[I] took off alone, by express train, on the railroad, from Dresden to Berlin—and it was a trip of life and death," Schreiber said. And that was all he was going to say about it.

Next, Schreiber began to lecture his audience on the Soviet threat. He singled out a former colleague, Vincenz Müller, to blame, not unlike what he had done with Dr. Blome before the International Military Tribunal at Nuremberg. Lieutenant General Vincenz Müller was a dangerous man, exclaimed Schreiber. Now that he'd gone over to the Russian side, he was a threat to world peace. Lieutenant General Müller had recently been installed by the Soviet government as the new police leader in Berlin, Schreiber said. "He is a fanatical communist," promised Schreiber, "completely devoted to the Russians. This is all the more astonishing as Müller comes from a very devout Catholic family." The Russians had plans to arm Müller's new Berlin police force with "heavy weapons, tanks, [and] artillery." The Soviets had only one goal, Schreiber promised, and that was world domination. It was beginning right now with the rearmament of East Berlin.

"Can you give us the names of the four other generals, outside of General Vincenz Müller?" a news reporter asked.

"I don't think it is necessary in the scope of this press conference to give those names," said Schreiber.

"Could the Russians be selling you a bill of goods?" asked another reporter.

"The Russians are animated by the idea of world revolution," Schreiber said. He explained that in Russia, most people believe "the revolution is coming."

Another news reporter asked, "Were you wearing your [Soviet] uniform" when you escaped? It was a good question. If Schreiber had been wearing his Soviet uniform, then clearly he would have been noticed by border patrol guards, stopped, and questioned as he passed from the Soviet zone to the American zone of occupation. If Schreiber had not been wearing his Soviet uniform, then the obvious next question was, Why not? Schreiber's answer was convoluted. His Soviet uniform happened to be at the tailors' shop on the day of his escape, he said, getting new shoulder straps and embroidery on the collar. To emphasize his point, Schreiber even went so far as to re-create a conversation between himself and his Soviet handler — a man named "Fisher" — regarding the missing uniform. "Fisher said [to me], 'You are going to get [your uniform] later. For the time being, this is not yet possible.'"

The explanation seemed implausible to at least one newsman. "Why didn't you get your uniform tailored [earlier]?" the reporter asked.

Schreiber said that his measurements had been taken for the new uniform, but the tailoring was delayed.

When Schreiber's American handler moved to change the subject, another reporter asked for more information about the human experiments Dr. Schreiber had spoken of during the Nuremberg trial. "How did the Doctor obtain knowledge of experiments on human beings?" he queried.

Schreiber insisted he had "never taken part in any such research work.... The knowledge [I] gained about it, [I] either gained through documents [I] ran across in [my] position or in medical conventions, where intellectuals could see that something like that was being conducted in the background."

Every aspect of Schreiber's escape story seemed unreasonable, which made it difficult for the reporters to take seriously almost anything else he said. Yet the press conference went on for more than thirty minutes, with Schreiber standing his ground.

As it turned out, Schreiber's press conference was not impromptu but rehearsed. He had been discussing his testimony with officers from the U.S. Counter Intelligence Corps for two weeks — since October 18, 1948, the day he had walked into the CIC's Berlin office. CIC special agent Severin F. Wallach was Schreiber's handler. Wallach had heard a much longer version of what had allegedly been going on with Schreiber since his capture by the Red Army during the fall of Berlin.

According to the thirteen-page report by Wallach in Schreiber's intelligence dossier, "On the 5th of May Dr. Schreiber was sent, together with other captured

German Generals, back to Berlin. The Generals were put in a cellar of the Reich Chancellery in Berlin and received orders to emerge from this cellar under strong Soviet guard. This whole scene was photographed by the Soviets, who were engaged in putting together an 'authentic' documentary film of the capture of Berlin." On May 9, with the Reich's surrender complete, Dr. Schreiber was sent with other officers to a much larger prisoner of war camp, in Posen, where he stayed until August 12, 1945. A transport of generals to Moscow had been organized; Schreiber said he arrived there on August 29. "The transport was very badly organized," Schreiber said, according to the dossier report. "There was a food shortage because the cooks on the transport sold the food on the black market or kept it for themselves." Schreiber's testimony was resplendent with details. "All generals were sent to the PW camp No. 7027 in Krasnogorsk, near Moscow," he recalled. Here, the food tasted wonderful because it came from the United States, in cans. Schreiber repeatedly told Wallach how much he loved everything about the United States.

On March 12, 1946, Schreiber said he was transferred to the Lubyanka prison in Moscow. "Treatment not bad." On March 20 he claimed to have been interrogated by the Russians for the first time: "Subject was The German Preparation for Biological Warfare." Wallach had to have known that this was highly improbable. Schreiber was one of the Third Reich's highest-ranking medical doctors, and he was a major general in the army. On March 20, 1946, he would have been in Soviet custody for ten months. That this was his first interrogation was absurd. Schreiber told Wallach he was questioned by a lieutenant general named Kabulow for three days. Kabulow didn't believe his testimony, Schreiber said, and so he was told, "Soviet interrogators are going to use now physical violence to break [you] and get the whole truth out of [you]." The next interrogation, recalled Schreiber, took place at three o'clock that same morning. "[I was] beaten by a Soviet officer who me knows as Lt. Smirnow [Smirnov]. Together with a Col. Walter Stern, who speaks German without the slightest accent and who [was] an excellent interrogator."

Schreiber said he withstood three weeks of rough interrogation, at which point he finally broke down and "wrote the statement which later on was submitted by the Soviet Government to the International War Crimes Tribunal in Nuremberg." He was flown from Moscow to Berlin, then down to Nuremberg to testify at the war crimes trial. During one of the flights Schreiber said his German-speaking Soviet interrogator, Colonel Stern, leaned over and whispered a warning to him. If Schreiber were to go off-book and say "anything detrimental to the interests of the Soviet Union, he would be hanged on his return to Russia."

After testifying at Nuremberg, Schreiber said he was taken back to the Soviet Union, where he and three generals were set up in a two-story country house in Tomilino, sixteen miles southeast of Moscow. One of the three generals was Field Marshal Friedrich Paulus—the highest-ranking Nazi general to ever have surrendered to the Soviets, which Paulus did during the Battle of Stalingrad. Paulus's own story, of the events leading up to his capture and his final communication with Hitler, was remarkable. The Soviets had also brought Paulus to testify at Nuremberg.

As William Shirer explained in *The Rise and Fall of the Third Reich,* the last days of Paulus's command during the battle for Stalingrad were cataclysmic. "Paulus, torn between his duty to obey the mad Fuehrer and his obligation to save his own surviving troops from annihilation, appealed to Hitler." Paulus sent an urgent message to the Führer that read, "Troops without ammunition or food...Effective command no longer possible...18,000 wounded without any supplies or dressings or drugs...Further defense senseless. Collapse inevitable. Army requests immediate permission to surrender in order to save lives of remaining troops." But Hitler refused to allow Paulus to surrender. "Surrender is forbidden," Hitler wrote in return, "Sixth Army will hold their positions to the last man and the last round and by their heroic endurance will make an unforgettable contribution toward the establishment of a defensive front and the salvation of the Western world."

"Heroic endurance" was a euphemism for suicide. Paulus was now supposed to kill himself. Hitler nudged him further in this direction by making Paulus a field marshal in what he hoped would be the last hour of the general's life. "There is no record in military history of a German Field Marshal being taken prisoner," Hitler told Alfred Jodl, who was standing next to him at the time. Instead, at 7:45 the following morning, Field Marshal Paulus surrendered. His last message to Hitler: "The Russians are at the door of our bunker. We are destroying our equipment." He was taken prisoner shortly thereafter. What Paulus left behind was, as described by Shirer, a terrifying scene: "91,000 German soldiers, including twenty-four generals, half-starved, frostbitten, many of them wounded, all of them dazed and broken, were hobbling over the ice and snow, clutching their blood-caked blankets over their heads against the 24-degrees-below-zero cold toward the dreary, frozen prisoner-of-war camps of Siberia." Of the ninety-one thousand Germans taken prisoner by the Soviets, only five thousand would come out of the prison camps alive. Paulus was one of them.

By 1947, he was living comfortably in this two-story country house with Major General Dr. Schreiber, outside Berlin. Actually, explained Schreiber, there were a total of four former Nazi generals living together under one roof. In addition to Field Marshal Friedrich Paulus, there was Lieutenant General Vincenz Müller, captured outside Minsk in 1944, and General Erich Buschenhagen, captured in eastern Romania in August 1944. For what purpose? Wallach asked. "Subject [Schreiber] is convinced Lt. Gen Vincenz Mueller was ordered by the Soviets to indoctrinate Professor Dr. Schreiber with communistic ideas." Whatever the real reason, Schreiber said he and his fellow generals lived a relatively enjoyable life full of Soviet perks. At one point General Schreiber and General Buschenhagen were taken to live "in Moscow in a nicely furnished private house." Their Soviet handler, with them constantly, "acted as a guide and took them to the museum, opera and to play-houses stressing the fact that Soviet Russia has a highly developed culture." For Schreiber, the motive was clear. "This, too, was of course part of the planned indoctrination program," he told Special Agent Wallach. All the while, Schreiber feigned that he was a happy Communist.

In July 1947, Field Marshal Paulus became sick and the group was "taken to a summer resort, Livadia, on the Crimea." There was no shortage of irony here. This was the same palatial resort at which the Yalta Conference took place, in February 1945. The dangerous Lieutenant General Vincenz Müller was with Paulus and Schreiber at the resort. The group stayed through the summer and returned to Moscow, by private jet, when the summer weather passed. For the next year, the former Nazi generals resided again at the country house in Tomilino, Schreiber said. Only now they were heavily engaged in antifascist courses that the Soviets required them to take. Studying kept the generals occupied until September 7, 1948. That's when Schreiber said he learned that he and twenty-five other former Nazi generals would be leaving for East Germany at once. After Schreiber said no to the police job, he was brought to Dresden and put up in the Hotel Weisser Hirsch, at Bergbahnstrasse 12. His handler, the man called Fisher, agreed to release him from police work, Schreiber said. Fisher stepped away to work on arrangements regarding Schreiber's teaching position. According to Schreiber, that was when he got away.

Special Agent Wallach summarized the details. "Subject remained alone without anybody looking after him...Subject simply took a train in Dresden on the 17th and arrived in Berlin on the same day. After contacting his family in Berlin...subject established contact with this agent...and was since then under the protection of U.S. authorities in Berlin. At the end of October subject was evacuated with his family to the U.S. Zone for detailed exploitation by ECIC [European Command Intelligence Center]," Camp King.

Was Schreiber a double agent? Was he a true-to-life James Bond? How was he able to resist the Soviets' notoriously brutal interrogation techniques when so many others—from hardened generals to civilians to spies—were beaten into human ruin? Was he a charlatan? Or a weasel of a man, uniquely skilled at saving his own hide? What was he really doing in Soviet Russia for three and a half years? Special Agent Wallach drew his own conclusion. "Subject made an excellent impression on the undersigned agent. It is not believed that subject is a Soviet plant," wrote Wallach. He signed his name in black ink.

Wallach's interrogation report was sent to the director of the Intelligence Division of the U.S. Army, EUCOM, with a memo written by Wallach's CIC superior marked "Secret-Confidential." It read: "Subject [Schreiber] claims to know everybody in his transport, their background, political attitude and new job assignments...Will be ready for transfer to your headquarters for detailed interrogation in about six days." Schreiber had told Wallach he had information on all the high-ranking Nazis now working for the East German police.

On November 3, 1948, the director of the Intelligence Division of the U.S. Army sent a telegram marked "Secret" to JIOA headquarters at the Pentagon, with a copy also sent to the Joint Chiefs of Staff. "If the Surgeon General replies that Schreiber is of importance to national security, his case should be processed under JCS procedure for immigration to the U.S." Major General Dr. Prof. Walter Schreiber, the former surgeon general of the Third Reich, was about to become part of Operation Paperclip. In the meantime, he and his family were taken to Camp King and put up in a safe house there. When General Charles E. Loucks learned that Schreiber was in U.S. custody, he traveled to Camp King to interview him. Loucks was the man who had welcomed Hitler's chemists into his home in Heidelberg to work on the secret formula for sarin production. He was particularly interested in learning from Dr. Schreiber about vaccines or serums produced by the Reich to defend against nerve agents.

Loucks found Schreiber to be "cooperative in all respects" and hired him to work for the U.S. Chemical Corps "in compiling data concerning the Nazi Chemical Corps." To oversee the project, General Loucks traveled back and forth from Heidelberg to Camp King. Next, Schreiber was hired to write a monograph for the U.S. Army about his experiences in Russia. When Loucks was finished working with Schreiber, he was asked by Camp King's commanding officer, Lieutenant Colonel Gordon D. Ingraham, if he would testify as to Schreiber's character for the doctor's OMGUS security report. Given Schreiber's position as a general in the Nazi high command, it was going to take serious effort on the part of JIOA to bring Schreiber into the United States. Loucks agreed but was uncharacteristically skeptical of the Nazi general's motivations. "Loucks stated subject was energetic and a good organizer of work projects.... Schreiber had apparently given accurate information [to Loucks] on all occasions [which has] been checked and confirmed by Technical Research experts in the United States.... However, Loucks stated that Schreiber may also have given this same information 'to the Russians.'" Loucks told Lieutenant Colonel Ingraham that he "believed that Schreiber could be persuaded by any attractive offer." In other words, Schreiber's loyalty could be bought.

At Camp King, Dr. Schreiber and his family were moved into a nice home provided by the U.S. Army. Despite General Loucks's concern about Schreiber's trustworthiness, in November 1949 Major General Dr. Walter Schreiber was hired by army intelligence to serve as post physician at the clandestine interrogation facility that was Camp King. According to Schreiber's declassified OMGUS security report, his new job involved "handling all the medical problems at Camp King [and] caring for internees." This meant Schreiber was in charge of the health and well-being of the Soviet prisoners held here, some of whom were being subjected to "special interrogation methods" by the CIA. Given the army's obsession with Soviet spies and the possibility of double agents, hiring a Nazi general turned Soviet *starshina* was an unusual choice when one considered the real possibility that Schreiber had not escaped from the Russians but was working for them. If Schreiber was a Soviet spy, it would have been very easy for him to learn everything that the CIA and military intelligence were doing at Camp King.

On the other hand, if Schreiber really had escaped from the Russians, then there was a lot to be exploited from his Soviet experience. Having been a prisoner of the Russians for the past three and a half years, he was familiar with at least some of the Soviets' interrogation techniques. He spoke Russian fluently as well. Lieutenant Colonel Ingraham was confident that Dr. Schreiber was a truth-teller. Ingraham kept him on as post doctor until August 1951. Colonel Ingraham also hired Schreiber's twenty-three-year-old daughter, Dorothea Schreiber, to serve as his personal secretary.

While employed at Camp King, Schreiber told his Army handler that the Russians were trying to capture and kill him and he asked to use the cover name of "Doc Fischer," to hide his identity. It was a cryptic choice for an alias. "Fisher" had been the name of the Soviet handler from whom Schreiber had allegedly escaped, in Dresden, and it was also the name of an SS doctor who served as one of Schreiber's wartime subordinates at the Ravensbrück concentration camp.

Dr. Fritz Fischer had performed medical experiments on Polish women and girls at Ravensbrück, crimes for which he had been tried and convicted of murder at the Nuremberg doctors' trial. Fritz Fischer was one of the few doctors who had accepted his guilt over the course of the trial. After hearing some particularly shocking witness testimony against him, Dr. Fischer confided in war crimes investigator Dr. Alexander about how he felt. "I would have liked to stand up and say hang me immediately," Fischer told Alexander.

Looking at the whole scenario — Dr. Schreiber, Doc Fischer, the Soviet Mr. Fisher, and the SS doctor Fritz Fischer — was like seeing a man standing in a hall of mirrors. But then again Operation Paperclip was a world marked by duplicity and deception. It was impossible to know who was telling the truth.

In September 1949, John J. McCloy became U.S. high commissioner of Allied Germany, marking the end of more than four years of military rule of Germany by the Allies. The day also marked the beginning of the end of the time Dr. Otto Ambros would spend in prison for war crimes. Soon he would be placed on the Operation Paperclip target list.

Ambros, Hitler's favorite chemist, had been incarcerated for roughly one year of an eight-year prison sentence. On July 30, 1948, Ambros had been convicted of mass murder and slavery in Case No. VI of the subsequent Nuremberg proceedings, the IG Farben trial, and sent to Landsberg Prison, also called War Criminal Prison No. 1. Located thirty-eight miles west of Munich, Landsberg was home to 1,526 convicted Nazi war criminals. The men were housed in a central prison barracks inside individual cells, but the facility itself was situated on a boarding school–like campus, with nineteenth-century buildings, leafy parks, and a grand, wood-paneled Catholic church. Adolf Hitler had been a prisoner here for eight months in 1924. Landsberg Prison was where Hitler wrote *Mein Kampf.*

In prison Ambros taught "chemical technology" to other inmates as part of a prisoner education program. He penned letters to his mother expressing how unfair it all was. "Politic[s] is a bitter disease and it is grotesque that I, as a non-political person, should suffer for something I have not done," he wrote. "But one day, all this suffering will cease and then it will not be long before I have forgotten all this bitterness." Ambros's nineteen-year-old son, Dieter Ambros, wrote clemency appeals on his father's behalf. "My father is innocent as you know," began one letter to Bishop Theophil Wurm, a Protestant leader who regularly advocated for the war criminals' release. "Thank you for supporting our efforts...my father is [being] illegally held." Ambros was a model prisoner. Only once was he written up for disciplinary action: "Inmate

Ambros, Otto, WCPL No. 1442 was standing and looking out the window at the women's exercise yard [and] this is against the prison regulations," reads a note in his prison file.

Otto Ambros had many lawyers working for him to secure an early release. He also wrote petitions himself, requesting small items. In 1948, he asked the prison board for an extra pillow, softer than the one provided. In 1949, he requested permission to keep his accordion in his cell. Each year, Ambros saw the Landsberg Prison doctor for a checkup. Convicted Nazi war criminal Dr. Oskar Schröder, former chief of the Medical Corps Services of the Luftwaffe, checked Ambros's vitals and wrote up his annual health report. Schröder had been employed by the U.S. Army Air Forces at the Aero Medical Center in Heidelberg before the doctors' trial and was now serving a life term at Landsberg. Also in the prison were Hermann Becker-Freyseng and Wilhelm Beiglböck, serving twenty and fifteen year sentences, respectively.

The twelve subsequent war crimes trials at Nuremberg had ended just a few months prior to John J. McCloy's becoming high commissioner of Germany. Most Americans had long since lost interest in following any of the trials. The majority of Germans disagreed with the whole war crimes trial premise, and many saw those convicted as having been singled out by American and British victors and given "victors' justice" as punishment. At war's end, U.S. occupation authorities had determined that 3.6 million Nazis in the American zone alone were "indictable" for political or war crimes. This enormous number was eventually whittled down to a more manageable 930,000 individuals, who were then processed through 169,282 denazification trials. More than 50,000 Germans had been convicted of various Nazi-era crimes, most in the Spruchkammern courts but also in Allied military tribunals. The majority of those convicted served some time in postwar detention camps or paid nominal fines. When McCloy took office, 806 Nazis had been sentenced to death and sent to Landsberg Prison, with 486 executions carried out to date. By the fall of 1949, the German press had begun referring to the convicted criminals held at Landsberg as the "so-called prisoners of war." This was just one of the sensitive issues that Commissioner McCloy was faced with when he arrived. Another was Operation Paperclip.

McCloy had been a champion of the Nazi scientist program from its very first days, back in the late spring of 1945, when he served as assistant secretary of war. He was also the chairman of the State-War-Navy Coordinating Committee at the time, which put him in charge of making some of the first decisions regarding the fate of the program. McCloy was a statesman and a lawyer but he was also an economist. In between his tenure as assistant secretary of war and high commissioner of Germany, he was president of the World Bank. His service there came at a critical time in the bank's early history. McCloy is credited in World Bank literature as "defining the relationship between the Bank and the United Nations and the Bank and the United States." Now he was back in government service as a diplomat, having come to Germany to fill the shoes of General Lucius D. Clay, exiting OMGUS chief. John J. McCloy was a short, plump man, balding, with a banker's bravado. When in public he almost always wore a crisp suit. As high commissioner he traveled around Germany in the private diesel train that had belonged to Adolf Hitler. While power had been officially transferred to a new West German civilian government, run by Chancellor Konrad Adenauer, McCloy remained in charge of many aspects of Germany's law and order as West Germany transitioned into becoming its own sovereign nation once again. One area that Chancellor Adenauer had absolutely no jurisdiction over was the Landsberg prisoners. Several hundred of these convicted war criminals had already been hanged in the Landsberg courtyard. Eighty-six others faced death. When McCloy took office as high commissioner, the rhetoric around the Landsberg prisoners was at an all-time high. Many Germans wanted the prisoners released.

In November of 1949, a group of German lawyers

linked to the Farben industrialists, including Otto Ambros, requested a meeting with John J. McCloy at his office in the former IG Farben building, in Frankfurt. The IG Farben building had been taken over by the U.S. Army when troops entered Frankfurt in March of 1945 and had served as a home for the U.S. Army and various U.S. government organizations ever since. The massive complex—the largest office building in Europe until the 1950s—had panoramic views of Frankfurt, as well as parklands, a sports field, and a pond. In August 1949, OMGUS moved its headquarters from Berlin to Frankfurt, and shortly thereafter the U.S. high commissioner's office headquarters were set up in the IG Farben complex. In September 1949, McCloy settled in. The CIA maintained an office in the IG Farben building throughout the Cold War. It was located just a few floors and a few doors down from McCloy's office.

It was a precarious time for an American civilian to be governing occupied Germany. The Soviets had just detonated their first atomic bomb, years ahead of what had been predicted by the CIA. The U.S. military was on high alert, perhaps nowhere more so than in West Germany. During the November meeting in McCloy's office at the IG Farben complex, German lawyers told McCloy that if West Germany and the United States were going to move forward together in a united front against the Communist threat, something had to be done about the men incarcerated at Landsberg. These prisoners were viewed unanimously by Germans as "political prisoners," the lawyers said, and they told McCloy that he should grant all of them clemency.

After the meeting, McCloy sent a memo to the legal department of the Allied High Commission, inquiring if "after sentences were imposed by military tribunal," he, as U.S. high commissioner, had any authority to review the sentences. The legal department told him that as far as the Landsberg war criminals were concerned, he had the authority to do whatever he thought appropriate. In America, Telford Taylor, the former Nuremberg prosecutor general, caught wind of what was going on in the high commissioner's office in Frankfurt and was outraged. He wrote to McCloy to remind him that the Nazi war criminals at Landsberg "are without any question among the most deliberate, shameless murderers of the entire Nuremberg List, and any idea of further clemency in their cases seems to me out of the question." McCloy never responded, according to McCloy's biographer, Kai Bird.

McCloy established an official review board to examine the war criminals' sentences—the Advisory Board on Clemency for War Criminals, known as the Peck Panel, after its chairman, David W. Peck. A powerful former Nazi lieutenant general, Hans Speidel, appealed personally to McCloy. Speidel was one of

Chancellor Konrad Adenauer's chief advisers on rearmament, a highly controversial subject but one being discussed nonetheless. Hans Speidel's younger brother, Wilhelm Speidel, was a convicted war criminal at Landsberg. Speidel told McCloy's adjunct in Bonn, "[If] the prisoners at Landsberg were hanged, Germany as an armed ally against the East was an illusion." In a similarly bullish manner, Chancellor Konrad Adenauer told McCloy the same thing, advising him that he should grant "the widest possible clemency for persons sentenced to confinement."

In June 1950, North Korean forces, supported by Communist benefactors, moved across the 38th parallel, marking the start of the Korean War. The idea that the Communists were also about to invade Western Europe took hold in the Pentagon. On July 14, 1950, the commander at Wright-Patterson Air Force Base sent an urgent memo to Operation Paperclip's in-house champion, Colonel Donald Putt: "Due to the threat of impending hostilities in Europe and the possibility that forces of the USSR may rapidly overrun the continent, this command is concerned with the problem of the immediate implementation of an evacuation program for German and Austrian scientists." Were these scientists to "fall into enemy hands... they would constitute a threat to our national security." Air force intelligence recommended to JIOA that it initiate a "mass procurement effort" in Germany. JIOA agreed and began making formal plans with the high commissioner's office to effect this.

The Korean War sparked a new fire under Operation Paperclip. Inside the high commissioner's office, McCloy maintained a group called the Scientific Research Division that was specifically dedicated to the issue of German scientists. The head of the division was Dr. Carl Nordstrom, and ever since McCloy had taken office Nordstrom had been trying to expedite the procession of German scientists to America. Dr. Nordstrom·maintained a thick file labeled "Allocation of German Scientists and Technicians" and had sent many eyes-only memos to McCloy in "support of certain research projects" he foresaw as valuable to national interest. Now, in light of the Korean War, Nordstrom got a new job from JIOA. He was assigned to be the German liaison to a new JIOA program being fast-tracked out of the Pentagon, named Accelerated Paperclip but called Project 63 in the field: A number of Germans had soured on the name Paperclip. The premise of the Accelerated Paperclip program was to move "especially dangerous top level scientists" out of Germany in a "modified Denial Program" that needed to be kept away from the Soviets at all costs. The high commissioner's office began working with army intelligence to "evacuate" 150 of these scientists, code-named the "K" list, from Germany to the United States. A group of American officers called the Special Projects Team would be dispatched to recruit the "K" list scientists. The Joint Chiefs of Staff approved a prodigious $1 million procurement budget to help entice these "especially dangerous top level scientists" to come to America, the equivalent of approximately $10 million in 2013.

Accelerated Paperclip, or Project 63, meetings were held at the high commissioner's offices in Wiesbaden and Frankfurt, with Dr. Carl Nordstrom keeping notes. Representatives from JIOA, the army, the air force, EUCOM, and the CIA attended. Because many on the "K" list did not have a job offer already in place, the JIOA decided to set up a clandestine office in New York City, at the Alamac Hotel, where the scientists could live while they waited for assignments. The Accelerated Paperclip project director in America, Colonel William H. Speidel (no known relation to the Wilhelm Speidel war criminal at Landsberg Prison or his lawyer brother), maintained an office there. An entire block of rooms was set aside in the nineteen-story hotel, on Seventy-first Street and Broadway, for a yet-unnamed group of German scientists scheduled to arrive at a future date. A welcome brochure was printed up and kept on file at the high commissioner's office. "To insure your comfort, convenience and interest in general," it read, "a competent officer, assisted by a carefully selected staff...will serve your interests from the time of your arrival until the time of your departure to enter employment." The officer "will maintain an office at the hotel in which you reside and be prepared to complete, or make provisions for, all arrangements incident to housing, restaurant facilities, securing medical services, and the administrative details of the project." The U.S. Army's "primary interest," the scientists were told, "is in providing for your comfort, contentment, happiness and security [and] efforts will be directed to help you in attaining these goals with a minimum [of] friction, distraction, and delay."

But the program did not take off like fire in dry grass, as Dr. Nordstrom had hoped it would. Much to everyone's surprise, the offers made under Accelerated Paperclip were rejected by many of the German scientists who were approached. When JIOA requested an explanation from the high commissioner's office as to why, Nordstrom reported that some on the "K" list were simply "too old, too rich, too busy and too thoroughly disgruntled with past experiences with Americans," to see a free room at the Alamac Hotel in New York City as a career move. Besides, Germany had its own chancellor now, and for the first time in five years, many German scientists saw that a prosperous scientific future was possible in their own country.

Others could not wait to come to America. With Accelerated Paperclip's newest policy in place, Class I offenders could now be put on a JIOA list. This included Dr. Schreiber, still serving as post physician at Camp King. Another Class I offender was Dr. Kurt Blome, former deputy surgeon general of the Third Reich and Hitler's biological weapons expert. The sword and the shield.

Finally, there was Dr. Otto Ambros, the war criminal convicted at Nuremberg of slavery and mass murder. In the winter of 1951, Otto Ambros was placed on the JIOA list for Accelerated Paperclip even though he was still incarcerated at Landsberg Prison.

In January of 1951, John J. McCloy's office announced that he had come to a decision regarding the war criminals incarcerated at Landsberg Prison. The Peck Panel had finished its review process and recommended "substantial reductions of sentences" in the majority of cases involving lengthy prison terms. As for those who had been handed death sentences, the panel advised McCloy to consider each case individually. Also at issue was a financial matter. At Nuremberg, the judges had ordered the confiscation of property of convicted war criminals whose money was so often earned on the backs of slave laborers, tens of thousands of whom had been worked to death. Now, the Peck Panel suggested that this confiscation order be rescinded. For Otto Ambros, this would mean that he could keep what remained of the gift, from Adolf Hitler, of 1 million reichsmarks, a figure that has never been revealed before. McCloy spent several months considering the panel's recommendations. During this time he was deluged with letters from religious groups and activists in Germany urging for the war· criminals' release. McCloy sent a cable from Frankfurt to Washington asking for counsel from the White House. The White House advised McCloy that the decision was his to make.

John J. McCloy commuted ten of the fifteen death sentences. This meant that ten men condemned by International Military Tribunal judges—including the commander of the Malmédy Massacre, considered one of the war's worst atrocities against prisoners of war, and several SS officers who had overseen the mobile killing units called Einsatzgruppen—would be released back into society within one and seven years. Among the death sentences McCloy chose to uphold were those of Otto Ohlendorf, commander of Einsatzgruppe D, responsible for ninety thousand deaths in Ukraine; Paul Blobel, commander of Einsatzgruppe C, responsible for thirty-three thousand deaths at Babi Yar, in Kiev; and Oswald Pohl, chief administrator of the concentration camps. McCloy also drastically reduced the sentences of sixty-four out of seventy-four remaining war criminals, which meant that one-third of the inmates tried at Nuremberg were freed. On February 3, 1951, Otto Ambros traded in his red-striped denim prison uniform for the tailored suit he had arrived in.

He walked out of the gates of Landsberg Prison a free man, his finances fully restored.

General Telford Taylor was indignant. In a press release he stated, "Wittingly or not, McCloy has dealt a blow to the principles of international law and concepts of humanity for which we fought the war." Eleanor Roosevelt asked in her newspaper column, "Why are we freeing so many Nazis?"

The will and wherewithal to punish Nazi war criminals had faded with the passage of time. "Doctors who had participated in the murder of patients continued to practice medicine, Nazi judges continued to preside over courtrooms, and former members of the SS, SD and Gestapo found positions in the intelligence services," explains Andreas Nachama, curator of the Nazi Documentation Center in Berlin. "Even some leaders of the special mobile commandos ("Einsatzkommandos") [paramilitary extermination squads] tried to pursue careers in the public service."

The following month, on March 27, 1951, Dr. Carl Nordstrom dispatched Charles McPherson, an officer with the Special Projects Team, to go locate and hire Dr. Kurt Blome. The Special Projects Team was now composed of a group of twenty agents, each with his own "K" list of scientists to find. McPherson learned that Blome lived at 34 Kielstrasse, in Dortmund, and he traveled there to interview the doctor.

During his first visit, Charles McPherson learned that Dr. Blome lived in the apartment adjacent to Blome's private physician's practice in Dortmund during the week. On weekends he returned to his home in Hagen, twelve miles away, to be with his family. "His English is excellent and no interpreter is necessary to carry on a conversation," McPherson wrote in his report. The reason for the visit, McPherson told Dr. Blome, was to offer Blome a contract with Operation Paperclip. "He stated he would definitely be interested." Dr. Blome requested more details. "He feels he is too old to begin a new type of work and would prefer to return to biological research or cancer research." Blome alluded to the fact that he had already worked on Top Secret germ warfare research for the British, under Operation Matchbox, the British equivalent of Operation Paperclip. Blome said that the British had helped secure his house in Hagen for him. McPherson left Dortmund with the impression that the fifty-seven-year-old was "very interested but would need a definite offer before he could make up his mind."

Approximately three months later, on Thursday, June 21, 1951, McPherson again interviewed Dr. Blome. "I presented him with a copy of our contract form and informed him that we were willing to pay him about $6400 per year for the duration of the contract." Blome had additional questions. He asked

McPherson about the buying power of this salary and the amount of taxes he would have to pay. "He then had another request which I informed him that I could do nothing about," McPherson wrote. Blome said he had "some money which was tied up in a professional account because it had been determined that these are funds of the Nazi Party." Blome asked McPherson for his help in trying to release the money back to him, and to look into "the possibilities of transferring [the money] from Marks into Dollars," in order to bring it to the United States. McPherson explained, "I informed him that there was no legal means at present of doing this but that this could be transacted through Switzerland."

Blome said he needed some time to read over the contract and to discuss the matter with his wife. He said he'd get in touch with McPherson in about two weeks. In August it was official: "Professor Kurt Blome was contracted under Project 63 on 21 August. Will be ready for shipment 15 November," McPherson wrote. The Blomes took their boys out of school and began teaching them English. Dr. Blome turned his practice over to another doctor in Dortmund. The couple traveled to the Berlin Document Center and provided sworn testimony regarding their Nazi past. The documents were reviewed by McCloy's office. Per Accelerated Paperclip, a key document that would be used for a visa application, the Revised Security Report on German (or Austrian) Scientist or Important Technician, was drawn up.

The single most important element governing justification of Accelerated Paperclip/Project 63 was now stated on page one: "Based on available records... Subjects have not been in the past and are not at the present time members of the Communist Party." The issue of being an ardent Nazi had lost first position and was relegated down to section six. There, the issue of Blome's Nazi Party record was addressed: "Kurt Blome entered the Party on 1 July 1931 with Party number 590233. He is also listed as a member of the SA since 1941 and is a holder of the Golden Party Badge since 1943. His wife, Dr. Bettina Blome, entered the Party on 1 April 1940 with Party number 8,257,157." It was also noted, "The 66th CIC Central Registry contains a Secret dossier on Dr. Kurt Blome." Those details were separately classified.

"Based on available records, Subjects were not war criminals but undoubtedly were ardent Nazis," wrote Lieutenant Colonel Harry R. Smith, an authorized representative of John J. McCloy. Blome was a Nazi ideologue and Smith stated so. He also wrote that Blome was not likely to become a security threat. "It is the opinion of the United States High Commissioner for Germany that they are not likely to become security threats to the United States." The report was signed and dated September 27, 1951. Two weeks later, on October 10, 1951, Blome's secret Accelerated Paperclip contract was approved. But on October 4, 1951, the chief of Army Intelligence, G-2, a colonel by the name of Garrison B. Coverdale, read the high commissioner's security report on Dr. Blome and rejected Blome's admission to Operation Paperclip. McCloy's representative, Lieutenant Colonel Harry R. Smith, had failed to do the most important thing necessary when it came to drawing up a Paperclip contract: to lie by omission about Nazi party loyalty. All kinds of phrases could be used to allow the State Department to turn a blind eye to its legal obligation to keep Nazis out of the United States. Most OMGUS security officers knew to write "not a security threat" or "merely an opportunist" in the space that asked about the scientist's Nazi Party record. After reading the report, Colonel Garrison B. Coverdale sent a confidential cable to the director of JIOA in Washington, stating, "Attention is invited to paragraph 6 of subject report." In stating the truth, Lieutenant Colonel Harry R. Smith made it impossible for Dr. Blome's Paperclip contract to be approved.

On October 12, 1951, the JIOA and Army Intelligence read the high commissioner's report. A confidential memo from the Department of the Army in

Heidelberg was sent back marked urgent: "Suspend shpmt Dr. Kurt Blome appears inadmissible in view of HICOG [High Commissioner of Germany]."

McPherson would not readily accept this setback. "Blome contract signed and approved Commander in Chief. Subject completing preparations for shipment late November. Has already turned over private practice Dortmund to another doctor. In view of adverse publicity which might ensue and which may destroy entire program this theatre recommend[s] subject be shipped for completion 6 months portion contract," he wrote. The case was sent to the consulate for an opinion. On October 24, 1951, the consulate in Frankfurt agreed with army intelligence: "Frankfurt Consul states Blome inadmissible."

Charles McPherson saw Blome's rejection as calamitous. It interfered with the core mission of the Special Projects Team and jeopardized the success of the entire Accelerated Paperclip program, he believed. "Recommend Blome be shipped and ostensibly occupied on inconsequential activities in the United States for 6 months," McPherson suggested. But that idea was rejected, too.

McPherson contacted Blome by telephone. They agreed to meet at the Burgtor Hotel in Dortmund. McPherson brought a colleague from the Special Projects Team along with him this time, Philip Park. McPherson's job was "to explain to [Blome] the reasons why we were unable to send him to the United States at the present time." Agent Park could back him up if uncomfortable questions arose.

"Professor Blome was alone when we came in. I commenced by saying that I had some bad news to tell him," McPherson wrote in a memo. "I then stated that due to another change in our security laws we were forced to suspend our shipments to the United States."

Blome was no fool. This was a man who skillfully argued at Nuremberg that intent was not a crime and was acquitted on these grounds. Now, at the Burgtor Hotel, Blome made clear to McPherson that he did not believe McPherson's lies. Blome had recently been in communication with his colleague and former professor Pasqual Jordan and was entirely aware that Jordan was actively being prepared for Accelerated Paperclip work. Blome told McPherson that he had recently been in contact with one of his deputies in biological warfare research, the rinderpest expert Professor Erich Traub. Blome told McPherson that he was well aware that Traub had recently gone to America "under the auspices of the Paperclip Project." Blome was particularly upset because Traub had worked under him during the war.

McPherson tried to placate Dr. Blome. "I then said that we were still ready to put his contract in effect and that we had a position for him at a military post in Frankfurt, probably as a post doctor." McPherson suggested that Blome come have a look at the facility, that it was a pleasant place and offered a well-paying job. "He and his wife agreed," McPherson wrote in his report. Still, McPherson saw the situation with Dr. Blome as potentially disastrous. "The undersigned wishes to point out that he did not tell the Blomes that their opportunity to go to the United States is apparently nil and apparently will remain that way, but took the slant that our shipments have been suspended for an indefinite duration. This will leave the way open in the event something can be done."

McPherson worried that the Blome affair was going to have an unfavorable effect "upon our project not only by the immediate individuals concerned but by the chain reaction produced by one individual telling the other that the Americans have broken their word." McPherson was actively recruiting other German scientists at the same time. He believed that he had a handle on how the scientists communicated among themselves. "The professional class of Germany is tightly enough knit so that this word will be widely disseminated and the future effectiveness of our program will be greatly curtailed."

McPherson was wrong. Not all the former Nazis in the "professional class" talked among themselves. Dr. Blome was not aware that the reason a job offer as a military post doctor near Frankfurt was available to him was because the previous post doctor, also a former high-ranking Nazi in the Reich's medical chain of command, had just been shipped to America under Accelerated Paperclip. There was irony in the fact that the shoes Dr. Blome was about to fill had, for the past two years and four months, belonged to Major General Dr. Walter Schreiber, the man who had betrayed Dr. Blome in his testimony during the trial of the major war criminals at Nuremberg. The army post outside Frankfurt where Blome would soon start working was Camp King. McPherson made arrangements with its new commanding officer, Colonel Howard Rupert, to meet the Blomes and arrange for them to have a nice house.

Blome had been to Camp King before. He had worked for the U.S. government there on a "special matter" during the Reinhard Gehlen era. When Dr. Blome's wife, Bettina, learned more about Camp King, she declined the invitation to live there. She was not interested in having her children live at an American military facility. The couple separated. Dr. Kurt Blome moved to Camp King alone. On November 30, 1951, McPherson reported: "Dr. Blome employed by ECIC effective 3 Dec for 6 months. Contract placed in effect."

Meanwhile, halfway across the world, in the suburbs of San Antonio, Texas, Camp King's former post physician, Major General Dr. Walter Schreiber, had arrived in the United States and was working at the School of Aviation Medicine at Randolph Air Force Base (formerly Randolph Field). In addition to his U.S. government salary, Schreiber had recently received a check from the U.S. government in the amount of $16,000 — roughly $150,000 in 2013 — as a settlement for the alleged lost contents of his former Berlin home. (Schreiber claimed that the Russians had stolen all his property in retaliation for his working for the Americans.) With the money, Schreiber bought a home in San Antonio and a car, and he enrolled his son in the local high school. Even Schreiber's eighty-four-year-old mother-in-law had been brought along to live in Texas, courtesy of the United States Air Force. It could be said that Dr. Schreiber was living the American dream. But in the fall of 1951 the dream was unexpectedly interrupted by a former war crimes investigator named Dr. Leopold Alexander, and a concentration camp survivor named Janina Iwanska.

CHAPTER EIGHTEEN

Downfall

In the fall of 1951, Dr. Leopold Alexander's life in Boston had returned to normal. It had been four years since he had served as an expert consultant to the secretary of war during the Nuremberg doctors' trial. Since returning home, Dr. Alexander continued to speak out against medical crimes, nonconsenting human experiments, and medicine under totalitarian regimes. He coauthored the Nuremberg Code, the set of research principles that now guided physicians around the world. The first principle of the Nuremberg Code was that informed voluntary consent was required in absolute terms. Dr. Alexander gave lectures, wrote papers, and practiced medicine. The doctors' trial had affected him deeply, as evidenced in nearly five hundred pages of journal entries.

Whenever the occasion arose, he provided pro bono service to victims of the Nazi regime.

One day in the fall of 1951, Dr. Alexander was contacted by an aid group called the International Rescue Office. The group was organizing medical assistance for several concentration camp survivors who had been experimented on by Nazi doctors during the war and asked if Dr. Alexander could help. Dr. Alexander in turn contacted his friend and colleague at Beth Israel Hospital in Boston, chief surgeon Dr. Jacob Fine, and the two men helped arrange for the camp survivors to come to Massachusetts for medical treatment. One of the women was twenty-seven-year-old Janina Iwanska, a former prisoner at the Ravensbrück concentration camp. On November 14, 1951, Iwanska arrived at the Port of New York aboard a Greek ocean liner, the SS *Neptunia*.

At the doctors' trial, Janina Iwanska had delivered much of her testimony with Dr. Alexander standing beside her, pointing to her injuries and providing the judges with a professional medical analysis of what had been done to her by Nazi doctors during the war. Iwanska's testimony was generally regarded as among the most powerful evidence presented at the trial. At Ravensbrück she had had her legs broken by Waffen-SS surgeon Dr. Karl Gebhardt and pieces of her shinbones removed. Dr. Gebhardt then ordered that Iwanska's surgical wounds be deliberately infected with bacteria to cause gangrene, so he could treat them with sulfa drugs to see if the drugs worked. It was nothing short of a miracle that Janina Iwanska survived. Now, nine years later, she continued to suffer great physical pain. She walked with a limp because of the decimation of her shinbones. The purpose of the trip to the United States was to allow Iwanska to undergo surgery, at Beth Israel Hospital in Boston, to help alleviate this pain.

Dr. Gebhardt had been Major General Dr. Schreiber's direct subordinate at Ravensbrück. Gebhardt had been one of the twenty-three defendants tried at the doctors' trial. He was convicted, sentenced to death, and hanged in the courtyard at Landsberg Prison in June, 1948. Dr. Schreiber was working for the U.S. Air Force in Texas.

The same month that Janina Iwanska arrived in the United States, a brief note appeared in a medical journal stating that a doctor from Germany named Walter Schreiber had just joined the staff of the U.S. Air Force School of Aviation Medicine in Texas. As circumstance would have it, Dr. Leopold Alexander was a regular reader of this journal. When he came across Schreiber's name—familiar as he was with Dr. Schreiber's testimony at the Nuremberg trial of the major war criminals—Alexander was appalled. He wrote to the director of the Massachusetts Medical Society at once. "I regard it as my duty to inform you that the record shows that Dr. Schreiber is a thoroughly undesirable addition to American Medicine—in fact, an intolerable one," Alexander explained. "He has been involved as an accessory before and after the fact in the worst of the Medical War Crimes which were carried out by the Nazi Government during the war, and was a key person in perverting the ethical standards of the members of the medical profession in Germany during the war." Dr. Alexander demanded that Dr. Walter Schreiber be barred from practicing medicine in the United States. When he did not get the response he'd expected, he took the matter to the *Boston Globe*.

It was shortly after 10:30 p.m. on December 8, 1951, in San Antonio, Texas, when Dr. Walter Schreiber heard the telephone ring. He had been in America for three months, and it was an unusual hour for him to receive a phone call. It was not as if Dr. Schreiber was a hospital physician on call waiting to hear about a sick patient. In his capacity as a research doctor at Randolph Air Force Base, he spent most of his time lecturing about classified matters to a small group of other doctors.

Sometimes, he boasted to other doctors about how his area of expertise was extremely rare. He would tell colleagues that he was particularly valuable to the U.S. military because his knowledge was so esoteric. Dr.

Schreiber knew everything there was to know about winter warfare and desert warfare. About hygiene and vaccines and bubonic plague. At the officers' club at the School of Aviation Medicine, where he gave lectures, he could be loud and boastful, delivering a highly sanitized version of his colorful life. He enjoyed telling long-winded stories about himself: how he had been a prisoner of the Soviets after the fall of Berlin; how he'd spent years in the notorious Lubyanka prison, in Moscow; how he'd doctored Field Marshal Paulus in a Russian safe house when Paulus got sick. But what Dr. Schreiber never discussed with anyone at Randolph Air Force Base was what he did before the fall of Berlin—from 1933 to 1945.

Schreiber answered the telephone and was greeted by a man who identified himself as Mr. Brown.

"I am calling from the *Boston Globe*," Brown said.

Mr. Brown did not ask Schreiber his name. Instead, Brown asked if he'd reached "telephone number, 61-210 in San Antonio, Texas." Dr. Schreiber told Brown the number was correct.

There was a pause. Later, Dr. Schreiber recalled to military intelligence that he'd asked Mr. Brown what it was that he wanted at this time of night.

"Are you the individual who performed experiments on the bodies of live Polish girls who were interned in German concentration camps during World War II?" Brown asked.

Schreiber told Brown he had never been connected in any way with experiments of that nature. "I [have] never worked in a concentration camp," Schreiber said. "I have never in my entire life conducted, ordered, or condoned experiments on humans of any nationality."

Brown told Schreiber he assumed he'd passed investigations before being brought to America to conduct secret U.S. Air Force work.

"I [have] been thoroughly investigated," Schreiber said.

Brown thanked Schreiber. He said he was just checking up on a story that had been relayed to him by a physician in Boston, Dr. Leo Alexander.

Dr. Schreiber hung up the phone. He did not tell anyone that night about the call.

The following morning, the *Boston Globe* published an explosive story with an eye-catching headline: "Ex-Nazi High Post with United States Air Force, says Medical Man Here." News reached Texas quickly. Dr. Schreiber was called into the Office of Special Investigations (OSI) at Randolph Air Force Base to provide details of the phone call with Mr. Brown the previous evening. Schreiber confirmed that "during World War II he had held a position in the Wermacht [*sic*] similar to the Surgeon General of the U.S. Army." But his story, his struggle, he told the investigating officer, was so much more than that. Schreiber relayed his capture in Berlin, his life as a prisoner of war in Soviet Russia, and how he'd "served as a prosecution witness during the Nuremberg War Crimes Trials." Schreiber explained that owing to "feigned compliance" with the Russians he'd been "awarded a high position with the East German Politzei [*sic*]." Instead of taking that job he'd escaped. Schreiber lied to the Office of Special Investigations and said that he had met Dr. Leo Alexander at Nuremberg. That he had been "given a clean bill of health by [Doctor] Alexander" himself.

Schreiber told the air force investigator he was certain that the Russians were behind all of this. He had recently written a manuscript called "Behind the Iron Curtain," Schreiber said, and while he hadn't found a publisher yet, he believed that the Russians had gotten hold of a copy of it. "He was of the opinion that [Soviet Intelligence] is perhaps attempting to slander him by implying that he was engaged in atrocious experiments on human subjects," the investigating officer wrote in his report. Schreiber's full statement was forwarded to air force headquarters. The security officers at Randolph Air Force Base did not have the kind of clearance that allowed them access to Dr. Schreiber's JIOA file or his OMGUS security report. They had no idea who Schreiber really was. They most certainly had not been made aware that he was a Nazi ideologue and the former surgeon general of the Third

Reich. All anyone at the School of Aviation Medicine would have known was that he was a German scientist who was part of Operation Paperclip. The facility had already employed thirty-four German scientists.

On December 14, the FBI got involved in the case. Dr. Leopold Alexander, Schreiber's accuser in Boston, was an internationally renowned authority on medical crimes. His allegations had to be taken seriously. But the air force was unwilling to give up Dr. Schreiber right away. To garner support for him, a memo classified Secret was circulated among those involved, heralding the "Professional and Personal Qualifications of Dr. Walter Schreiber." The memo stated, "He possesses an analytical mind, critical judgment, objectivity, and a wealth of well detailed, exact information." Schreiber possessed "know how on preventative health measures, military and civilian, under conditions of total war." He had "detailed information of medical problems in connection with desert and 'Arctic' warfare," and had "contributed to Zeiss Atlas of Epidemology" [sic]. He'd been "a 'key' prisoner of war in Russia for three and a half years [and] he is in a position to provide authoritative information and serve as a consultant on vitally important medical matters in Russia."

With public outrage brewing over the *Globe* article, the situation escalated quickly. The Pentagon became involved, and the matter was sent to the office of the surgeon general of the United States Air Force. That position was now held by none other than Harry Armstrong, recently promoted to major general. Armstrong knew the situation could very quickly get out of control. And hardly anyone had more to lose, personally and professionally, than he did. With the assistance of Dr. Hubertus Strughold, Harry Armstrong had personally recruited fifty-eight former Nazi doctors for work at the U.S. Army Air Forces Aero Medical Center in Heidelberg, five of whom had been arrested for war crimes, four of whom were tried at Nuremberg, two of whom were convicted at Nuremberg, and one of whom was acquitted and then rehired by the U.S. Air Force to work in America before being revealed as incompetent and fired. That said nothing of the thirty-four doctors who had since been hired to work at the School of Aviation Medicine, many of whom were Nazi ideologues as well as former members of the SS and the SA. The Schreiber scandal could trigger a domino effect, shining an unwanted spotlight on the highly suspicious backgrounds of Dr. Strughold, Dr. Benzinger, Dr. Konrad Schäfer, Dr. Becker-Freyseng, Dr. Schröder, Dr. Ruff, and so many others.

Major General Harry Armstrong wrote to the director of intelligence of the air force regarding Major General Dr. Schreiber. "I have been advised by the Commandant, USAF School of Aviation Medicine,

Randolph Air Force Base, Texas, that they recently forwarded to your office a request for a new contract [for] Doctor Walter Schreiber until June 1952," Armstrong wrote. "Recent information indicated that Doctor Schreiber may have been implicated in the medical war crimes in Germany during World War II, and his presence in this country has aroused a considerable amount of criticism. As a consequence of this, it is the firm opinion of this office that the Air Force Medical Service cannot associate itself with Doctor Schreiber beyond the six months' contract under which he is now employed." Further, said Armstrong, "it may be advisable to terminate this contract even prior to its expiration." Armstrong promised that the commandant of the School of Aviation Medicine, Major General Otis Benson, "concurs with this recommendation." General Benson had equal reason to want the Schreiber scandal to quietly disappear. After the war, Benson had served as technical supervisor of the German scientists working at the Army Air Forces Aero Medical Center in Heidelberg.

Two weeks after the scandal broke, Dr. Schreiber was informed that his contract would not be renewed. General Benson delivered the news in person. According to Schreiber, Benson also proposed a secret, alternative plan involving Schreiber's future career in America. In an affidavit, Schreiber swore that General Benson "stated that he felt sure my services could be utilized in other branches of the government or in medical schools, and offered his assistance in soliciting for a position for me."

In Boston, the matter continued to generate press. The Janina Iwanska story was news people were interested in reading about. Despite what had been done to her during the war, Janina Iwanska was a vibrant, beautiful, credible young woman; it was almost impossible not to marvel at her resilience. After being liberated from Ravensbrück, she had moved to Paris, where she had been working as a journalist for Radio Free Europe. She also worked as the Paris-based correspondent for several Polish newspapers in Western Europe.

When Iwanska was experimented on at Ravensbrück she was only seventeen years old. While imprisoned there, she and several other female prisoners had taken remarkable steps to get word outside the camp about what was being done by Nazi doctors at Ravensbrück. With the goal of getting their message to the Vatican, the BBC, and the International Red Cross, Iwanska and four other women sent secret messages to their relatives outside the camp. Remarkably, a French prisoner named Germaine Tillion took photographs of the women's wounds, then smuggled a roll of film out of the camp. The story was printed in the Polish underground press during the war, notifying the world about the Ravensbrück medical experiments.

The story was eventually picked up by the BBC, as the women had hoped.

Now, with the Schreiber story gaining momentum, in January 1951, FBI agents arranged to interview Janina Iwanska in Boston. Under oath and from a photograph, she identified Dr. Schreiber as a high-ranking doctor who had overseen the medical experiments at Ravensbrück.

"How do you know that the Dr. Schreiber, whom you saw in the Concentration Camp in Germany in 1942 and 1943, is the same Dr. Walter Schreiber who is now in San Antonio?" an FBI agent asked her.

"Three weeks ago, the journalist was coming from the Boston Post [*sic*] and they showed me about fifty (50) pictures and they asked me if I know [which one] is Dr. Schreiber," Iwanska said. She described how she had no trouble picking Schreiber's photograph out of fifty presented to her. "I saw this face in the doctors' group in Ravensbrück," she said. "After, they asked me if I knew Schreiber's name. I told them, 'I know Schreiber's name.'"

Dr. Schreiber, in a separate interview, claimed that he had had nothing to do with the Ravensbrück experiments, that he had never even visited a concentration camp, and that he had never met Janina Iwanska, who was accusing him of outrageous acts. In response, Janina Iwanska had this to say: "I had an operation done on my legs by Dr. Gebhart [*sic*]....During the first dressing after the operations I spoke to him. I asked Dr. Gebhart why they did the operation and they gave the answer, 'We can do the experiment because you are condemned to die.' The number [tattooed on] my legs are T. K. M. III. If Dr. Schreiber cannot remember my name, I am sure perhaps he can remember the experiment [number]."

Janina Iwanska said she was sure that Dr. Schreiber was at the concentration camp; she had seen him with her own eyes. After operations were performed on seventy-four women, she explained, there was a doctors' conference at the concentration camp.

"Were you present at the Doctors' Conference?" the FBI agent asked.

"Yes, because every woman who had the experiment was taken in the [conference] room and Gebhart explained to the other doctors what he did."

"Do you remember the date of this conference by Dr. Gebhart and Dr. Schreiber about your experiment?"

"I think it was about three weeks after the [operation of] 15th of August 1942," Janina Iwanska said.

"Did any of the people who were experimented on die as a result of the experiments?"

"Yes. Five died 48 hours after the operation, and six were shot after the operation," Iwanska said.

"Do you know the names of those who died as a result of the operation?"

"I have the list at my house. I sent the names to the United Nations. They have all the documents. It was seventy-four who had the experiment."

"Do you have any knowledge as to whether Dr. Schreiber ordered these experiments done?"

"I don't know if he gave the orders. Only I know he was very interested in it," she said.

Back in Texas, Dr. Schreiber began plotting his escape to Argentina. He wrote to his married daughter, Elisabeth van der Fecht, who was living in San Isidro, in Buenos Aires. He asked her to get information about how he could get a visa fast. The FBI intercepted Schreiber's mail.

"I will be able to get a visa for you all within fourteen days," promised Schreiber's daughter. "To be honest, we always have feared something similar to come. . . . Everything started so well and seemed to be wonderful. Your nice house, your new furniture. . . ." She assured her father that she could help get the paperwork necessary to enter Argentina. "If, father, you wish us to take steps, send us the necessary documents," Elisabeth wrote. "Anyhow the risk would be much less than if you must go to Germany. For Heaven's sake do not go back."

By mid-January, the JIOA weighed in. The Joint Chiefs decided that it was necessary to repatriate Dr. Schreiber. "Reason for Requesting Repatriation: Dr.

Schreiber is basically not a research scientist, and as such his usefulness is very limited to the School of Aviation Medicine," a memo read. "Recent criticism and adverse publicity [have] been directed against the School of Aviation Medicine in that it has been charged that Dr. Schreiber, as a high ranking Nazi Medical Officer, was connected with brutal experiments on concentration camp victims. The School of Aviation Medicine and the Surgeon General do not wish to assume responsibility for Schreiber...in view of the above criticism."

Army headquarters in Heidelberg, Germany, disagreed. On February 5, 1952, the army cabled a message marked "Secret" to JIOA, written in shorthand. "Recommend take act[ion] to retain Schreiber in US. Subj has invaluable info of intelligence nature regarding Russia and is outstanding [in] his professional field."

The Office of the U.S. High Commissioner weighed in: "All reparations of paper clip personnel unless voluntary have some adverse effect [on] JIOA programs," a memo stated. "Request coordinate with G2 Army prior to final decision on repatriation." At air force headquarters, another idea was gaining support. Why not help Dr. Schreiber "move" to Argentina? To this end, a major named D. A. Roe, from Army Intelligence, G-2, contacted Argentinean General Aristobulo Fidel Reyes to discuss "the utilization

of the services of Dr. Walter Paul Schreiber, M.D."
Major Roe queried if "in any way possible...his talents could be used in Argentina." The army had originally intended to allow the talented Dr. Schreiber to immigrate to the United States, Major Roe explained, but, unfortunately, that had changed. "His admissibility, under current law, as an immigrate [*sic*] of the U.S. is questionable because of this close affiliation with the Nazi Army," Major Roe clarified. Argentina did not have the same kinds of immigration laws prohibiting entry of former high-ranking Nazis. It would be great if they could help.

Faced with pushback from High Commissioner John J. McCloy, JIOA retooled its repatriation position and instead made a case for extending Dr. Schreiber's stay. "Schreiber family may be subjected to reprisals due to his previously reported escape from Russian control," JIOA wrote. "If repatriation is inadvisable believe Schreiber may be retained here by issuance of visa. Above is additional reason for desirability of entry with immigration visa." But then a new document emerged, marked JIOA eyes only, and to be kept in Schreiber's classified dossier. Schreiber's wife of forty years, Olga, was also an old-time Nazi. According to official NSDAP paperwork, she had joined the Nazi Party on October 1, 1931, years before Hitler came to power.

Each week that passed brought more focus to the subject. The public was growing increasingly outraged with the notion that an ex-Nazi general and alleged war criminal was still living in the United States, and that his employment by the United States Air Force was still up in the air.

The Physicians Forum, a group of doctors representing thirty-six states, wrote to the Senate. "Had Dr. Schreiber been found and apprehended at the time of the Nuremberg Military Tribunal," the doctors wrote, "it is virtually certain that he would have been brought to trial along with his associates many of whom were sentenced to life imprisonment or hanged for their crimes. Instead this individual is now in the United States working for our Air Force." The physicians unanimously recommended "[i]mmediate expulsion of Dr. Schreiber from the United States [and] a thorough investigation of the circumstances leading to Dr. Schreiber's entry into this country and his assignment to the Air Force."

The final demand was far more threatening to the air force than the Physicians Forum doctors could have known. Were the Senate to hold hearings to investigate "similar appointments of German physicians formerly in important positions in the German armed forces during World War II," the whole Paperclip program could be revealed. Heads would roll.

Schreiber's story could travel all the way up to the Joint Chiefs of Staff, the scandal all the way to the president of the United States.

With an astonishing degree of hubris, Dr. Schreiber gave one media interview after the next, professing innocence and calling all the charges against him "lies."

"I am not fighting for a renewal of my contract," Schreiber told the *Washington Post,* "I am fighting for justice, and I will continue to do so as long as I live." He issued an egregiously false statement, declaring, "I never was a member of the staff of the supreme command of the Wehrmacht." In another interview, he said, "I never worked in a concentration camp." Later in the interview he clarified that he had actually "visited" a camp once, in eastern Pomerania, without knowing it was a concentration camp. His very brief job there was to inspect "a report for the delousing treatment," for a group of girls whose "linen was disinfected with DDT," Schreiber said. "After I arrived there, four days later, the girls were found clean and free of lice," Schreiber said.

Dr. Schreiber had the air force and the Joint Chiefs of Staff over a barrel, and he likely knew it. If any one of these high-ranking U.S. government officials was forced to admit what was really known about Schreiber—what had been known all along—it would be a scandal of the first degree. Instead, the lie was allowed to expand. "Until we come up with some basic facts," JIOA director Colonel Benjamin W. Heckemeyer told *Time* magazine, in an exclusive sit-down interview with a reporter named Miss Moran, "a man should be given the all-American treatment here and not be given the bum's rush."

Behind the scenes, Harry Armstrong, surgeon general of the air force, had been spearheading Schreiber's removal. But in statements to the press, he maintained a façade of support. There is "no evidence he's guilty other than serving his country during war same as I did mine," General Armstrong told the Associated Press. Legendary newspaper columnist Drew Pearson did not see things in the same light. He pulled transcripts from the Nuremberg doctors' trial and quoted from them in "Washington Merry-Go-Round." "Here are the facts regarding the Nazi doctor who escaped the Nuremberg war crimes trials and is now working for the Air Force at Randolph Field, Texas," Pearson wrote, "kicking, screaming young Polish girls were held down by SS troops and forcibly operated on.... Nuremberg document No. 619 also shows that Schreiber was second on a list of prominent medical officers detached to the SS for two days.... Human victims were also used in typhus experiments at Buchenwald and Natzweiler concentration camps. Deadly virus was transferred from men to mice and back in an attempt to produce live vaccine." Schreiber responded by saying, "The Nuremberg Military Tribunal has prosecuted and held those responsible for the crimes."

That statement further incensed Dr. Leo Alexander and Boston attorney Alexander Hardy, the former chief prosecutor at the doctors' trial. The two men drafted a ten-page letter to President Truman. "He was not a defendant [but] beyond a doubt he is responsible for medical crimes," Hardy and Alexander explained. "He certainly had full knowledge that concentration camp inmates were being systematically experimented on by doctors of the Medical Service in which he was a General.... Schreiber's subordinates performed experiments encouraged by Schreiber," who in turn made "materials and funds available [and] the holding of conferences."

Hardy and Alexander cited testimony from five of Schreiber's physician-colleagues who testified at the Nuremberg doctors' trial that Schreiber had overseen a host of "hygiene"-related Reich medical programs in which countless humans were sacrificed in the name of research. Included were details of yellow fever experiments, epidemic jaundice experiments, sulfanilamide experiments, euthanasia by phenol experiments and the notorious typhus vaccine program "with its 90% death rate." Alexander Hardy and Dr. Alexander's letter to President Truman portrayed Major

General Dr. Walter Schreiber not only as a war criminal of the worst order but as a sadist and a liar. Parts of the letter were made public. "Truman Is Urged to Expel Physician," read a headline in the *New York Times*. Harry Armstrong wrote to the Physicians Forum assuring them that Schreiber would be returned to Germany at once.

JIOA director Colonel Heckemeyer was asked by *Time* magazine what the air force was going to do if, after completing its investigation, it found Dr. Schreiber guilty of war crimes. "That I will have to get guidance from the Office of the Secretary of Defense on," Heckemeyer said. Could Schreiber be prosecuted, the *Time* reporter asked?

"We are not going to make a Nuernberg [*sic*] trial three years after the trials are closed," Heckemeyer said. Finally, the secretary of the air force, Thomas Finletter, made a public announcement stating that Dr. Schreiber would be dropped from his contract and put under military custody. He would leave the United States at once.

But Dr. Schreiber refused to leave the United States. Instead, he packed up his family, left Texas, and drove to San Francisco. There, the Schreibers moved in with Dorothea, the couple's married daughter, and her husband, William Fry, in their home at 35 Ridge Road, in San Anselmo. Just as Dorothea had worked at Camp King when her father was post surgeon there, her new husband, William Fry, had served at Camp King, as an army intelligence investigator. Having moved to America in July 1951, the two had lived in California since.

More than a month passed. The Physicians Forum received new information about Schreiber, which they submitted to President Truman on April 24, 1952, along with a telegram marked "Urgent." They attached a document that showed that Brigadier General Otis Benson, commanding general at Randolph Field School of Aviation Medicine, "sought continued employment for Dr. Schreiber in the United States, preferably a 'University Appointment,'" after the air force had already promised that Schreiber would soon be leaving the United States.

The outrageous part, said the physicians, was that the U.S. Air Force was colluding to keep Schreiber in the United States. In their letter to President Truman, they quoted General Benson from a letter he'd written to the dean of the Minnesota School of Public Health, seeking a new job for Schreiber in the private sector. "I like and respect the man [but] he is too hot for me to keep here using public funds," General Benson said of Schreiber. He said that the bad press had been little more than "an organized medical movement against him emanating from Boston by medical men of Jewish ancestry." The Physicians Forum's board of directors demanded that President Truman order the attorney general to open an investigation into the case.

A few days later, a lieutenant colonel named G. A. Little, representing the Joint Chiefs of Staff, flew to California to visit with Dr. Schreiber at his daughter's house, "for the purpose of attempting to persuade Dr. Schreiber to go to Buenos Aires at once, regardless of job opportunities which might or might not be developed through [our] office." According to Little, the visit went well.

"We were very cordially received by Dr. Schreiber and spent approximately two hours in conversation with him," Lieutenant Colonel G. A. Little wrote in a confidential report for the Joint Chiefs of Staff. In exchange for "transportation to South America" and "funds for travel," and "travel allowance for Schreiber Family Group," Dr. Schreiber agreed to leave the United States at once. "Schreiber's voluntary and immediate departure for Buenos Aires will provide a satisfactory solution for all concerned." The amount Schreiber was paid to leave was never disclosed and is not mentioned in his declassified Operation Paperclip file.

On May 22, 1952, Dr. Schreiber and his family were flown by military aircraft from Travis Air Force Base, in California, to New Orleans, Louisiana. There, they boarded a ship bound for Argentina.

When they arrived in Buenos Aires, they were taken by car to the American consulate and given documents that allowed them to stay. The arrangements were made by General Aristobulo Fidel Reyes. The air force paid for police protection for Dr. Schreiber and his family during the transition. It was important that the Schreibers' resettlement in South America go smoothly. There were too many American officials whose reputations would never survive if the facts were revealed.

The Senate hearings never happened, and the attorney general never opened a case. In Argentina, Dr. Schreiber bought a home and named it Sans Souci ("without a care"), the name of the summer palace in Potsdam, outside Berlin, of Frederick the Great, king of Prussia. According to personal family documents, Schreiber's final quest was to prove that he had been born a baron and was descended from Prussian royalty. "As a consequence of what the official investigation turned up, in addition to his own extraordinary meritorious achievements," say the family documents, "Oberstabsarzt [Military Medical Doctor] Walter P. Schreiber was given 'official' permission to add 'von' to his name." The personal papers do not say who gave this official permission. Schreiber died in September 1970 in San Carlos de Bariloche, Río Negro, Argentina.

CHAPTER NINETEEN

Truth Serum

As Dr. Schreiber sailed for Argentina, the Operation Bluebird interrogation program at Camp King expanded to include "the use of drugs and chemicals in unconventional interrogation." Dr. Kurt Blome was Camp King's post doctor during this period. According to a memo in his declassified foreign scientist case file, Blome worked on "Army, 1952, Project 1975," a Top Secret project that itself has never been declassified. Blome's file becomes empty after that.

"Bluebird was rechristened Artichoke," writes John Marks, a former State Department official and authority on the CIA's mind control programs. The goal of the Artichoke interrogation program, Marks explains, was "modifying behavior through covert means." According to the program's administrator, Richard

Helms—the future director of the CIA—using drugs was a means to that end. "We felt that it was our responsibility not to lag behind the Russians or the Chinese in this field, and the only way to find out what the risks were was to test things such as LSD and other drugs that could be used to control human behavior," Helms told journalist David Frost, in an interview in 1978. Other U.S. intelligence agencies were brought on board to help conduct these controversial interrogation experiments. "In 1951 the CIA Director approved the liaison with Army, Navy, and Air Force intelligence to avoid duplication of effort," writes Marks. "The Army and Navy were both looking for truth drugs while the prime concern of the Air Force was interrogation techniques used on downed pilots." Since the end of the war, the various U.S. military branches had developed advanced air, land, and sea rescue programs, based in part on research conducted by Nazi doctors. But the Soviets had also made great advances in rescue programs and this presented a serious new concern. If a downed U.S. pilot or soldier was rescued by the Russians, that person would almost certainly be subjected to unconventional interrogation techniques, according to the CIA. The purpose of Operation Bluebird was to try to predict what kinds of methods the Soviets might use against American soldiers and airmen. One of the so-called truth serum drugs the CIA believed the Soviets were most heavily involved in researching was LSD. One CIA report, later shared with Congress, stated that "the Soviets purchased a large quantity of LSD-25 from the Sandoz Company...reputed to be sufficient for 50 million doses." Or so the CIA thought. A later analysis of the information determined that the CIA analyst working on the report made a decimal point error while performing dosage calculations. The Soviets had in fact purchased enough LSD from Sandoz for a few thousand tests—a far cry from fifty million.

For its Operation Bluebird experiments involving LSD and other drugs, the CIA teamed up with the Army Chemical Corps. The initial research and development was conducted by officers with the Special Operations Division who worked inside a classified facility designated Building No. 439, a one-story concrete-block building set among similar-looking buildings at Camp Detrick so as to blend in. Almost no one outside the SO Division knew about the Top Secret work going on inside. The SO Division was paid for by the CIA's Technical Services Staff (TSS), a unit within the CIA's Clandestine Service; many of its field agents were culled from a pool of senior bacteriologists at Detrick. One of these SO Division field agents was Dr. Harold Batchelor, the man behind the Eight Ball who consulted with Dr. Kurt Blome in Heidelberg in 1947. Another SO Division agent was Dr. Frank Olson, a former army officer and bacteriologist turned Agency operative whose sudden demise in 1953 would nearly bring down the CIA. The two men were assigned to the program at Camp King involving unconventional interrogation techniques.

In April 1950, Frank Olson was issued a diplomatic passport. Olson was not a diplomat; the passport allowed him to carry items in a diplomatic pouch that would not be subject to searches by customs officials. Frank Olson began taking trips to Germany, flying to Frankfurt and making the short drive out to Camp King. In one of the rare surviving official documents from the program, Deputy Director of Central Intelligence Allen Dulles sent a secret memo to Richard Helms and CIA Deputy Director for Plans Frank Wisner regarding the specific kinds of interrogation techniques that would be used. "In our conversation of 9 February 1951, I outlined to you the possibilities of augmenting the usual interrogation methods by the use of drugs, hypnosis, shock, etc., and emphasized the defensive aspects as well as the offensive opportunities in this field of applied medical science," wrote Dulles. "The enclosed folder, 'Interrogation Techniques,' was prepared in my Medical Division to provide you with a suitable background." Camp King was the perfect location to conduct these trials. Overseas locations were preferred for Artichoke interrogations, since foreign governments "permitted certain activities which were not permitted by the United States government (i.e. anthrax etc.)."

The next trip on record made by Frank Olson occurred on June 12, 1952. Olson arrived at Frankfurt from the Hendon military airport in England and made the short drive west into Oberursel. There, Artichoke interrogation experiments were taking place at a safe house called Haus Waldorf. "Between 4 June 1952 and 18 June 1952, an IS&O [CIA Inspection and Security Office] team…applied Artichoke techniques to two operational cases in a safe house," explains an Artichoke memorandum, written for CIA director Dulles, and one of the few action memos on record not destroyed by Richard Helms when he was CIA director. The two individuals being interrogated at the Camp King safe house "could be classed as experienced, professional type agents and suspected of working for Soviet Intelligence." These were Soviet spies captured by the Gehlen Organization, now being run by the CIA. "In the first case, light dosages of drugs coupled with hypnosis were used to induce a complete hypnotic trance," the memo reveals. "This trance was held for approximately one hour and forty minutes of interrogation with a subsequent total amnesia produced." The plan was straightforward: drug the spies, interrogate the spies, and give them amnesia to make them forget.

Another surviving memo from this otherwise

unreported chapter of Cold War history is from Dr. Henry Knowles Beecher, chief anesthetist at Massachusetts General Hospital in Boston and a CIA-army-navy adviser on Artichoke techniques. Beecher traveled to Germany to observe what was happening at Camp King. He was a colleague of Dr. Leopold Alexander's in Boston and, like Alexander, was an outspoken advocate of the Nuremberg Code, the first principle of which is informed consent. And yet in one of the stranger Cold War cases of dissimulation, Dr. Beecher was a participant in secret, government-sponsored medical experiments that did not involve consent. Beecher was paid by the CIA and the navy to consult on how best to produce amnesia in Soviet spies after they were drugged and interrogated so they would forget what had been done to them.

Dr. Frank Olson returned from Germany to his office in Detrick in a moral bind, according to his Detrick colleague the bacteriologist Norman Cournoyer. "He had a tough time after Germany... [d]rugs, torture, brainwashing," Cournoyer explained decades later, for a documentary for German television made in 2001. Cournoyer said that Olson felt ashamed about what he had witnessed, and that the experiments at Camp King reminded him of what had been done to people in concentration camps. Back in America, Olson contemplated leaving his job. He told family members he was considering a new career, as a dentist. Instead, he stayed on at the bioweapons facility, becoming chief of the SO Division for a while. He continued to work on Top Secret biological and chemical weapons programs in the CIA's office at Detrick, Building No. 439.

Unknown to Frank Olson, the CIA was expanding its Artichoke program in new ways, including expanding its use of LSD in "unconventional interrogations" through covert means. Strapping a suspected Soviet spy to a chair and dosing him with drugs, as was done at Camp King's Haus Waldorf, was one approach to getting a spy to spill his secrets. But the CIA wondered what would happen if an enemy agent were to be given an incapacitating agent like LSD on the sly, without knowing he had been drugged. Would this kind of amnesia be effective? Could it produce loyalty? How much, if any, of the experience would be remembered? These were questions the CIA wanted answered. The director of the Technical Services Staff, Dr. Sidney Gottlieb, a stutterer with a club foot, decided that the first field tests should be conducted on men from the SO Division without them knowing about it. One of those targeted for experimentation was Dr. Frank Olson.

A week before Thanksgiving, in November 1953, six SO Division agents including Frank Olson and the new SO Division chief, Vincent Ruwet, were invited to a weekend retreat at a CIA safe house in western

Maryland called Deep Creek Lake. There, the men from Detrick were met by TSS director Sidney Gottlieb and his deputy, Robert Lashbrook, a chemist. One part of the agenda was to discuss the latest covert means of poisoning people with biological agents and toxic substances, including those that had been acquired from around the globe by Fritz Hoffmann, the Operation Paperclip scientist with the Chemical Corps. The second goal was to covertly dose the six SO Division men with LSD and record what happened. After dinner on the second night, Robert Lashbrook secretly added LSD to a bottle of Cointreau. The unwitting test subjects were all offered a drink, and all but two of the SO Division officers drank the aperitif; one had a heart condition and the other abstained from alcohol. Frank Olson had a terrible reaction to having been drugged with LSD. He became psychologically unstable and could not sleep. His boss, Vincent Ruwet, who had also been dosed, described his own LSD poisoning experience as "the most frightening experience I ever had or hope to have." The CIA had at least one of their questions answered now; covert LSD poisoning did not produce amnesia.

When Monday morning came around, Ruwet arrived for work at 7:30, as usual. Inside Detrick's Building No. 439 he found a very agitated Dr. Frank Olson waiting for him. Olson told Ruwet that he was devastated by what had happened at Deep Creek Lake. He wanted to quit or be fired. Ruwet told him to give the issue some time and to get back to work. But when Ruwet arrived for work the following morning at 7:30, he again found Frank Olson waiting for him. Olson's mental state had deteriorated considerably, and Ruwet decided that he needed medical help. He called Agency headquarters, in Washington, D.C., and told Robert Lashbrook what was going on. Frank Olson was privy to the CIA's most controversial behavior modification and mind control programs. If he had a psychotic break in public he could inadvertently talk—terribly ironic, since getting a man to spill his secrets was what this LSD poisoning program was all about. Were Frank Olson to talk, the Agency would have a nightmare on its hands.

"Dr. Olson was in serious trouble and needed immediate professional attention," Lashbrook wrote in an after-incident CIA report. Lashbrook told Vincent Ruwet to bring Olson to headquarters immediately. From there, the two men took Olson to a townhouse in New York City, at 133 East Fifty-eighth Street. There, Olson met with a doctor on the CIA's payroll named Harold Abramson. Abramson was not a psychiatrist. He was an allergist and immunologist who worked on LSD tolerance experiments for the CIA, and he carried a Secret clearance of his own. Frank Olson told Dr. Abramson that he was suffering from memory loss, confusion, feelings of inadequacy, and terrible guilt. Dr. Abramson noted that Olson seemed to have a perfectly good memory and could remember people, places, and events easily on demand. In other words, Frank Olson's problems were all in his mind.

The next morning, Robert Lashbrook and Vincent Ruwet took Frank Olson on another visit, to another CIA contract employee, John Mulholland, a semifamous New York City magician. Like Dr. Abramson, magician John Mulholland had a Secret security clearance with the TSS. Mulholland taught CIA agents how to apply "the magician's art to covert activities." One of his specialties was "the delivery of various materials to unwitting subjects." During the visit, Olson became suspicious of what was going on and asked to leave. By the next day, Frank Olson was hearing voices. He told Dr. Abramson that the CIA was trying to poison him, which they already had done at least once. Vincent Ruwet returned to Maryland to be with his family for Thanksgiving; Abramson and Lashbrook decided to have Frank Olson committed to the Chestnut Lodge sanitarium, in Rockville, Maryland. Conveniently, the CIA had doctors on staff there.

Robert Lashbrook and Frank Olson spent one last night in New York City, in the Statler Hotel, on

Seventh Avenue and Thirty-third Street. They were given room 1018A, on the tenth floor. After eating in the hotel restaurant, Olson and Lashbrook returned to their room to have a drink and watch television. Olson called his wife for the first time since he had left home and told her not to worry — that he'd be home soon. Then he went to sleep.

At approximately 2:30 a.m. Olson crashed through the hotel window and fell more than one hundred feet to his death on the street below. According to the coroner's report, Olson hit the ground feet first, as if standing up, fell backward, and broke his skull. The Statler Hotel night manager, Armand Pastore, heard the impact and ran outside. He found Olson lying on the pavement, still alive. His eyes were open, Pastore told the police, and he tried to say something. But no words came out, and after a few moments, Frank Olson took his last breath.

Pastore looked up to see which room Olson had come out of. He could see that one of the window shades in a room high above was sticking out, as if the shade had been down when Olson crashed through it. Pastore noted which room it might be. When the police arrived, he took them up to the tenth floor and into room 1018A, using the manager's passkey to get inside. There, inside the bathroom, CIA agent Robert Lashbrook sat on the toilet seat in his underwear, holding his head in his hands. Lashbrook had already made two telephone calls. The first call was to the CIA's director of the Technical Services Staff, Sidney Gottlieb, the man who had, with Lashbrook, poisoned Frank Olson with LSD a little over a week before. The second call was to the hotel's front desk, to report Olson's suicide. When New York City Police detective James W. Ward arrived on the scene, he asked Robert Lashbrook several questions, to which Lashbrook responded using only the words "yes" and "no." Lashbrook did not identify himself as a CIA agent.

Dr. Lashbrook was taken in for questioning; homicide had not yet been ruled out. At the precinct, Detective Ward asked Lashbrook to empty his pockets. Among the contents were papers with addresses for Dr. Harold Abramson, in New York City, and the Chestnut Lodge sanitarium, in Rockville, Maryland. When asked about his profession, Lashbrook told Detective Ward that he was a chemist with the War Department and that Frank Olson was a scientist at Camp Detrick—and that Olson was mentally ill. Detective Ward called Dr. Abramson, who verified Lashbrook's story, leaving out that he also worked for the CIA.

Two days later, Detective Ward submitted Case Number 125124 to his station chief. The death of Frank Olson was determined to be a suicide. The case was closed.

*   *   *

If the Greek philosopher Heraclitus is right and war is the father of all things, then America's Nazi scientist program was a nefarious child of the Second World War. Operation Paperclip in turn created a host of monstrous offspring, including Operations Bluebird, Artichoke, and MKUltra. Before Frank Olson became involved in the CIA's poisoning and interrogation programs, he conducted research and development of the airborne delivery of biological weapons. Olson had been working in the field of biological weapons research since 1943. He was recruited by Detrick's first director, Ira Baldwin, during the war. After the war, Dr. Olson became a civilian scientist at Detrick. He joined the Special Operations Division in 1950 and was part of a team that covertly tested how a weaponized biological agent might disperse if used against Americans.

In the late 1940s and early 1950s, Olson had traveled across the United States overseeing field tests that dispersed biological agents from aircraft and crop dusters in San Francisco, the Midwest, and Alaska. Some field tests involved harmless simulants and others involved dangerous pathogens, as Senate hearings later revealed. One such dangerous experiment was conducted by Olson and his Detrick colleague Norman Cournoyer. The two men went to Alaska and oversaw bacteria being sprayed out of airplanes to see how the pathogens would disperse in an environment similar to that of a harsh Russian winter. "We used a spore," Cournoyer explained, "which is very similar [to] anthrax, so to that extent we did something that was not kosher. Because we picked it up all over [the United States] months after we did the tests." A third man involved in the covert tests with Cournoyer and Olson was Dr. Harold Batchelor, the bacteriologist who learned airborne spray techniques from Dr. Kurt Blome, whom Batchelor consulted with in Heidelberg. Olson and Batchelor also conducted covert field tests in closed spaces across America, including in subways and in the Pentagon. For these tests, the Special Operations Division used a relatively harmless pathogen that simulated how a deadly pathogen would disperse. A congressional inquiry into these covert tests found them "appalling" in their deception.

By being part of a team of covert poisoners, be it out in the Alaskan tundra or inside a safe house at Camp King in Germany, Dr. Frank Olson and his colleagues violated the Nuremberg Code, which requires informed consent. As circumstances would have it, the great tragedy of Frank Olson's life and death was that his own inalienable right to be protected from harm from his government and his doctor was violated on orders from the very same people to whom he had dedicated his life's work.

This new war, the Cold War, was now the father of its own dark events.

PART V

"War is the father of all things."

— Heraclitus

CHAPTER TWENTY

In the Dark Shadows

The Cold War became a battlefield marked by
doublespeak. Disguise, distortion, and deception were accepted as reality. Truth was promised in a serum. And Operation Paperclip, born of the
ashes of World War II, was the inciting incident in
this hall of mirrors.

But in 1952, the heedless momentum of Operation
Paperclip began to slow as conflicts emerged between
the JIOA and the CIA over policies with the new West
German government. German officials warned High
Commissioner John J. McCloy that Operation Paperclip violated NATO regulations and even America's
own policies for governance in Germany. On February 21, 1952, McCloy sent a memo to the U.S. Secretary of State expressing his concern that if Paperclip
was not curtailed, it could result in a "violent reaction"

from officials in West Germany. With McCloy no longer expressing unbridled enthusiasm for Operation Paperclip, the JIOA began to lose its once indomitable grip on the program. But the CIA was not bound to the same NATO policies as were the Joint Chiefs of Staff, and so the CIA continued to do what it had been doing—namely, recruiting Nazi scientists and intelligence officers to act as advisers at Camp King. The five-year partnership between the two agencies began to unravel.

JIOA officials became furious as they watched the CIA poach German scientists and technicians from the Accelerated Paperclip lists. In response, and in spite of McCloy's requests otherwise, in the winter of 1952 the JIOA prepared to send a twenty-man team to Frankfurt on a recruiting trip. Delegates included JIOA's new deputy director, Colonel Gerold Crabbe; General Walter Dornberger; and five unnamed Paperclip scientists who were already working in America. When McCloy learned of the trip, he asked the State Department to intervene and cancel it, fearing it would draw the ire of German officials, which it did. The trip happened anyway.

A compromise was reached between U.S. officials and Chancellor Konrad Adenauer's office whereby the JIOA and the CIA agreed to stop recruiting new scientists but could continue to work with the scientists who remained on the original, President Truman–approved, thousand-person list. Official numbers vary dramatically in different declassified record groups, but there were approximately six hundred Paperclip scientists in the United States at this time, meaning some four hundred German scientists were still on the target list. JIOA renamed Paperclip the Defense Scientists Immigration Program, or DEFSIP, and the CIA renamed one vein of its involvement the National Interest program, but most parties still referred to it all as Operation Paperclip.

It was a dwindling empire. In 1956 the CIA ceded control of the Gehlen Organization to the West German government, which renamed it the BND (Bundesnachrichtendienst). The former Nazi general and his men were spies for Chancellor Adenauer's government now. Then, in 1957, JIOA got a new officer in Lieutenant Colonel Henry Whalen, a man whose actions would have a profound effect on the legacy of the Paperclip program. By 1959 Whalen was promoted to deputy director of JIOA, which meant he had access to highly classified intelligence reports from scientists working on atomic, biological, and chemical weapons. Whalen had an office in the "E" Ring of the Pentagon, reserved for senior officials, and enjoyed direct access to the Joint Chiefs of Staff. During his year-long tenure as JIOA deputy director, no one had any idea that Whalen was working as a Soviet spy. It wasn't until 1963 that the FBI learned that he had been passing military secrets to Colonel Sergei Edemski, a GRU intelligence agent posing as a military attaché in the Soviet embassy in Washington. By then Henry Whalen had already left the JIOA.

When the Justice Department began investigating Whalen, they seized all of the JIOA records that he had been working with. The FBI learned that Whalen had destroyed or given away thousands of Paperclip files. In 1966, a grand jury was presented with evidence against Whalen behind closed doors. He was indicted and the trial was conducted under a gag order, with the press denied access to what the FBI had learned and to Whalen's confession. Journalists were prohibited from reporting on the trial, and no one made the connection between Whalen, the JIOA, and Operation Paperclip. The Nazi scientist program had long since faded from public discourse. Whalen was sentenced to fifteen years at a federal penitentiary but was paroled after six years. Most of the FBI's investigation of Whalen remains classified, which likely explains why so few Paperclip files from that time frame are housed at the National Archives.

In 1962, the JIOA was officially disbanded. What remained of the Paperclip program was taken over by the Research and Engineering Department at the Pentagon. Under the DoD Reorganization Act of 1958, this new office had been created to handle the military's scientific and engineering needs under a scientific director who reported to the Secretary of Defense. The act also gave a home to the Pentagon's new in-house, cutting-edge science agency — the Advanced Research Projects Agency, or ARPA, later renamed DARPA — with a *D* for defense. The first director of the Research and Engineering Department was the nuclear physicist Herbert York. York also served as the first scientific director of ARPA. He was one of the nation's leading experts on nuclear weapons and on Intercontinental Ballistic Missiles, or ICBMs.

The ICBM is the "truly revolutionary military offspring" of Hitler's V-2 rocket, says Michael J. Neufeld, curator of the Department of Space History at the Smithsonian National Air and Space Museum and author of several books and monographs on German rocket scientists. The ICBM is capable of carrying, in its nose cone, a weapon of mass destruction and delivering it to a target almost anywhere in the world. The ICBM became the centerpiece of the Cold War — the ultimate sword. It also became the ultimate shield. "Total war" with the Soviets never happened. The Cold War never became a shooting war. Deterrence prevailed.

Today, the Research and Engineering Department at the Pentagon, renamed the Department of Defense Research and Engineering Enterprise, develops all next-generation weapons and counterweapons of mass destruction — the twenty-first century's swords and shields. It is as true today as it was when World War II ended that America relies upon the advancement of science and technology—and industry—to prepare for the next war. This relationship is understood as America's military-industrial complex. It was President Eisenhower who, in his Farewell Address to the nation in 1961, coined this phrase. Eisenhower cautioned Americans to be wary of "the acquisition of unwarranted influence, whether sought or unsought, by the military-industrial complex." Eisenhower's famous warning is well known and often paraphrased. But he also delivered a second warning in his farewell speech, not nearly as well known. Eisenhower told the American people that, indeed, science and research played a crucial role in national security, "[y]et, in holding scientific research and discovery in respect, as we should, we must also be alert to the equal and opposite danger that public policy could itself become the captive of a scientific-technological elite."

Herbert York, as both ARPA chief and director of the Research and Engineering Department at the Pentagon, worked closely with President Eisenhower on matters of military science during the last three years of Eisenhower's presidency. He was deeply troubled by Eisenhower's words in his Farewell Address. "Scientists and technologists had acquired the reputation of being magicians who had access to a special source of information and wisdom out of reach of the rest of mankind," said York. In the mid-1960s, York went to visit Eisenhower at the former president's winter home, in the California desert. "I asked him to explain more fully what he meant by the warnings, but he declined to do so," York said. "I pressed this line of questions further by asking him whether he had any particular people in mind when he warned us about 'the danger that public policy could itself become the captive of a scientific-technological elite.'" York was surprised when Eisenhower "answered without hesitation: '(Wernher) von Braun and (Edward) Teller [father of the hydrogen bomb].'"

York spent decades considering what the President had told him. "Eisenhower's warnings[,] which were based largely on his intuition, pointed up very real and extremely serious problems. If we forget or downgrade his warnings, it will be to our peril," York wrote in his memoir, *Arms and the Physicist,* in 1995.

The legacy of some of the Operation Paperclip scientists as individuals parallels the heritage of many of the Cold War weapons programs they participated in. The biological and chemical weapons programs can now be looked back upon as distinct failures and the product of vague and often wrong intelligence. So it was with the Chemical Corps' relationship with former SS–Brigadier General Dr. Walter Schieber. Declassified files reveal that Schieber was double-crossing the

Americans from the moment he began working for the U.S. Army in Germany, including the entire time he worked for Brigadier General Charles Loucks on the sarin gas project at Loucks's private home in Heidelberg. It took military intelligence until 1950 to determine that something about Hitler's trusted servant was untoward, and even longer to fully realize the extent to which Speer's Armaments Supply Office chief was deceiving them. Almost immediately after Schieber was released from Nuremberg, he began using his old Nazi contacts to sell heavy weapons to at least one enemy nation through a Swiss intermediary. In 1950, military intelligence intercepted a four-page letter addressed to Schieber, sent from Switzerland. There was no return address on the envelope and only an illegible signature inside. The letter writer discussed with Schieber the sale of weapons to a third party, describing the buyer as "A Prince of the Royal House," which military intelligence surmised was a code name. "They are looking for everything, tanks, aircraft, etc., weapons, ammunitions in short everything that pertains to armament. I can't write all that in a telegraph," stated the author, who requested "PAK anti-tank guns...willing to pay $5,000 a piece; 75 mm weapons at $3,000-$4,000 a piece, and 50mm (about $2,000.00)." This was a "first class business deal if we could arrange it," the intermediary promised.

Special Agent Carlton F. Maxwell, commander of

a CIC Team in Heidelberg, was assigned the task of analyzing the letter and the situation. Maxwell quickly determined that the letter was legitimate and its implications dangerous. "Dr. Schieber, described as an extremely valuable scientist in the employ of the Chemical Division, EUCOM, where he possibly has access to Chemical Corps information of a highly classified nature, appears to be one person whom the United States can ill afford to have involved in any sort of international scheme of the nature which the letter seems to imply," Maxwell wrote. "Subject's connections during the Nazi regime with the armament industry, in his capacity as Armaments Supply Chief under Dr. Albert Speer, would make him valuable as a consultant and intermediary in any sort of plan involving illegal sales and shipments of arms." It is remarkable that despite many warning signs, the U.S. Army placed so much trust in Schieber in the first place, and that General Loucks was allowed to make back-door deals with him.

Special Agent Maxwell worried that "in light of recent indication of revival of right-wing activity it is not impossible to image that these arms might be used in a possible German monarchist coup." Schieber needed to be watched closely, Maxwell advised. The Counter Intelligence Corps put a tail on him and followed him into the Soviet zone. In four months' time, CIC confirmed, "Schieber has contacts with a Swiss

Import-Export Agency in a dubious transaction involving the sale and shipment of arms to a foreign power." But there was even worse news for U.S. national security. The CIC also learned that Schieber was working for Soviet intelligence. "Subject is in some way involved with the MGB [Ministerstvo Gosudarstvennoi Bezopasnosti, or the Ministry of State Security, a forerunner of Komitet Gosudarstvennoi Bezopasnosti, or the KGB] in Weimar," Maxwell learned.

Maxwell advised his superiors that it appeared Schieber was a Soviet mole and summarized the dangers involved. Schieber had been cleared to work with highly classified material for the U.S. Chemical Corps, including but not limited to tabun and sarin nerve gas. He had worked on classified design projects involving underground bunker systems for the U.S. Air Force, ones that would supposedly protect the U.S. government's greatest military assets in the event of a nuclear attack. And now he was meeting with Soviet agents who were connected to the intelligence service. Maxwell was told to notify the CIA. The response from the Agency was surprising. "Just forget it," the CIA told Maxwell, and he was instructed to share this message with his colleagues inside the Counter Intelligence Corps. According to declassified memos in Schieber's file, in addition to working for the Chemical Corps, Schieber was also working for the CIA. The Agency told Special Agent Maxwell that they had Schieber under their control. Whether that meant that Schieber was spying on the Soviets as part of an Agency plan or that he was double-crossing both nations remains a mystery. Either way, he was also working on an illegal arms deal. According to a declassified memo in Schieber's foreign scientist case file, he remained on the Operation Paperclip payroll until 1956.

In America, Dr. Walter Schieber left an indelible mark on the future of the U.S. Chemical Corps. After he and his team of Farben chemists provided General Loucks with the secret of sarin gas, the United States began stockpiling the deadly nerve agent for use in the event of total war. The program was fast-tracked at the start of the Korean War. On October 31, 1950, Secretary of Defense George C. Marshall authorized $50 million (roughly $500 million in 2013) for the design, engineering, and construction of two separate sarin production facilities, one in Muscle Shoals, Alabama, and another inside the Rocky Mountain Arsenal. Most information on biological and chemical weapons was classified, and no one outside a core group of senior congressmen who portioned out money for these programs had a need to know about them. "Only five members of the House Appropriations Committee, and no more than 5 percent of the entire House of Representatives, were cleared for information on chemical and biological weapons," writes chemical weapons expert Jonathan Tucker. "As a result, a small clique of senior congressmen was able to allocate money for these programs in secret session and then bury the line items in massive appropriations bills that were brought to the floor for a vote with little advance notice, so that few members had time to read them."

Sarin production took off at a frenzied pace. Twenty-four hours a day, seven days a week, 365 days a year, the two plants cranked out thousands of tons of sarin nerve agent each year. The facility at Rocky Mountain Arsenal, code-named Building 1501, was a windowless five-story blockhouse designed to withstand a 6.0 earthquake and 100-mph winds. It was the largest poured concrete structure in the United States, and for years, no one in America without a need to know had any idea what went on inside. The Chemical Corps also fast-tracked its chemical warfare munitions program, developing state-of-the-art weapons with which to deliver deadly nerve agents in battle. At the Rocky Flats munitions loading plant, sarin was fitted into artillery shells, aerial bombs, rockets, and warheads for missiles, with the preferred method of delivery being the M34 cluster bomb, a 1,000-pound metal cylinder with 75 sarin-filled mini-bombs sealed inside.

Shortly after the Korean War ended with an

armistice on July 27, 1953, the Chemical Corps began releasing public service announcements to educate Americans about chemical warfare. In November 1953, when *Collier's* magazine published "G-Gas: A New Weapon of Chilling Terror. We Have It—So Does Russia," the public learned for the first time that World War III with the Soviets would most likely involve nerve agents. Journalist Cornelius Ryan presented the information about sarin in the starkest of terms. "Right now, you and your family—all of us—are unprotected against the threat of a terror weapon which could prove more deadly than an atomic bomb." The army described sarin as "an odorless, colorless, tasteless nerve gas designed to destroy people with paralyzing suddenness" and warned Americans of a possible "Pearl Harbor–type attack." Ryan described for readers a scenario wherein a Russian strategic bomber, the Tu-4, carrying seven tons of sarin bombs, would drop its load on an American city, killing every unprotected human in a 100-square-mile radius within four minutes of the attack. The chief of the Chemical Corps, Major General E. F. Bullene, promised the nation that the only defense was offense. "At this time the only safe course is to be prepared to defend ourselves and ready to use gas in overpowering quantities." Brinkmanship—the practice of pushing dangerous events to the edge, or brink, of disaster—was the new Cold War mentality.

In 1957, after years of producing sarin and filling munitions around the clock, the Chemical Corps finally fulfilled the Defense Department's stockpile requirement. Later that same year, Edgewood chemists found an even more lethal killer, "a toxic insecticide that penetrated the skin like snake venom," explains Jonathan Tucker. This nerve agent was code-named VX (the V stood for venomous) — a battlefield killer that was three times more toxic than sarin when inhaled and one thousand times more lethal when it came in contact with the skin. Ten milligrams of VX could kill a man in fifteen minutes. VX would be much more effective on the battlefield than sarin ever would be; sarin dissipated within fifteen or so minutes, but when VX was sprayed, it stayed on the ground for up to twenty-one days. Now, in 1957, the Chemical Corps began producing VX by the thousands of tons. Operation Paperclip scientist Fritz Hoffmann moved over from synthesizing tabun at Edgewood to working on VX munitions. But Fritz Hoffmann's more haunting legacy lies in the work he performed for the CIA's Special Operations Division and the Chemical Corps' antiplant division. Antiplant agents include chemical or biological pathogens, as well as insects, that are then used as part of a program to harm crops, foliage, or other plant life.

After the death of Frank Olson, the SO Division continued its LSD mind control schemes. But Sidney

Gottlieb, the man who had suggested poisoning Frank Olson at the CIA safe house in Deep Creek Lake, Maryland, was assigned to also work on the CIA's assassination-by-poison program. Fritz Hoffmann was one of the chemists at the locus of this program. "He was our searcher," Edgewood laboratory director Dr. Seymour Silver told journalist Linda Hunt. "He was the guy who brought to our attention any discoveries that happened around the world and then said, 'Here's a new chemical, you better test it.'"

Hoffmann's daughter, Gabriella, remembers her unusually tall, soft-spoken father regularly traveling the globe in the late 1950s and early 1960s, always with a military escort, gathering obscure poisons from exotic toads, fish, and plants. "He would send me postcards from places like Japan, Australia and Hawaii," she recalls. "He always flew military and he was always escorted by military staff. They would pick him up at our house and bring him back home on a Sunday night." A teenager at the time, Gabriella Hoffmann remembers the unconventional items her father brought home. "He'd unpack his luggage before he would go back to Edgewood on Monday and in his suitcase he'd have all these little jars. They were filled with sea urchins and things. It all seemed very exotic to me."

The poisons Fritz Hoffmann sought for the CIA included substances like curare, a South American blowpipe poison that paralyzes and kills people. Curare was the poison that the CIA's U-2 pilots carried in their flight suit pockets, hidden inside a tiny sheath inserted into an American coin. The SO Division had a Device Branch, which was run by Herb Tanner, co-designer of the Eight Ball. The Device Branch was responsible for the hardware behind the delivery systems, including fountain pens filled with poison projectiles, briefcases that spread bacterial aerosols, poisoned candies, invisible powders, and the "non discernible microbioinoculator," a high-tech dart gun that injected a tiny, poison-tipped dart into the bloodstream without leaving a mark on the body.

In other situations it behooved the CIA to locate and weaponize a poison where death came after a delay, sometimes with an incubation period of about eight or twelve hours, sometimes much longer. The SO Division's Agent Branch worked to find poisons that could make a target mildly ill for a short or long period of time followed by death, very ill for a short or long time followed by death, or any number of combinations, including mild to extreme illness followed by death. The rationale behind assassinating someone with a built-in time delay was to allow the assassin to get away and to deflect suspicion. SO Division targets included Fidel Castro, whose favorite drink, a milkshake, the CIA tried to poison several times.

Another CIA target was Patrice Lumumba, the

first legally elected prime minister of Congo, who the CIA believed was a Soviet puppet. Sidney Gottlieb, the man who poisoned Frank Olson, was assigned the job of assassin. Gottlieb later told congressional investigators that for this job, he needed to locate a poison that was "indigenous to that area [Congo] and that could be fatal." Gottlieb decided on botulinum toxin. Armed with this toxin concealed inside a glass jar within a diplomatic pouch, Sidney Gottlieb traveled to Congo with the intention of killing Prime Minister Lumumba himself. On September 26, 1960, Gottlieb arrived in the capital, Léopoldville, and headed to the U.S. embassy. There, Ambassador Lawrence Devlin was expecting him.

Two days prior, Ambassador Devlin had received a Top Secret cable from CIA director Allen Dulles. "We wish give every possible support in eliminating Lumumba from any possiblity [sic] resuming govermental position," Dulles wrote. Ambassador Devlin knew to be on the lookout for a visitor who would introduce himself as "Joe from Paris." This was Sidney Gottlieb. Gottlieb's plan was to inject the botulinum toxin into Lumumba's toothpaste tube with a hypodermic syringe. Ideally, Lumumba would brush his teeth and eight hours later he'd be dead. But while in Léopoldville, Gottlieb could get nowhere near Prime Minister Lumumba, who was living in a house on a cliff high above the Congo River. Lumumba was constantly surrounded by bodyguards. After several days, the botulinum toxin lost its potency. Gottlieb mixed it with chlorine, tossed it into the Congo River, and left Africa. Patrice Lumumba died in January of the following year, beaten to death — allegedly by Belgian mercenaries.

"My father understood the risks [of] being a chemist at Edgewood," Gabriella Hoffmann says. "Because he never spoke of anything he did there is so much that is unknown." For Gabriella Hoffmann, memories of an unusual childhood include trips to military bases. "Whenever we traveled, me and my mom and dad, it was always to a military installation. We went to White Sands in New Mexico. We went to military bases in California, Arizona, North Dakota, and Dugway, Utah. I remember dad giving a lecture at Dugway Proving Ground." Hoffmann was a central player in some of the most mysterious and controversial government programs of the 1950s and 1960s, but the record of most of his work was destroyed or, as of 2013, remains classified. Gabriella Hoffmann is in the dark about her father's legacy, as is most of the rest of the world.

What remains are Gabriella Hoffmann's memories of the company her father kept. Dr. L. Wilson Greene, the man who coined the term "psychochemical warfare," lived down the street and was a colleague and family friend. During the late 1950s and early 1960s,

Greene continued his LSD research for the army and the CIA's psychochemical warfare programs. LSD and other incapacitating agents were tested on thousands of U.S. soldiers and sailors with a questionable degree of informed consent. To Gabriella Hoffmann, Greene was simply an eccentric neighbor. "I recall [going] to the Greenes' house to see the amazing train garden, his hobby, that he built in the basement. It [had] hills and towns and little people. Ponds and lights and trains that whistled and chugged along with wisps of smoke. My attention and interest was solely on the trains." Another unusual neighbor on the same street was Maurice Weeks, from the Directorate of Medical Research at Edgewood. Weeks was chief of the Vapor Toxicity Branch, with expertise regarding "the inhalation toxicity of combustion products," and spent his time researching how biological and chemical agents became even more deadly when smoke and gas were involved. This picked up where Dr. Kurt Blome's research left off; it had been discussed by Blome in Operation Paperclip consultations in Heidelberg. "Maurice Weeks was a neighbor of ours," recalls Gabriella Hoffmann. "His son, Christopher, and I were the best of friends. There were all these monkeys in cages in Christopher's backyard. Christopher and I would amuse ourselves watching these monkeys for hours on end. Obviously it never dawned on me back then what they were for." Gabriella only learned the true nature of her father's job during interviews for this book.

The strange, tragic thing about Fritz Hoffmann and his legacy in Operation Paperclip is that during the war he was anti-Nazi—at least according to the affidavit written by the wartime American diplomat Sam Woods. And yet here in America, working for the Army Chemical Corps and for the CIA, Fritz Hoffmann's science projects took on a monstrous life of their own. This includes what his daughter believes was a role in the development of Agent Orange, the antiplant weapon, or defoliant, used by the U.S. military during the Vietnam War.

"During the Vietnam War, I remember one evening we were at the dinner table and the war was on the news," Gabriella Hoffmann explains. The family was watching TV. "Dad was usually a quiet man, so when he spoke up you remembered it. He pointed to the news—you could see the jungles of Vietnam, and he said, 'Wouldn't it be easier to defoliate the trees so you could see the enemies?' That's what he said. I remember it very clearly. Years later I learned one of Dad's projects was the development of Agent Orange."

The army's herbicidal warfare program during the Vietnam War started in August 1961 and lasted until February 1971. More than 11.4 million gallons of Agent Orange were sprayed over approximately 24 percent of South Vietnam, destroying 5 million acres of uplands and forests and 500,000 acres of food crops—an area about the size of the state of Massachusetts. An additional 8 million gallons of other anticrop agents, code-named Agents White, Blue, Purple, Pink, and Green, were also sprayed, mostly from C-123 cargo planes. Fritz Hoffmann was one of the earliest known U.S. Army Chemical Corps scientists to research the toxic effects of dioxin—possibly in the mid-1950s but for certain in 1959—as indicated in what has become known as the Hoffmann Trip Report. This document is used in almost every legal record pertaining to litigation by U.S. military veterans against the U.S. government and chemical manufacturers for its usage of herbicides and defoliants in the Vietnam War.

It is the long-term effects of the Agent Orange program that Gabriella Hoffmann believes would have ruined her father, had he known. "Agent Orange turned out not only to defoliate trees but to cause great harm in children," Gabriella Hoffmann says. "Dad was dead by then and I remember thinking, Thank God. It would have killed him to learn that. He was a gentle man. He wouldn't hurt a fly."

Fritz Hoffmann's untimely death came like something out of a Special Operations Division's Agent Branch playbook. He suffered a serious illness that came on quickly, lasted for a relatively short time, and was followed by death. On Christmas Eve 1966, Fritz

Hoffmann was diagnosed with cancer. Racked with pain, he lay in bed watching his favorite television shows—"Cowboy westerns and Rod Serling in *The Twilight Zone*," Gabriella Hoffmann recalls. One hundred days later, Fritz Hoffmann was dead. He was fifty-six years old.

"He did try, periodically, to go to work," Gabriella Hoffmann remembers, "but he was in too much pain. I do have a recollection of a lot of different men in dark suits always coming by the house to talk to him. At the time, Demerol as a painkiller had just come on the market. He had a prescription for it, from a doctor at Edgewood. During the time that [he was dying] there were a lot of questions asked of my mother and [me] that led me to think the FBI or the CIA, or whoever the men in the dark suits were, were worried that my father would start talking about whatever it was that he did. He never did say anything. He was silent until the end."

Hoffmann's antiplant work in herbicides was one element of Detrick's three-part biological weapons division, the other two being antiman and antianimal. Antianimal weapons were aimed at killing entire animal populations, with the goal of starving to death the people who relied on those animals for food. At the locus of the U.S. antianimal program was Operation Paperclip's Dr. Erich Traub, Dr. Kurt Blome's deputy.

Traub was recruited into the Accelerated Paperclip program by Dr. Blome's handler, Charles McPherson, and arrived in America on April 4, 1949.

Traub worked on virological research at the Naval Medical Research Institute, in Bethesda, Maryland. Almost all of Traub's work remains classified as of 2013. At the Naval Medical Research Institute, Traub became friendly with the Luftwaffe physiologist and explosive decompression expert Dr. Theodor Benzinger, whose early work for the navy also remains classified. In addition to working for the navy, Traub worked at Camp Detrick on antianimal research. The agents and diseases being studied by Detrick researchers at this time, meant to decimate a specific animal population, included rinderpest, hoof-and-mouth disease (also called foot-and-mouth disease), Virus III disease of swine (likely African swine fever), fowl plague, Newcastle disease, and fowl malaria. All of Traub's Camp Detrick work remains classified as of 2013.

In 1948, Congress had approved a $30 million budget for antianimal weapons research (roughly $300 million in 2013), but because this work was so dangerous, Congress mandated that it needed to take place outside the continental United States, on an island and not connected to the nearest mainland by a bridge. Plans moved forward and the army chose Plum Island, a 1.3-square-mile land parcel located off the coast of Connecticut in the Long Island Sound. The obvious choice for a director was the world's expert on antianimal research, Dr. Erich Traub. But the plan to activate Plum Island for biological weapons research languished for several years.

Traub received his immigrant visa on September 7, 1951, and he worked on classified programs for three more years. Then, under mysterious circumstances, in 1954, he resigned his position as medical supervisory bacteriologist at the Naval Medical Research Institute and asked to be repatriated to Germany. He told the navy he'd accepted a position with the West German government, as director of the Federal Institute for Virus Research. This move alarmed the JIOA. "Dr. Traub is a recognized authority in certain fields of virology, particularly in hoof and mouth disease of cattle and in Newcastle disease of poultry," read a declassified report. "It can be anticipated that this institution in Germany [where Traub was going] will become one of the leading research laboratories of the world in virological research." In view of the "recognizable military potentialities in possible application of his specialty, it is recommended that future surveillance in appropriate measure be maintained after the specialist's return to Germany." Traub needed to be kept under surveillance, likely for the rest of his life.

In Germany, throughout the late 1950s and early 1960s, U.S. military intelligence spied on Dr. Traub at his new home on Paul-Ehrlich-Strasse, in Tübingen. Agents stopped by the home of Traub's colleague and friend Dr. Theodor Benzinger, in Maryland, to ask whether Benzinger knew "of any associations retained by Dr. Traub outside the United States." Benzinger said he did not. Whenever Traub traveled outside Germany, military intelligence kept a watchful eye on him.

Dr. Traub was a man experienced in the illegal trafficking of deadly pathogens. During World War II, he was the trustworthy scientist chosen by Heinrich Himmler to travel to Turkey to obtain samples of rinderpest to weaponize on Riems. And after the war, when Traub fled the Russian zone at great risk of personal harm, he managed to smuggle deadly cultures with him out of the Eastern bloc, which he then stashed at an intermediary laboratory in West Germany until he was able to locate an appropriate buyer. In the mid-1960s, according to Traub's FBI file, Traub moved from Germany to Iran, with a new permanent address at the Razi Institute for Serums and Vaccines, in Hesarak, a suburb of the city of Karaj. When Traub traveled from Iran to the Shoreham Hotel in Washington, D.C., for a meeting with unknown persons, FBI agents watched him. What they learned about the mysterious Dr. Traub remains classified as of 2013.

Through the lens of history, it is remarkable to think that U.S. biological warfare and chemical warfare programs grew so quickly to the size they did. But the Pentagon was able to keep the scope and cost of these weapons programs secret from Congress in much the same way that it was able to keep the damaging details of Operation Paperclip secret from the public. Everything was classified.

It took President Richard Nixon to realize that playing chicken with the Russians, using a huge arsenal of biological and chemical weapons, was pure madness. On November 25, 1969, Nixon announced the end of all U.S. offensive biological warfare research and ordered that America's arsenal was to be destroyed. "I have decided that the United States of America will renounce the use of any form of deadly biological weapons that either kill or incapacitate," Nixon said. His reasons were simple and self-evident. The use of biological weapons could have "uncontrollable consequences" for the world. "Mankind already carries in its hands too many of the seeds of its own destruction," said Nixon. After twenty-six years of research and development, America's biological weapons programs came to an end. For the first time during the Cold War, a president decided that an entire group of weapons was going to be unilaterally destroyed.

The end of America's chemical weapons program was not far behind. Nixon reinstated the "retaliation-only" policy, which meant no new chemical weapons would be developed and produced. Over the next few years, Congress worked with the military to determine the best way to destroy this entire group of weapons. The original plan was to dispose of some twenty-seven thousand tons of chemical-filled weapons in the deep sea. But upon investigation, it turned out that many of the sarin- and VX-filled bombs stored at the Rocky Mountain Arsenal were already leaking nerve agent. These munitions needed to be encased in steel-and-concrete "coffins" before they could be dumped in the ocean. The Pentagon also had thirteen thousand tons of nerve agent and mustard gas stored in secret on its military base in Okinawa, Japan, which now needed to be disposed of. In 1971, these munitions were brought to an American-owned atoll in the South Pacific called Johnston Island in an operation called Red Hat. The plan was to store the sarin- and VX-filled bombs in bunkers on the atoll until scientists figured out how best to destroy them. But as it turned out, the sarin and VX bombs were not made to ever be dismantled. So the army had a massive new scientific endeavor on its hands, for which it created the Johnston Atoll Chemical Agent Disposal System, the world's first "full-scale chemical weapons disposal facility."

It took another thirty-four years for America's arsenal of chemical weapons to be destroyed. "The numbers speak volumes," says the army. "More than 412,000 obsolete chemical weapons—bombs, land mines, rockets and projectiles—all destroyed." The elaborate destruction process involves the robotic separation of the chemical agent from the munitions, followed by incineration of the separated parts in three separate special types of furnaces. The army says it is "proud [of its] accomplishment," which cost an estimated $25.8 billion as of 2006, or approximately $30 billion in 2013.

CHAPTER TWENTY-ONE

Limelight

America's Cold War biological and chemical weapons programs existed in the shadows, and the majority of the Nazi scientists who worked on them maintained anonymity for decades. Their JIOA case files and OMGUS security reports were classified, as were the programs they worked on. But some of the Operation Paperclip scientists enjoyed the limelight for their work, notably in instances where their work crossed over from weapons projects into space-related endeavors. In this manner, Walter Dornberger, Wernher von Braun, and Hubertus Strughold attained varying degrees of prominence and prestige in the 1950s and 1960s and onward.

Within two years of his arrival in the United States, Dornberger had transformed from public menace to American celebrity. In 1950 he left military custody at

Wright-Patterson Air Force Base to work for Bell Aircraft Corporation in Niagara Falls, New York, quickly becoming vice president and chief scientist. His vocation was now to serve as America's mouthpiece for the urgent need to weaponize space. Dornberger was given a Top Secret security clearance and a job consulting with the military on rockets, missiles, and the future of space-based weapons. In his desk diary, housed in the archives at the Deutsches Museum in Munich, he kept track of his cross-country business trips with an engineer's precision. He attended "classified meetings" at U.S. Air Force bases including Wright-Patterson, Elgin, Randolph, Maxwell, and Holloman, as well as at Strategic Air Command headquarters, in Omaha, Nebraska, and the Pentagon. He also became a consultant to the Joint Chiefs on Operation Paperclip, visiting the inner circle in the Pentagon to discuss "clearance procedures" and the "hiring of German Scientists." As a Paperclip scout, in 1952 Dornberger traveled with what he called "Pentagon Brass" to Germany to "interview German scientists and engineers [in] Frankfurt, Heidelberg, Wiesbaden, Stuttgart, Darmstadt, Witzenhausen."

In his desk diary Dornberger also detailed an ambitious schedule of public appearances, carefully noting the places he traveled and the people he met with. They were the kinds of engagements usually reserved for congressmen. Throughout the 1950s, he jetted from one event to the next, lecturing at dinners and luncheons and sometimes weeklong events. His speeches were always about conquest, with titles like "Rockets — Guided Missiles: Key to the Conquest of Space," "Intercontinental Weapons Systems," and "A Realistic Approach to the Conquest of Space." He orated to anyone who would listen: the Men's Club of St. Mark's Episcopal Church, the Boy Scouts of America, the Society of Automotive Engineers. When the Rochester Junior Chamber of Commerce hosted General Dornberger for a women's luncheon in the spring of 1953, the local press covered the event with the headline, "Buzz Bomb Mastermind to Address Jaycees Today."

Dornberger became so popular that his memoir about the V-2, originally published in West Germany in 1952, was published in America in 1954. In these pages, Dornberger was able to reengineer his professional history from that of warmongering Nazi general to beneficent science pioneer. According to Dornberger, the research and development that had gone into the V-2 at Peenemünde was a romantic, science-laboratory-by-the-sea affair. There was no mention made of the slave labor facility at Nordhausen or the slaves at Peenemünde. The book was originally titled *V-2: The Shot into Space (V-2: Der Schuss ins All)*. "It would be nice to know who invented the subtitle," says Michael J. Neufeld, "which so neatly captures the reinvention of a Nazi terror weapon as the space rocket it most certainly was not."

In 1957, Dornberger seemed to have found his true post-Nazi calling, attempting to sell Bell Aircraft's BoMi (bomber-missile) to the Pentagon. BoMi was a rocket-powered manned spacecraft designed for nuclear combat in space. Occasionally, and behind closed doors, usually at the Pentagon, Dornberger faced challenges. He was once pitching the benefits of BoMi to an audience of air force officials when "abusive and insulting remarks" were shouted at him, according to air force historian Roy F. Houchin II. In that instance, Dornberger is said to have turned on his audience and insisted that BoMi would receive a lot more respect if Dornberger had had a chance to fly it against the United States during a war. There was "deafening silence," in the room, Houchin noted.

In 1958 the FBI opened an investigation into General Dornberger based on an insider's tip that he might be engaged in secret discussions with Communist spies. The special agent who interviewed Dornberger did not believe he was spying for the Soviets but honed in on Dornberger's duplicitous nature: "It is believed that subject [Dornberger] could carry on satisfactorily in the role of a double agent." Dornberger was a cunning man, and this quality, coupled with his scientific acumen, served him. No matter what the circumstances, Dornberger always seemed to come out on top.

* * *

Wernher von Braun became the biggest celebrity of the Operation Paperclip scientists. By 1950, the army decided that it required a much larger facility to research and develop longer-range rockets, so the Fort Bliss rocket team moved from Texas to the Redstone Arsenal, near Huntsville, Alabama. There, the German scientists began work on the army's Jupiter ballistic missile. The group had launched sixty-four V-2s from White Sands. At the same time, von Braun was ambitiously developing a persona for himself as America's prophet of space travel. Rockets, outer space, and interplanetary travel had gained a foothold in American culture. It was the dream of many 1950s American children to fly into outer space, and von Braun and Dornberger became national spokesmen on this issue. They promised the nation that the army's development of a ballistic missile was the necessary first step in reaching outer space.

In 1952, von Braun experienced a breakthrough in his role of national space advocate when *Collier's* magazine paid him $4,000 (approximately $36,000 in 2013) to write the lead article in what would eventually become an eight-part series on future space travel, edited by Cornelius Ryan. "Within the next 10 or 15 years," von Braun wrote, "the earth will have a new companion in the skies, a man-made satellite that could be either the greatest force for peace ever devised, or one of the most terrible weapons of war—depending on who makes and controls it." This space satellite, or station, would also be a "terribly effective atomic bomb carrier," von Braun added. From its earliest days, space travel would be intertwined with war making. It still is.

In addition to earning von Braun a small fortune, the *Collier's* magazine series propelled him into the national spotlight, increasing his fame and affording him additional writing opportunities. Most important, this newfound limelight provided von Braun with a platform to recast himself as a patriotic American. In the summer of 1952, *American Magazine* published a piece with von Braun's byline entitled "Why I Chose America," in which he professed a deep love for America, Christianity, and democracy. In the article, von Braun claimed that during the war he opposed Nazism and was never in a position to do anything but follow orders. The piece earned him an award for "patriotic writing," even though it had been written by a ghostwriter. The article was reprinted in at least one book and, explains Michael J. Neufeld, would thereafter be "taken as a fundamental source by many later journalists and authors."

The national attention caught the eye of Walt Disney Studios, in Burbank, California, and an executive called von Braun to see if he wanted to film a couple of space-related shows for a new television series that Disney was working on. For over a year, von Braun had been trying to find a New York publisher for the science fiction novel he had written while living in the Texas desert. By this time, the novel had been rejected by eighteen publishers. The Disney contract offered a wider road to fame and von Braun signed on. The first Disneyland TV broadcast in which von Braun appeared, in 1955, called *Man in Space,* had an estimated 42 million viewers and was reported to be the second-highest-rated television show in American history at the time. The following month, on April 15, 1955, von Braun and many of his fellow German rocket scientists became U.S. citizens, in a public ceremony held in the Huntsville High School auditorium. In 1958, von Braun and his team launched America's first successful space satellite, *Explorer I,* as a quick response to the Soviets' *Sputnik.* Kurt Debus was in charge of the launch.

In 1960, von Braun and a group of approximately 120 Operation Paperclip scientists, engineers, and technicians were transferred from the army to the newly established National Aeronautics and Space Administration, or NASA, with a mandate to build the Saturn rockets designed to take man to the moon. Von Braun was made director of the new NASA facility, the Marshall Space Flight Center, also located at the Redstone Arsenal, as well as chief architect of the Saturn V launch vehicle, or "superbooster" rocket, as it would become known. Von Braun's deputy developer on the Saturn program was Arthur Rudolph, the former operations director at the Mittelwerk slave labor facility.

The Saturn V rocket would need its own launch complex and hangar. Cape Canaveral, on Florida's east coast, was chosen as the perfect site. On July 1, 1962, NASA activated its Launch Operations Center there, naming Kurt Debus as director. Debus was the ardent Nazi who, during the war and on his own volition, had turned an engineering colleague over to the Gestapo for making anti-Hitler remarks. To house the giant Saturn rocket, NASA constructed the Vertical Assembly Building on nearby Merritt Island. The structure would soon become the most voluminous building in the world—larger than the Pentagon and almost as tall as the Washington Monument. It was designed by Bernhard Tessmann, former facilities designer at Peenemünde and Nordhausen, and one of the two men who, at war's end, stashed the V-2 documents in the Dörnten mine.

As had Dornberger, von Braun worked carefully to whitewash his Nazi past. He knew never to speak of the fact that he had become a Nazi Party member in 1937. When a reporter once asked him, incorrectly, about his joining the party "in 1942," von Braun put his scientist's precision aside and chose not to correct the newsman. Instead von Braun did what he always did—he said that he'd been coerced into joining the

Nazi Party. Never did von Braun speak of the SS cavalry unit he joined, in 1933, or that he was made an SS officer in 1944 and wore the SS officer's uniform, with its swastika on the armband and its cap with the death's-head image. With a revelation like that, he could have been deported under the Internal Security Act of 1950, which was meant to keep Communists out of the country but also covered anyone who had held membership in a "totalitarian dictatorship." Instead, the fact that he was an officer with the SS remained a jealously guarded secret by all parties — von Braun, the U.S. military, and NASA until CNN journalist Linda Hunt broke the story in 1985.

But the dark shadow of von Braun's complicity in Nazi war crimes followed him around. Sometimes it snuck up on him from behind. One day, while visiting the *Collier's* offices, he was riding in the elevator with fellow Paperclip Scientist Heinz Haber, from the School of Aviation Medicine at Randolph Air Force Base, and Cornelius Ryan, his editor. Also in the elevator were a few *Collier's* magazine staffers, one of whom reached out to Haber, rubbed a piece of the scientist's leather coat between his fingers, and said wryly, "Human skin, of course?"

As man got closer and closer to the moon, Wernher von Braun enjoyed a parallel ascent in fortune and in fame. Because von Braun was a public figure, his Nazi past was always there, but in shadow. By the 1960s, it was sometimes treated as a joke. One night, before an Apollo mission, von Braun stormed out of a press conference after a reporter asked him if he could guarantee that the rocket would not hit London. Tom Lehrer wrote a song famously satirizing von Braun: " 'Once the rockets are up, who cares where they come down? That's not my department,' says Wernher von Braun." Filmmaker Stanley Kubrick created a von Braun–inspired character in his black comedy *Dr. Strangelove,* in which a mad scientist famously gets out of a wheelchair and cries, "Mein Führer, I can walk!" But in 1963, in East Berlin, a popular author and lawyer named Julius Mader wrote a book called *The Secret of Huntsville: The True Career of Rocket Baron Wernher von Braun.* On the dust jacket there was a drawing of von Braun wearing a black SS-Sturmbannführer's uniform. Around von Braun's neck was the Knight's Cross, bestowed upon him by Albert Speer at the Castle Varlar event. The book, published by Deutscher Militärverlag, portrayed von Braun as an ardent Nazi and included detailed pages about the murderous conditions at the slave labor facility at Nordhausen, where von Braun oversaw work, and at the Dora-Nordhausen concentration camp, which supplied the slave laborers.

Several German-speaking U.S. citizens alerted NASA about the allegations against von Braun, as if NASA did not know. The space agency's three top officials, James E. Webb, Hugh Dryden, and Robert Seamans, discussed the situation with von Braun and counseled him on how to handle it. If anyone were to ask him about this matter, NASA administrator James Webb told von Braun pointedly, his answer was to be, "Everything related to my past activity in Germany... is well known to the U.S. Government." In the end, nothing much came of the book in the Western world. The narrative mixed "damaging facts" with "completely fabricated scenes," explains von Braun's biographer Michael J. Neufeld, which made the book easy to discount. When NASA learned that the book's author, Julius Mader, worked as an intelligence agent for the East German secret police, the revelations lost their potential power. If need be, Mader's book could be discounted as Soviet-engineered lies, meant only to damage the prestige of the U.S. space program.

The Secret of Huntsville: The True Career of Rocket Baron Wernher von Braun was a success in the Eastern bloc. It was also translated into Russian and sold half a million copies in the Soviet Union. The book inspired a Soviet-sponsored film, *Die Gefrorenen Blitze* (Frozen Lightning), also written by Mader and released by the East German state film studio, DEFA (Deutsche Film-Aktiengesellschaft), in 1967. The same year, West German prosecutors opened a new Dora-Nordhausen trial, this one called the Essen-Dora trial. An SS guard, a Gestapo official, and the chief of security for the V-2 were all charged with war crimes that took place while V-weapons were being built in the Mittelwerk tunnel complex. Mittelwerk general manager and former Paperclip scientist Georg Rickhey was called as a witness, as was Wernher von Braun. Rickhey, who lived in Germany, took the stand. He testified under oath that he did not know about the Dora-Nordhausen concentration camp; that he learned about it after the war. "I don't know much more because they were secret [concentration] camps," Rickhey told the court. "It was a secret commando group in charge. Eyes only. There were bad catastrophic conditions [at Nordhausen] but I learned about this only after the war." This was an absurd misrepresentation of the truth.

Back in America, NASA general counsel refused to send the star of their space program to Germany; this would open Pandora's box. Instead, they allowed von Braun to deliver oral testimony at a West German consulate in the United States. As a venue, NASA lawyers suggested New Orleans, Louisiana—a place about as far away from the media glare as there was. Representing the Soviet-bloc survivors of Dora-Nordhausen was an East German lawyer by the name of Dr. Friedrich Kaul. Upon learning that von Braun would not be coming to Germany, Dr. Kaul arranged to travel to New Orleans to take von Braun's testimony himself.

Kaul was a supremely skillful lawyer, and he had information on von Braun that was damaging.

According to Michael J. Neufeld, the U.S. government knew that Dr. Kaul had served as legal adviser on *Die Gefrorenen Blitze,* and that Kaul's goal in interviewing von Braun was "to broadcast the connection between the rocket engineer, the SS and the concentration camps." NASA's moon program would never survive that kind of publicity, and the State Department denied Dr. Friedrich Kaul a visa.

Instead, at the courthouse in Louisiana, von Braun answered questions put to him by a German judge. As the U.S. Army had done with the 1947 Nordhausen trial, the government sealed von Braun's testimony. Word leaked to the news media that von Braun had been deposed for a war crimes trial, causing von Braun to issue a brief statement on the matter. He said he had "nothing to hide, and I am not implicated." When a reporter asked if there had been any concentration camp prisoners used as slave laborers at Peenemünde, von Braun said no.

The judge prosecuting the Essen-Dora trial also took the testimony of General Dornberger, in Mexico, where the now-retired Dornberger spent winters with his wife. Some months later, von Braun and Dornberger corresponded about the matter. "In regards to the testimony, fortunately I too have heard nothing more," wrote von Braun. No news was good news for both men.

After the Apollo 11 moon landing in July 1969, columnist Drew Pearson—whose fierce exposés of Major General Dr. Walter Schreiber had helped to banish the man from America—wrote in his column that von Braun had been a member of the SS. But von Braun's glory had reached epic proportions, and Pearson's article went by relatively unnoticed. Von Braun was an American hero. Citizens all across the nation showered him with praise, glory, and the confetti of ticker-tape parades.

After the Apollo space program ended, von Braun moved into the private sector. In his new life as a defense contractor, he traveled the world and met its leaders, including Indira Gandhi, the Shah of Iran, and Crown Prince Juan Carlos of Spain. In 1973, he decided to take up flying, and in June of that year he applied to get his pilot's license with the Federal Aviation Administration. This required a physical and a body X-ray, which revealed a dark shadow on his kidney. Von Braun had terminal cancer but would live for another four years.

The year before he died, there was a motion inside the Ford White House to award Wernher von Braun the Presidential Medal of Freedom. The idea almost passed until one of President Ford's senior advisers,

David Gergen, famously wrote, in a note passed to colleagues, "Sorry, but I can't support the idea of giving [the] medal of freedom to [a] former Nazi whose V-2 was fired into over [*sic*] 3000 British and Belgian cities. He has given valuable service to the US since, but frankly he has gotten as good as he has given." Von Braun was awarded the Medal of Science instead. He died on June 16, 1977. His tombstone, in Alexandria, Virginia, cites Psalm 19:1, invoking God, glory, heaven, and earth.

It would be another eight years before the intrepid CNN reporter Linda Hunt became the first person to crack the pretense of Wernher von Braun. It took a Freedom of Information Act (FOIA) record release to reveal Wernher von Braun's Nazi past.

Dr. Hubertus Strughold, though never as famous as Wernher von Braun, played an equally vital role in the U.S. space program—in the field of medicine. In November 1948, fifteen months after Strughold arrived at the School of Aviation Medicine at Randolph Air Force Base, he and Commandant Harry Armstrong hosted the first ever U.S. military panel discussion on biology in space. Strughold served as professional adviser to Armstrong at the SAM, and he oversaw the work of approximately thirty-four German colleagues under Operation Paperclip contracts there. Strughold's broader vision, which he shared with Armstrong, was to create a space medicine program for the U.S. Air Force.

In 1948 the notion of human space flight was still considered science fiction by most. But Strughold's team had recently conducted a groundbreaking experiment with von Braun's rocket team at White Sands, the results of which they desired to make public. On June 11, 1948, a nine-pound rhesus monkey named Albert was strapped into a harness inside the nose cone of a V-2 rocket and jettisoned into space. Albert's pressurized space capsule, its harness and its cage, had been designed by Dr. Strughold and his team. The V-2 rocket carrying Albert traveled to an altitude of 39 miles. Albert died of suffocation during the six-minute flight, but for Dr. Strughold, the monkey's voyage signified the momentous first step toward human space flight.

Armstrong and Strughold's biology in space panel was cosponsored by the air surgeon, the National Research Council, and the medical research laboratory at Wright-Patterson Air Force Base, Wright Lab. For the first time ever in America, space medicine was now being looked at as a legitimate military science. Two months later, in January 1949, Armstrong and Strughold decided it was time to seek funding for a new department inside the SAM dedicated solely to researching space medicine. For this, Armstrong later explained, "we needed much larger accommodations, more space and facilities and much more sophisticated research equipment."

Armstrong traveled to Washington, D.C., to sell the idea to Congress and to explain that he and Dr. Strughold needed between fifteen and twenty years' lead time to do the necessary research to prepare humans for space travel. Congress approved Armstrong's idea. "There were no wild headlines," Armstrong later explained. The Department of Space Medicine at the SAM opened with little fanfare on February 9, 1949. "I appointed myself Director of the new aerospace laboratory...not Dr. Strughold since there were [*sic*] still some lingering enmity toward the Germans," Armstrong told an air force historian in 1976. But just a few months after the department officially opened, Armstrong was transferred to the Pentagon to serve as surgeon general of the U.S. Air Force. The man he was replacing was his long-time mentor, Major General Malcolm Grow. Back in Texas, Dr. Strughold was promoted to the position of scientific director of the Department of Space Medicine at SAM.

Strughold continued to oversee the professional activities of the Paperclip scientists. He also played an active role in further recruiting endeavors. In the fall of 1949, he traveled to Germany to try to facilitate the hiring of Dr. Siegfried Ruff, Strughold's long-time colleague and one of the doctors tried—and acquitted—at the Nuremberg doctors' trial. Siegfried Ruff was the Luftwaffe doctor who had overseen Rascher's medical murder experiments at Dachau. Information about Ruff's recruitment was leaked to journalist Drew Pearson, who allegedly threatened to tell the president if Dr. Ruff came to the United States as part of Operation Paperclip. Ruff's contract never materialized.

A second defendant acquitted at the Nuremberg doctors' trial was able to secure a Paperclip contract with Dr. Strughold's help. Konrad Schäfer, the "thirst and thirst quenching" expert who had "participated in the planning and execution of the saltwater experiments" at Dachau, arrived at the School of Aviation Medicine in August 1949. In Texas, Schäfer's military research was supervised by Dr. Strughold and an air force captain named Seymour Schwartz. During Schäfer's time at SAM, he worked in three departments: internal medicine, radiobiology and pharmacology. One of his research projects involved an effort to make Mississippi River water drinkable. Konrad Schäfer's American supervisor found him inept. "The scientist [Schäfer] has been singularly unsuccessful in producing any finished work and has displayed very little real scientific acumen," Captain Schwartz wrote. "The experience of this Headquarters indicates that this man is a most ineffective research worker and on the basis of his performance here his future worth to the U.S. Armed Forces is nil."

In a letter to the director of intelligence at U.S. Air Force headquarters in Washington, D.C., in March 1951, Captain Schwartz requested that Schäfer's Paperclip contract be terminated and that he be sent back to Germany. The air force terminated Schäfer's contract, but the German doctor refused to leave the United States. Thanks to expedited protocols put in place by the JIOA, Schäfer was already in possession of a valid U.S. immigration visa by the time it was determined that he was incompetent. He moved to New York City to work in the private sector.

In the early 1950s, the School of Aviation Medicine and its Department of Space Medicine swiftly gained momentum as formidable medical research laboratories. The *Journal of Aviation Medicine* began publishing articles on space medicine, including ones written by Paperclip scientists and covering topics like weightlessness, urinating in zero gravity, and the effects of cosmic radiation on humans. There was also classified military research going on at the SAM, including experiments involving human test subjects. In January 1994, President Clinton created the Advisory Committee on Human Radiation Experiments to investigate and report on the use of human beings in medical research related to Cold War atomic tests. The Advisory Committee found many of these experiments to be criminal and to be in violation of the Nuremberg Code, including studies conducted by German doctors with the SAM.

Starting in 1951, the Atomic Energy Commission (AEC) and the Department of Defense conducted aboveground nuclear weapons tests in the Nevada desert, at a facility called the Nevada Proving Ground and later renamed the Nevada Test Site. The experiments were designed to determine how soldiers and airmen would perform on the nuclear battlefield. The AEC and the DoD agreed that subjecting soldiers to blast and radiation effects of various-sized atomic bombs was required to accurately prepare for a nuclear war.

Colonel John E. Pickering was the director of medical research at the SAM in the early 1950s. In this capacity, Pickering oversaw the research of two German doctors working on a group of AEC studies involving "flashblindness" — the temporary or permanent blindness caused by looking at a nuclear weapon during detonation. According to testimony Colonel Pickering gave to the Advisory Committee in 1994, one set of flashblindness studies had been suggested by Dr. Strughold. Pickering said Strughold was interested in learning about permanent eye damage based on the fact that he himself had been partially blinded in one eye as a boy after watching a solar eclipse with a faulty viewing glass. "That's the thing that gave us curiosity," Pickering recalled of Strughold's accident.

Among the Paperclip scientists who carried out the initial flashblindness studies were Heinrich Rose and Konrad Büttner. Both men had been working for the U.S. government since the fall of 1945, when Dr. Strughold first recommended them for employment at the Army Air Forces Aero Medical Center in Heidelberg. And like so many of Strughold's close Luftwaffe colleagues, Rose and Büttner had been Nazi ideologues during the war. Both men were long-term members of the Nazi Party and also members of the SA.

For the first set of flashblindness studies, Rose and Büttner's test subjects were large pigs. The two doctors calculated that the flash of a 20-kiloton atomic bomb could produce retinal burns as far as forty miles from ground zero. As the atomic tests progressed, the Atomic Energy Commission sought more specific data regarding humans. In the fall of 1951, for an atomic test series called Operation Buster-Jangle, soldier volunteers were asked to stare at a nuclear explosion from varying distances. Crew flying in C-54 aircraft approximately nine miles from ground zero were told to look out the windows of the airplane when the bomb went off. Some wore protective eye goggles while others did not. "No visual handicaps" were reported, according to Pickering. In the spring of

1952, for an atomic test series called Operation Tumbler-Snapper, a trailer was set up on the Nevada desert floor, with porthole-like windows at viewing range. Soldier volunteers were asked to sit in front of the windows and stare at the nuclear explosion. The distance from ground zero was also approximately nine miles, only this time the soldiers were at blast level. One of the volunteers in the viewing trailer was Colonel Pickering, who said that some of the human test subjects were given protective eye goggles to wear but others were asked to observe the fireball with the naked eye. Two soldiers had their eyeballs burned and the trailer tests were suspended.

In the spring of 1953, for an atomic test series called Operation Upshot-Knothole, soldier volunteers were asked to perform duck-and-cover drills inside five-foot-deep trenches that had been dug into the desert floor several miles from ground zero. The men were specifically instructed not to look at the atomic blast. But curiosity got the better of at least one young officer, a twenty-two-year-old lieutenant identified in declassified records as "S.H." Instead of facing away from the blast, the lieutenant looked over his left shoulder in the direction of the atomic bomb when it detonated with a force of 43 kilotons.

"He didn't wear his goggles and he looked," Colonel Pickering told President Clinton's Advisory Committee in 1994. Because light enters the eye through the cornea and is refracted when it hits the lens, Pickering explained, images are flipped upside down. As a result, the image of an inverted nuclear fireball was seared on the lieutenant's retina "forever," leaving what Pickering described as "probably one of the most beautiful images of a fireball you'd ever see in your life." Pickering said that doctors at the SAM kept a photograph of the man's eyeball for their collection. The Advisory Committee determined that the SAM continued its classified flashblindness studies until at least 1962 but that most of the records had been lost or destroyed.

Dr. Strughold's personal research efforts remained focused on space medicine. Monkey astronaut rocket tests at White Sands progressed. On June 19, 1949, a monkey named Albert II was blasted off into space. During that flight, Strughold and his team monitored Albert II's vitals as the V-2 carried him past the Kármán line — the point at which outer space begins — and reached an altitude of 83 miles. Albert II died back on Earth, on impact, when his parachute failed to open. By 1952, Strughold had succeeded in convincing the air force to fund the construction of a sealed chamber, or capsule, to be used for space medicine research on humans. The cabin was designed by fellow Paperclip scientist Fritz Haber, brother of Heinz Haber, and completed in 1954. The sealed chamber was one hundred cubic square feet, with a single seat and an instrument panel, and was meant to duplicate the conditions that an astronaut would experience during a voyage to the moon. The first human test was conducted in March 1956, when an airman named D. F. Smith spent twenty-four hours inside the chamber, performing various tasks while being monitored by Strughold and his team. Approximately two years later, in February 1958, pilot Donald G. Farrell, a twenty-three-year-old native of the Bronx chosen from a pool of rigorously screened airmen, stepped into the chamber for seven consecutive days and nights. This time period, Strughold explained, was inspired by Jules Verne's prediction of how long it would take for a spaceship to get to the moon. During the test, Strughold and Hans-Georg Clamann, Strughold's former assistant during the Luftwaffe years, as well as two air force colleagues, monitored Donald Farrell's vitals and his ability to perform tasks.

The space capsule simulator test with airman Donald Farrell attracted all kinds of media attention. On the seventh and final day, Texas senator Lyndon B. Johnson stopped by the School of Aviation Medicine to personally escort Farrell out of the chamber and join him at a press conference. Dr. Strughold stood by Farrell's side, as did General Otis Benson—the man who had tried, and failed, to find Major General Dr. Walter Schreiber a teaching position so he could remain in America. Benson singled out Strughold for his excellent work "in the name of medical science."

So excited was Lyndon Johnson by the event that he flew Dr. Strughold and his team to Washington, D.C., to attend a luncheon with seventy congressmen, the secretary of the air force, and a half-dozen four-star generals. Strughold later recalled the event: "After the soup, [Lyndon] Johnson asked me to give a five minute talk about the scope of space medicine and the meaning of the experiment." Hubertus Strughold, like Wernher von Braun and Walter Dornberger, was now an accepted member of the U.S. military scientific elite.

But there was a bump in the road. In May 1958, *Time* magazine featured a piece about Strughold, showering him with praise as America's space medicine research pioneer. In response, the widely read *Saturday Review* published an editorial by Julian Bach Jr., a former war correspondent for the *Army Talks* series of pamphlets for GIs, and "the first American correspondent to report in the general press on medical experiments on human beings by Nazi doctors during World War II." Bach's editorial was called "Himmler the Scientist." In it, the war correspondent reminded the public of the human experiments that the Nazi doctors had conducted on prisoners in concentration camps. "The German doctors carving them up were medical men of stature in many cases," wrote Bach. "Only the fewest were quacks." Bach correctly linked Strughold to the freezing experiments at Dachau, stating that, at minimum, he "had knowledge of them." This was the first time the public had heard anything about Dr. Strughold having knowledge of the medical murder experiments, a fact he had previously been able to keep hidden.

The article prompted an investigation at the federal level. Because Strughold had become a U.S. citizen two years prior, in 1956, the Immigration and Naturalization Service (INS) was now compelled to investigate. But after checking with the air force, the INS released a statement saying that Strughold had been "appropriately investigated" before becoming a U.S. citizen. The INS had not been shown Strughold's OMGUS security report or his classified dossier. If it had been, INS would have learned that military intelligence had concluded that Strughold's "successful career under Hitler would seem to indicate that he must be in full accord with Nazism."

In October 1958, the twenty-ninth meeting of the Aero Medical Association convened in San Antonio, Texas, for a daylong symposium, "Aviation Medicine on the Threshold of Space." The group's first international convention had taken place at the Waldorf-Astoria hotel twenty-one years before. Dr. Strughold and Harry Armstrong had both been in attendance. Here they were together again: Armstrong and

Strughold were co-hosts of the event. General Dornberger delivered a speech. Eleven of the forty-seven scientists in attendance had been brought to America from Germany as part of Operation Paperclip. They were American citizens now.

A prolific writer, Strughold authored papers and journals, sometimes more than a dozen in a year. He had contributed a piece about space cabins to the *Collier's* magazine series. He created new, space-related nomenclature, including the words "bioastronautics," "gravisphere," "ecosphere," and "astrobiology." He studied jet lag—how the body responds to flight—and wrote a book about his findings, *Your Body Clock.* In 1964, he was interviewed by the space writer and journalist Shirley Thomas for her eight-volume series *Men of Space.* Now, with nearly two decades of distance between himself and his Nazi past, and with so many accolades to his name, Strughold began to construct a fictional past for himself, one in which he had actually been an opponent of the Nazis.

"I was against Hitler and his beliefs," Strughold told Shirley Thomas. "I sometimes tried to hide myself because my life was in danger from the Nazis," he said. This was, of course, absurd. Strughold was among the highest-ranking doctor-professors in the Luftwaffe.

"Were you ever forced to join the Nazi Party?" Thomas asked.

"It was tried," Strughold said. "They tried it" but failed. Strughold told Thomas that the same went for the staff at his Aviation Medical Research Institute in Berlin. "Only the janitor and the man who took care of the animals" were members of the Nazi Party, Strughold lied.

It was as if his close colleagues from the Nazi era—Doctors Siegfried Ruff, Konrad Schäfer, Hermann Becker-Freyseng, and Oskar Schröder—did not exist.

People began to look harder into the past. On September 3, 1973, Simon Wiesenthal, a freelance Nazi hunter who had played a role in the capture of Adolf Eichmann, wrote to Dr. Adalbert Rückerl, director of the Central Office of the State Justice Administration for the Investigation of National Socialist Crimes, in Ludwigsburg, Germany.

"I have information that Dr. Hubertus Strughold participated in experiments with human beings in Ravensbrück," Wiesenthal wrote. "We don't have the name in our archive. The information states that the experiments were in context of the German Air Force."

Dr. Adalbert Rückerl told Simon Wiesenthal that he would look into it. A few months later, Rückerl wrote back to Wiesenthal to say that a comprehensive investigation had been done and that the Central Office could not find any direct evidence of Strughold's personal participation in the medical crimes. Rückerl provided Wiesenthal with a copy of the documents pertaining to the conference, "Medical Problems of Sea Distress and Winter Distress," at the Hotel Deutscher Hof in Nuremberg in 1942, including the travel expenses that Strughold had submitted and the amount he was paid to attend. The Luftwaffe doctors in attendance had discussed the medical data of murdered people. But knowledge was not a crime, and there was no further action for the Central Office of the State Justice Administration to take, Rückerl said.

A few months later, toward the end of 1973, the U.S. Immigration and Naturalization Service announced that it was investigating Strughold and thirty-four other long-dormant cases of Nazi war criminal suspects living in the United States. That Strughold's name was on the list, which was published in the *New York Times,* came as a shock to him. Strughold rehearsed a succinct response for reporters, the same as von Braun had when allegations arose. "I was cleared before I came here, before I was hired," Strughold said.

Leaders of Jewish groups sprang into action, and Charles R. Allen Jr., the former senior editor of *The Nation*, writing for *Jewish Currents,* presented evidence about Strughold's knowledge of human experiments. Allen had acquired the first public copy of the proceedings of the Nuremberg medical conference at the Hotel Deutscher Hof in 1942, which included a list of those in attendance. On this list were five other Nazi doctors who had worked in America under Operation Paperclip—Major General Walter Schreiber, Konrad Schäfer, Hans-Georg Clamann, Konrad Büttner, and Theodor Benzinger—but neither Charles R. Allen Jr. nor anyone else involved in the Strughold exposé appears to have recognized any of these names. Twenty years prior, Major General Walter Schreiber had been fired from the SAM and banished to Argentina, and Konrad Schäfer had been fired from the SAM and moved to New York. But when this story broke in 1973, Hans-Georg Clamann, Konrad Büttner, and Theodor Benzinger had all been gainfully employed by the U.S. government for decades and were now living the American dream.

The cloud of suspicion hung over Strughold. Ralph Blumenthal wrote a follow-up article in the *New York Times,* in November of 1974, describing the freezing experiments in detail. "Victims were forced to remain naked outdoors or in tanks of ice water. Tests were made periodically as they were freezing to death... the victims suffered agonizing deaths, after which they were dissected for data." INS director Leonard Chapman made inquiries with the U.S. Air Force and was told that there was "no derogatory information" on Strughold. The case was closed, yet again.

Now more than ever Strughold felt the need to promote the fiction that he was anti-Nazi. In an air force oral history interview that took place two days after Blumenthal's piece ran in the *New York Times,* Strughold told interviewer James C. Hasdorff that during the war he constantly feared for his life. He said that he had been written up on Hitler's "so-called enemy list." Strughold insisted that at one point, shortly after the attempted assassination of Hitler by Count Claus von Stauffenberg, in July 1944, he, Strughold, had to go into hiding because he was receiving death threats—that he went from farmhouse to farmhouse until it was finally "safe to return to Berlin." This was an absurd distortion of reality. In the months after the Stauffenberg incident, in the winter of 1945, Hubertus Strughold had been promoted to full colonel in the Luftwaffe reserves.

Three years passed. The spotlight moved away. In a ceremony on January 19, 1977, a bronze plaque bearing a portrait of Dr. Hubertus Strughold was unveiled in San Antonio, Texas, in the foyer at the Aeromedical Library at the School of Aerospace Medicine. The library, which was the largest medical library in the U.S. Air Force, was being dedicated to Dr. Hubertus Strughold, the Father of Space Medicine. The air force promised that Strughold's bronze plaque "will be permanently displayed in the foyer of the library." But permanence is impossible to predict. Only time tells what lasts.

CHAPTER TWENTY-TWO

Legacy

In the 1960s, what Operation Paperclip left behind in the country it recruited from reveals contradictory truths about the program's legacy. A watershed event occurred in 1963 when the West German courts opened the Frankfurt Auschwitz trial. This was the first large-scale public trial of Holocaust perpetrators to take place inside Germany. There had been one Auschwitz trial previously, in Cracow, Poland, in 1947, in which a number of camp functionaries were tried, including Auschwitz commandant Rudolf Höss, who was sentenced to death and hanged. But before the Frankfurt Auschwitz trial, which ran through 1965, Germany had not conducted any major war crimes trials of its own. The Nuremberg trials, of course, were presided over by the Allied forces. At the Frankfurt Auschwitz trial, 360 witnesses from 19 countries, including 211 Auschwitz survivors, would confront the accused. The judges hearing the case were all Germans.

Throughout the 1950s, Germany's political and legal elite had opposed holding these kinds of trials for a variety of reasons, but mostly because it would likely generate additional trials. Just as it behooved U.S. military intelligence to keep the details of Operation Paperclip a secret, so it went for many German jurists. Thousand of their fellow citizens had committed heinous crimes during the war and were now living perfectly normal and sometimes very productive lives.

But in 1963, with the trial under way, for the first time since 1945 people in Germany were openly talking about gas chambers, death camps, and the Holocaust. "The trial was triggered by a letter I received," recalled Hermann Langbein, the secretary of the International Auschwitz Committee. "It was mailed to me by an Auschwitz survivor. A German [named] Rögner." At Auschwitz, where both Langbein and Rögner had been prisoners, the SS guards had made Rögner a kapo—someone who supervised slave laborers. "He was a kapo in an electrical detail," explained Langbein, "a decent kapo. There were some decent kapos there [at Auschwitz]."

In Rögner's letter to Langbein he wrote, "I know where Boger is."

Boger was a loaded name to anyone who had

survived Auschwitz. Survivors knew Boger as the Tiger of Auschwitz. Wilhelm Boger was a man of indisputable evil. Witnesses watched him murder children on the train ramps with his own hands. In Rögner's letter to Langbein, he said that he'd filed murder charges against Boger with the German authorities in Stuttgart. In the spring of 1958, in his capacity as the secretary of the International Auschwitz Committee, Hermann Langbein followed up on Rögner's tip and discovered that Boger was working as an accountant in a town outside Stuttgart. There, the Tiger of Auschwitz lived with his wife and children as a family man. "Bringing the man to justice was one long obstacle course," said Hermann Langbein. But finally, after several months of delays, Boger was arrested.

There were twenty-two codefendants in the Frankfurt Auschwitz trial. With the exception of Richard Baer, the last commandant of Auschwitz, all of the former Auschwitz functionaries had been leading normal lives using their own names. Commandant Baer had been living as a groundskeeper on an estate near Hamburg under the alias Karl Neumann. Auschwitz adjunct Robert Mulka was working as a merchant in Hamburg; at the death camp, Mulka had served as Rudolph Höss's right-hand man. Josef Klehr was living in Braunschweig, working at an auto body shop called Büssing & Sons; Klehr had been a medical orderly at Auschwitz and had killed thousands with lethal injections to the heart. Oswald Kaduk was living in Berlin working in a nursing home; at Auschwitz Kaduk beat people to death with his boot. SS pharmacist Dr. Victor Capesius was living in Göppingen, a wealthy pharmacy and beauty shop owner; at Auschwitz, Capesius had assisted the notorious Dr. Mengele in selecting individuals for medical experiments and had managed the death camp's supply of Zyklon B. With the actions of all these men under scrutiny at the Frankfurt Auschwitz trial, Germany faced its past through the lens of its own jurists, not American judges. Many young Germans paid attention and began to wonder how so many war criminals could have gotten away with so many war crimes for so long.

More important to the Operation Paperclip legacy is that when Rögner filed his civil suit, he named another name: Dr. Otto Ambros. Ambros was listed as a criminal in Rögner's complaint because he had been general manager of IG Farben's Buna factory at Auschwitz. German prosecutors looked into the Ambros case and saw that Ambros had already been tried and convicted of war crimes at Nuremberg and that he had served two years and five and a half months in Landsberg Prison. They decided not to press charges against him a second time. In December 1963, however, Ambros was called to testify at the trial, which was held in the Römer, Frankfurt's town hall. This brought Ambros into the limelight once again.

By 1964, Ambros had been a free man for thirteen years. He was an extremely wealthy, successful businessman. He socialized in Berlin among captains of industry and the professional elite. When the Frankfurt Auschwitz trial started, he was a board member of numerous major corporations in Germany, including AEG (Allgemeine Elekrizitats Gesellschaft), Germany's General Electric; Hibernia Mining Company; and SKW (Süddeutsche Kalkstickstoff-Werke AG), a chemical company. On the witness stand at the Frankfurt Auschwitz trial, Ambros gave testimony that contradicted statements he had made during his own trial at Nuremberg in 1947. He also said that the conditions at the Auschwitz Buna factory had been "cozy," and that the workers had enjoyed "good hospitality," which many Auschwitz survivors found appalling in its offensiveness. While contradictory statements might have gone unnoticed by Frankfurt's judges and jurors, there were a number of high-profile Israeli journalists in the courtroom who had become experts on the subject and caught Ambros's lies. These journalists began at once to investigate Otto Ambros's post-Nuremberg life. Here was a man who had been convicted of slavery and mass murder and had served very little time, considering the crimes. The Farben slaves at Auschwitz numbered sixty thousand, approximately thirty thousand of whom had been worked to death.

The journalists covering the trial were outraged by what they discovered. Otto Ambros now sat on the board of directors of numerous private corporations, but he was also on the board of directors of five companies that were owned by the Federal Republic of Germany. The exorbitant fees Ambros commanded in these positions were being paid by the German taxpayer. An Israeli female journalist, identified in Bundesarchiv documents only as "Frau Deutschkren," became so incensed by the arrogance and hubris she saw in Otto Ambros's postwar life that she wrote a letter to the state minister of finance, Ludger Westrick. That the Federal Republic of Germany was paying a "consulting fee" of 12,000 deutschmarks—about $120,000 in today's dollars—to a convicted war criminal was shameful, Frau Deutschkren said. She demanded that Finance Minister Ludger Westrick meet with her. He agreed.

Frau Deutschkren could not have known that Otto Ambros and Finance Minister Ludger Westrick were business colleagues and apparently on very friendly terms, as state archive correspondence reveals. After the meeting, Westrick promised to look into the matter. Instead, he told Ambros what was going on. In an effort to hold on to his lucrative and prestigious positions on company boards, Ambros produced a summary of the allegations against himself and his Nuremberg codefendants, written by his attorney, a Mr. Duvall. Ambros asked Finance Minister Westrick to circulate this apologia around the various boards on his behalf. "As a short summary of our case [shows] you will clearly find out we are innocent," Ambros explained, referring to charges of slavery and genocide that he and his Farben colleagues were convicted of at Nuremberg. "I and my colleagues are the victims of the Third Reich," Ambros insisted. "The former government utilized the success of synthetic rubber which they used to make a profit. If there had been anything against me, then I would have never been released by the American military." U.S. High Commissioner John J. McCloy had granted Ambros clemency under intense political pressure, a fact now being used by Ambros to suggest that he had been wrongly convicted at Nuremberg.

In a letter dated April 25, 1964, Ambros reminded Finance Minister Westrick that after he had been released from Landsberg Prison in 1951, "You helped me get back on the boards." For this, Ambros said he was grateful. "I see it as an honor and a duty to [remain] there. I do this for pure altruistic reasons. I appreciate anything you can do." In turn, Finance Minister Westrick wrote letters to the various boards on Ambros's behalf. "Ambros was chosen for the board because of exceptional talent," Westrick said. "In his field he is as wanted as Wernher von Braun. Everyone wants him. He can get a job anywhere he wants, anywhere in the world."

The Israeli journalists refused to let up on Otto Ambros. They continued to write news stories about him, making it increasingly difficult for both the Federal Republic of Germany and the publicly traded companies on whose boards Ambros sat to maintain business associations with him. "Former War Criminal Found Refuge in Switzerland," read a headline on June 6, 1964. The story detailed time that Ambros spent in the Swiss village of Pura. Ambros, furious, submitted a "statement of facts" to Labor Ministry state secretary, Ludwig Kattenstroth, in response. "I did not hide myself in Pura," Ambros wrote. "It is my holiday home. And I have to say that when I bought the parcel there [in 1956] I informed the Swiss government by handing over the judgment at Nuremberg. I am only there for holidays... my children go there, and my friends. After consulting with my lawyer, I will never go back." Then came the blaming. "The whole affair," wrote Ambros, "has to be seen in the shadow of the Frankfurt trial. A certain faction of the press is trying to blame me." The subtext was that "the Jews" were trying to blame him.

In late summer of 1964, AEG's board members met and decided that they could no longer retain convicted war criminal Otto Ambros. Ambros then also quietly left at least two of the five taxpayer-funded consulting positions he held on Federal Republic of Germany boards.

In separate letters to Finance Minister Ludger Westrick and Deputy Finance Minister Dr. Dollinger, a new secret was revealed, though Ambros promised not to make public a piece of the information they shared. "Concerning the firms abroad where I am a permanent co-worker advisor," Ambros wrote, "I won't name them [publicly] because I don't want to tip off any journalists who might cause trouble with my friends. You know about W. R. Grace in New York...and I hope I can stay with Hibernia Company. Concerning the firms in Israel," Ambros wrote, "stating their names publicly would be very embarrassing because they are [run by] very public, well-respected persons in public positions that have actually been at my home and are aware of my position, how I behaved during the Reich, and they accept this."

The "well-respected" public figures in Israel to whom Ambros referred have never been revealed. That Ambros also had worked for the American company W. R. Grace would take decades to come to light. When it did, in the early 1980s, the public would also learn that Otto Ambros worked as a consultant for the U.S. Department of Energy, formerly the Atomic Energy Commission, "to develop and operate a plant for the hydrogenation of coal in a scale of 4 million tons/year at the former IG Farben industrie." That a convicted war criminal had been hired by the Department of Energy sparked indignation, and congressmen and journalists sought further details about Ambros's U.S. government contract. In a statement to the press, the Department of Energy insisted that the paperwork had been lost.

The scandal was brought to the attention of President Ronald Reagan. Letters on White House stationery reveal that Deputy National Security Adviser James W. Nance briefed Reagan about how it was that the U.S. government could have hired Otto Ambros. Nance's argument to the president was that many others hired him. "Dr. Ambros had contacts with numerous officials from Allied countries," wrote Nance. "Dr. Ambros was a consultant to companies such as Distillers Limited of England; Pechiney, the French chemical giant; and Dow Europe of Switzerland. He was also the chairman of Knoll, a pharmaceutical subsidiary of the well known chemical corporation BASF." President Reagan requested further information from the Department of Energy on its Ambros contract. Nance told the president, "The DOE and/or ERDA [The Energy Research and Development Administration] do not have records that would answer the questions you asked in the detail you requested. However, with Dr. Ambros' involvement in the company shown and his special knowledge in hydrogenation of coal, we know there were productive contacts between Dr. Ambros and U.S. energy officials." Even the president of the United States could not get complete information about an Operation Paperclip legacy.

In the midst of the scandal, a reporter for the *San Francisco Chronicle* telephoned Ambros at his home in Mannheim, Germany, and asked Ambros about his 1948 conviction at Nuremberg for mass murder and slavery.

"This happened a very long time ago," Ambros told the reporter. "It involved Jews. We do not think about it anymore."

The third act of Dr. Kurt Blome's life contradicts the idea that the Soviets were desperately trying to hire Hitler's former biological weapons maker. In February 1962, a group of Communist physicians from Karl Marx University, in Leipzig, East Germany, sent an open letter "to all the doctors and dentists living in Dortmund," as well as many widely distributed newspapers "regarding the matter of the Dortmund physician, Dr. Kurt Blome." The letter listed the war crimes Blome had been accused of at the Nuremberg doctors' trial and stated that everyone in the town of Dortmund should "know who he is and what he did to the Jewish people." The Leipzig doctors identified the doctor in Dortmund as "the Deputy Surgeon General of the Third Reich." They wrote that Western German medical colleagues should distance themselves "from this man who dares to call himself a doctor." Blome was contacted by the East German radio station, Radio Berlin International, for comment. He agreed, and an interview was arranged for February 22, 1962.

"I have read the two pages you have presented to me," Blome told his interviewer on the appointed day. "During the war I got the task to prepare a serum, a vaccine, against a bacterial plague. This all started because [Major General] Dr. Schreiber told a lie. Schreiber is a pawn of the Russians. It's a sworn lie." Blome spent the rest of his radio time blaming Schreiber for his own misfortunes. He ended by pointing out his innocence; Blome reminded the radio announcer that he had endured a ten-month trial at Nuremberg. "I was acquitted on every point that there was," Blome said. "The U.S. judges were not easy. They handed out seven death penalties and five lifelong imprisonments. If I had been guilty, they would have convicted me."

The allegations opened a subsequent investigation into Blome. Two weeks later, on April 4, 1962, the state's attorney in Dortmund wrote to the Central Office of the State Justice Administration for the Investigation of National Socialist Crimes to see if there was anything on Dr. Kurt Blome. The state's attorney put together a dossier on Blome, with most of the focus on a letter Blome had written on November 18, 1942, to Reich Governor (Reichsstatthalter) Arthur Karl Greiser, mayor of a Polish area near Posen and subsequently executed for war crimes. The subject matter was *Sonderbehandlung,* or "special treatment," of Poles with tuberculosis. By now, in 1962, it had been established that *Sonderbehandlung* was a Nazi euphemism for extermination. In this 1942 correspondence, Blome and Greiser agreed that the best way to deal with this group of tubercular Poles being discussed was to give them "special treatment." But Blome also raises "a problem" that had to be dealt with — namely, that if these tubercular Poles were to be exterminated, "[t]he general perception by relatives would be that something unorthodox is going on." Blome reminds Reich Governor Greiser that the group they are proposing special treatment for totals approximately thirty-five thousand people.

Blome was called in to an interview with the state's attorney. "I am aware that there is an investigation," Blome said. "My position is that this has already been investigated by the Americans." Dr. Blome then made the same argument he had used at Nuremberg as a defendant in the doctors' trial: that he may have recommended the extermination of thirty-five thousand Poles as a course of action, but that there was no proof that that action ever took place. Intent, Blome maintained, is not a crime. "Furthermore," Blome told investigators, "I want to add that I put the blame on [Major General Dr. Walter] Schreiber based on what I learned at the trial." Blome said that Dr. Schreiber was the true murderer and that he had framed Dr. Blome to deflect the guilt away from himself. After several days of consideration, on May 21, 1962, the state's attorney placed the investigation of Dr. Blome on hold. Blome continued to practice medicine in Dortmund and the case against him was never reopened. Three years later, Blome was dead. "He died of emphysema," says his son, who also reports that, by the time of his father's death, Dr. Kurt Blome was alone and estranged from the world.

Albert Speer served all twenty years of his sentence at Spandau Prison outside Berlin, despite the efforts by John J. McCloy, as U.S. High Commissioner for Germany, to get Speer released earlier. In 1956 McCloy wrote to Speer's wife, Margarete, "I have a very strong conviction that your husband should be released and would be very happy if I could do anything to expedite such a release."

In prison Speer secretly worked on the memoirs he would publish upon his release. On pieces of toilet paper, cigarette wrappers, and paper scraps he jotted down recollections of his days working alongside

Hitler and the inner circle. It was illegal to write and send unscreened notes out of Spandau, so Speer had his writing smuggled out by two sympathetic Dutch Red Cross nurses. The notes were delivered to Speer's old friend and colleague Rudolf Wolters, a diehard Hitler loyalist living in Berlin. Over the course of twenty years Wolters painstakingly typed up tens of thousands of these individual paper scraps that Speer continued to send to him. After twenty years they amounted to a thousand-page manuscript, which Wolters turned over to Speer upon Speer's release. With the advance payments Speer received on his memoirs—he would write two, *Inside the Third Reich* and *Spandau: The Secret Diaries*—Speer became a wealthy man once again. He earned a reported 680,000 deutschmarks from *Die Welt* for serialization rights, and a reported $350,000 (roughly $2.4 million in 2013) advance for English-language book rights. Albert Speer bought a sports car and embarked on a new life as a successful author.

He never thanked or acknowledged Rudolf Wolters for his twenty years of work, at least not in his book and not publicly. A decade later, Speer defended this decision to his biographer, Gitta Sereny, saying "[I]t was for [Wolters's] own protection." Wolters's son reported that when Wolters died, the last word he uttered was "Speer."

After Speer sent Wernher von Braun a copy of

Spandau, von Braun wrote a letter to his old friend and former boss at the Reich's Ministry of Armaments and War Production, pointing out how divergent their lives had been over the past twenty years. While Speer had been at Spandau Prison, von Braun's star had risen steadily. He thanked Speer for the book. "How often I thought of you during those last twenty long years when so much was happening in my life," wrote von Braun.

When Siegfried Knemeyer learned that Albert Speer was out of prison, Knemeyer took a trip to visit his old friend in Germany. The last time the two men had seen each other was in Berlin in April 1945, when Knemeyer was helping Speer plot his escape to Greenland. In the years that Speer had been in prison, Knemeyer had been working for the United States—formerly the sworn enemy of both men. As an employee of the U.S. Air Force in the 1940s, 1950s, 1960s, and 1970s, Knemeyer worked on classified and unclassified projects, won awards, and rose up through the ranks. For the air force, he established the Pilot Factors Program, which coordinated technologies as aircraft went from subsonic to supersonic flight. Upon his retirement in 1977, Pentagon officials awarded Knemeyer their highest civilian award, the Distinguished Civilian Service Award. Knemeyer died of emphysema two years later, on April 11, 1979. His last wish was that his body be "immediately taken to a crematory" and that no funeral be held. "He served his native country, Germany, and his adopted country, America, with equal enthusiasm and dedication," said his son Sigurd Knemeyer.

To the end, Albert Speer denied that he had direct knowledge of the Holocaust. Sixty-seven years after interviewing Speer at the Ashcan interrogation facility in Luxembourg, John Dolibois still takes umbrage at this. "I asked [Speer] if he was at the Wannsee Conference at which Himmler announced the 'Final Solution' for eradicating Jews," Dolibois explains. "In my opinion, anyone who was at that meeting could not say he knew nothing about extermination camps — like Auschwitz, Sobibor, Birkenau. Speer first denied being at Wannsee, then admitted [to me] he was there, but left before lunch and missed the important announcements. Others also asserted [Speer] was not telling the truth. I think he should have been hanged."

Albert Speer died in a London hotel room in 1981. He was in town doing an interview for the BBC. "One seldom recognizes when the Devil puts his hand on your shoulder," Albert Speer told James P. O'Donnell in a *New York Times Magazine* interview shortly after his release from Spandau. He was referring to Hitler but might have been talking about himself.

CHAPTER TWENTY-THREE

What Lasts?

In the 1970s, several events led to a major shift in the American public's perception of the Holocaust, Nazis, and the United States. A series of congressional oversight hearings were convened as a result of many high-profile cases of Nazi war criminals found living in America. The hearings drew attention to the fact that the Immigration and Naturalization Service had been negligent in its investigations of these individuals and in turn led to the creation of the Justice Department's Office of Special Investigations. Then came the 1978 broadcast of NBC's four-part miniseries *Holocaust,* which broke records for family viewing and made Americans who had never contemplated the Holocaust think seriously about what had happened in Nazi Germany and what the state-sponsored murder of six million people really meant. These events set the stage for a simple, fortuitous occurrence for a young Harvard law student named Eli Rosenbaum, the consequences of which would profoundly impact the Paperclip scientists who had something to hide.

In 1980, Eli Rosenbaum was perusing book titles inside a Cambridge bookstore when he came across *Dora: The Nazi Concentration Camp Where Modern Space Technology Was Born and 30,000 Prisoners Died*, written by a former Dora prisoner, the French Resistance fighter Jean Michel. The narrative is detailed and revelatory in its brutal honesty. Jean Michel called his book an homage to the thirty thousand slave laborers who died building the V-2s. He comments on the memoirs written by Wernher von Braun and General Dornberger, and the countless interviews given by many of the other V-2 scientists who worked under Operation Paperclip. None of these Germans, notes Michel, utters a word about Nordhausen.

"I do not reproach these men with not having made public confessions after the war," Michel writes. "I do not hold it against the scientists that they did not choose to be martyrs when they discovered the truth about the [death] camps. No, mine is a more modest objective. I make my stand solely against the monstrous distortion of history which, in silencing certain facts and glorifying others, has given birth to false, foul and suspect myths."

In that same Cambridge bookstore, on that same

day, Eli Rosenbaum came across a second book about the V-2 rocket. This one was called *The Rocket Team* and was written by Frederick I. Ordway III and Mitchell R. Sharpe, with an introduction by Wernher von Braun. The book discusses and quotes the German scientists involved in the U.S. rocket program. In one part of the book, engineer Arthur Rudolph — praised as the developer of NASA's Saturn V rocket program — shares his thoughts with the authors. Rudolph is not identified as having once been operations director at the Mittelwerk slave labor facility.

Rudolph relates an anecdote from the war, concerning his dismay at being called away from a New Year's Eve party in 1943–44 because of a problem with some of the V-2 rockets. An accompanying photograph shows a POW in striped prisoner pajamas moving rocket parts. Eli Rosenbaum had spent the previous summer working for the Department of Justice in its Office of Special Investigations. He knew that the Geneva Convention forbade nations from forcing prisoners of war to work on munitions. He had also just read about what went on in the Nordhausen tunnels in Jean Michel's book *Dora*. Rosenbaum later recalled that he "was particularly offended by Rudolph's taking umbrage at missing a gala party while slave laborers toiled." The following year, after graduating from Harvard Law School, Rosenbaum started full-time work at the Department of Justice as a trial lawyer. He persuaded his boss, Neal M. Sher, to open an investigation into Arthur Rudolph.

In September of 1982, Marianne Rudolph, the daughter of Arthur Rudolph, received an unexpected telephone call at her home in San Jose, California. It was the Department of Justice calling. They explained that they had been trying to reach her father, Arthur Rudolph, who apparently lived nearby, but that they had been unable to do so. Marianne Rudolph, who worked as an artist for NASA, told the caller that her parents were vacationing in Germany and would be back at the end of the month.

The day after the Rudolphs returned, Arthur Rudolph received a registered letter in the mail from the Department of Justice, Office of Special Investigations. The letter stated that questions had been raised regarding Arthur Rudolph's activities during the Second World War. Rudolph was asked to meet with DOJ officials at the San Jose Hyatt on October 13 and to bring any documents with him that he owned covering the period between 1933 and 1945. During this first meeting, which included Arthur Rudolph, OSI director Allan A. Ryan Jr., Deputy Director Neal M. Sher, and trial attorney Eli Rosenbaum, the interview lasted five hours. There were two central questions the lawyers wanted answered: Owing to what set of principles had Rudolph decided to join the Nazi Party, which he did in 1933, and what exactly did he know about the Nordhausen executions, specifically about the prisoners who had been hanged from a crane? DOJ lawyers Allen, Sher, and Rosenbaum had documents with them including the previously sealed testimony from the 1947 Dora-Nordhausen trial. These documents included pretrial investigative material regarding former Paperclip specialist Georg Rickhey. Among the documents was the interview that Rudolph gave, on June 2, 1947, to Major Eugene Smith of the U.S. Army Air Force. In that testimony, Rudolph had first said that he never saw any prisoners beaten or hanged. Later in the interview with Major Smith, Rudolph changed his story to say that he had been forced to watch the hangings but had nothing to do with them.

During the San Jose Hyatt interview in 1982, the three Justice Department lawyers presented Rudolph with the drawing made by his former Nordhausen colleagues from Fort Bliss, the rocket engineers Günther Haukohl, Rudolph Schlidt, Hans Palaoro, and Erich Ball. This drawing illustrated the layout inside the tunnels and had been used as evidence in the Dora-Nordhausen trial. The Justice Department lawyers pointed out to Rudolph that there was a clear dotted line, labeled "Path of Overhead Crane Trolly [sic] On Which Men Were Hung." The dotted line ran right by Rudolph's office, which suggested that it would have been impossible for Rudolph not to have seen the hangings. The lawyers asked why he had lied. They also asked Rudolph about testimony from the 1947 Dora-Nordhausen trial that revealed that, at the Mittelwerk tunnel complex, he received daily "prisoner strength reports which showed the number of prisoners available for work, the number of 'new arrivals,' and the number of people lost through sickness or death.'" They said that, clearly, Rudolph knew people were being worked to death and were being replaced by fresh bodies from the Dora-Nordhausen concentration camp.

Back in Washington, D.C., the Justice Department put together its case. Four months later, on February 4, 1983, a second meeting took place at the San Jose Hyatt, and, shortly thereafter, Rudolph received a letter from the Justice Department in the mail. "Certain preliminary decisions have now been made," wrote Neal Sher. "I would be prepared to discuss these decisions, as well as the evidence amassed to date, with an attorney authorized to represent your interest."

The government stated intentions to pursue its case against Arthur Rudolph "showing that Mr. Rudolph enforced the slave labor system at Mittelbau [Mittelwerk] and aided in the transmission of sabotage reports to the SS." Arthur Rudolph had two choices, the Justice Department said. He could hire a lawyer and prepare to stand trial, or, alternatively, he could renounce his U.S. citizenship and leave the country at once.

Thirty-eight years after coming to America as part of Operation Paperclip, Arthur Rudolph left the United States, on March 27, 1984. Neal Sher met Arthur Rudolph at the San Francisco International Airport and made sure he got on the airplane.

It was another seven months before the Justice Department made public that Arthur Rudolph had renounced his citizenship so as to avoid facing a war crimes trial. When asked to comment, Eli Rosenbaum said that Arthur Rudolph had contributed to "the death of thousands of slave laborers." But Rosenbaum also said that it was Rudolph's "almost unbelievable callousness and disregard [for] human life" that surprised him most. The story became front-page news around the world.

Some individuals affiliated with NASA and other rocket-related government programs remained staunch supporters of Arthur Rudolph, calling the Justice Department's actions against him a "witchhunt." One of Rudolph's leading proponents, Hugh McInnish, an engineer with the U.S. Army Strategic Defense Command, helped promote the idea that Rudolph, Dornberger, and von Braun had all confronted members of the SS during the war and had tried to get better working conditions for the slave laborers. "Their defenders' assertions must be regarded with the greatest skepticism, especially as there is not a single document to back them up," says Michael J. Neufeld. "There is little doubt in my mind that Rudolph was deeply implicated." The Rudolph exposé triggered keen interest in how it was that Arthur Rudolph came to America in the first place. This was a turning point in Operation Paperclip's secret history.

There is a broad misconception in America that there exists some kind of automatic declassification system that requires the government to reveal its secret programs after thirty or fifty years. In reality, the most damaging programs often remain classified for as long as they can be kept secret. The Freedom of Information Act, signed into law by President Lyndon Johnson, allows for the full or partial disclosure of some documents, but a request must be initiated by an individual or a group and is by no means a guarantee that information will be obtained. After the Arthur Rudolph story broke, journalist Linda Hunt began reporting on Operation Paperclip for CNN and the *Bulletin of the Atomic Scientists*. She filed FOIA requests with the different military organizations and intelligence agencies involved and received varied responses. "I obtained six thousand Edgewood Arsenal documents in 1987 but it took more than a year, two attorneys and a threatened lawsuit to get the records [released]," says Hunt. When she arrived at the Washington National Records Center in the late 1980s to inspect the documents, she was told that Edgewood's own historian had checked out seven of the boxes and that another twelve were missing. The army later agreed to look for the missing records and sent her a bill for $239,680 in so-called search fees (the equivalent of $500,000 in 2013). Eventually Linda Hunt was granted access to the documents, and in the early 1990s she published a book that unveiled many of Operation Paperclip's seemingly impenetrable secrets. It was now no longer possible for the government to uphold the myth that Paperclip was a program peopled solely by benign German scientists, nominal Nazis, and moral men.

As if pushing back against the Office of Special Investigations, just a few months after Arthur Rudolph was expelled from the United States, the Texas Senate passed a resolution declaring that June 15 was Dr. Hubertus Strughold Day. Then Ohio State University's medical college unveiled a portrait of Strughold on its mural of medical heroes. But the attempts by one organization to perpetrate a false myth can serve as a great motivator for another organization, and to this end the World Jewish Congress discovered and released information on Dr. Strughold that army intelligence had managed to keep classified for nearly fifty years. The *New York Times* verified that Dr. Strughold had been listed as among those being sought on the 1945 Central Registry of War Criminals and Security Suspects, or CROWCASS, list. Strughold's portrait was removed from the Ohio State mural, where he appeared, ironically, alongside Hippocrates.

Citizens who were offended by the Strughold library at the Brooks Air Force Base asked the air force when Strughold's name was going to be removed from the U.S. government building. Library spokesman Larry Farlow told the Associated Press that there were no plans to remove the name from the Strughold Aeromedical Library. But growing pressure from the public forced the air force to reconsider its position. In 1995 General Ronald R. Fogleman, chief of staff of the Department of the Air Force, issued a terse statement stating that after reviewing Nazi-era documents, "the evidence of Dr. Strughold's wartime activities is sufficient to cause concern about retaining his name in an honored place on the library." The sign was removed from the building's exterior brick walls, and the Strughold bronze portrait was taken down. The permanence of the honor had come to an end. Strughold died the following year, at the age of eighty-eight. The Justice Department had been preparing a Nazi war criminal case against Strughold in his final years.

* * *

What does last? The desire to seek the truth? Or, in the words of Jean Michel, the ability to take a stand against "the monstrous distortion of history" when it gives birth to "false, foul and suspect myths"? In 1998, Congress passed the Nazi War Crimes Disclosure Act, which required various U.S. government agencies to identify and release federal records relating to Nazi war criminals that had been kept classified for decades. In accordance with the act, President Clinton established an Interagency Working Group—made up of federal agency representatives and members of the public—to oversee the interpretation of over eight million pages of U.S. government records and report its findings to Congress. The documentation revealed a vast web of profitable relationships between hundreds of Nazi war criminals and U.S. military and intelligence agencies.

In 2005, in a final report to Congress, *U.S. Intelligence and the Nazis,* the Interagency Working Group determined that "[t]he notion that they [the U.S. military and the CIA] employed only a few 'bad apples' will not stand up to the new documentation." In hindsight, wrote the Interagency Working Group, the government's use of Nazis was a very bad idea, and "there was no compelling reason to begin the postwar era with the assistance of some of those associated with the worst crimes of the war." And yet history now shows us that that is exactly what the American government did—and continued to do throughout the Cold War.

In the decades since Operation Paperclip ended, new facts continue to come to light. In 2008, previously unreported information about Otto Ambros emerged, serving as a reminder that the story of what lies hidden behind America's Nazi scientist program is not complete.

A group of medical doctors and researchers in England, working on behalf of an organization called the Thalidomide Trust, believe they have tied the wartime work of IG Farben and Otto Ambros to the thalidomide tragedy of the late 1950s and early 1960s. After Ambros was released from Landsberg Prison, he worked as an economic consultant to German chancellor Konrad Adenauer and to the industrial magnate Friedrich Flick, the richest person in Germany during the Cold War. Like Ambros, Flick had been tried and convicted at Nuremberg, then released early by John J. McCloy.

In the late 1950s, Ambros was also elected chairman of the advisory committee for a German company called Chemie Grünenthal. Grünenthal was about to market a new tranquilizer that promised pregnant women relief from morning sickness. The drug, called thalidomide, was going to be sold under the brand name Contergan. Otto Ambros served on the board of directors of Grünenthal. In the late 1950s, very few people knew that Grünenthal was a safe haven for many Nazis, including Dr. Ernst-Günther Schenck, the inspector of nutrition for the SS, and Dr. Heinz Baumkötter, an SS captain (Hauptsturmführer) and the chief concentration camp doctor in Mauthausen, Natzweiler-Struthof, and Sachsenhausen concentration camps.

Ten months before Grünenthal's public release of thalidomide, the wife of a Grünenthal employee, who took the drug to combat morning sickness, gave birth to a baby without ears. No one linked the birth defect to the drug, and thalidomide was released by the company. After several months on the market, in 1959, Grünenthal received its first reports that thalidomide caused polyneuropathy, or nerve damage, in the hands and feet of elderly people who took the drug. The drug's over-the-counter status was changed so that it now required a prescription. Still, thalidomide was marketed aggressively in forty-six countries with a label that stated it could be "given with complete safety to pregnant women and nursing mothers without any adverse effect on mother and child." Instead, the drug resulted in more than ten thousand mothers giving birth to babies with terrible deformities, creating the most horrific pharmaceutical disaster in the history of modern medicine. Many of the children were born without ears, arms, or legs and with reptilian, flipperlike appendages in place of healthy limbs.

The origins of thalidomide were never accounted for. Grünenthal had always maintained that it lost its documents that showed where and when the first human trials were conducted on the drug. Then, in 2008, the Thalidomide Trust, in England, headed by Dr. Martin Johnson, located a group of Nazi-era documents that produced a link between thalidomide and the drugs researched and developed by IG Farben chemists during the war. Dr. Johnson points out that Grünenthal's 1954 patents for thalidomide cryptically state that human trials had already been completed, but the company says it cannot offer that data because it was lost, ostensibly during the war. "The patents suggest that thalidomide was probably one of a number of products developed at Dyhernfurth or Auschwitz-Monowitz under the leadership of Otto Ambros in the course of nerve gas research," Dr. Johnson says.

The Thalidomide Trust also links Paperclip scientist Richard Kuhn to the medical tragedy. "Kuhn worked with a wide range of chemicals in his nerve gas research, and in his antidote research we know he used Antergan, which we are fairly sure was a 'sister drug' to Contergan," the brand name for thalidomide, Dr. Johnson explains.

In 2005, Kuhn experienced a posthumous fall from grace when the Society of German Chemists (Gesellschaft Deutscher Chemiker, GDCh) announced it would no longer award its once-prestigious Richard

Kuhn Medal in his name. Nazi-era documents on Kuhn had been brought to the society's attention, revealing that in "the spring of 1943 Kuhn asked the secretary-general of the KWS [Kaiser Wilhelm Society], Ernst Telschow, to support his search for the brains of 'young and healthy men,' presumably for nerve gas research." The Society of German Chemists maintains that "the sources indicate that these brains were most likely taken from execution victims," and that "[d]espite his scientific achievements, [Richard] Kuhn is not suitable to serve as a role model, and eponym for an important award, mainly due to his research on poison gas, but also due to his conduct towards Jewish colleagues."

It seems that the legacy of Hitler's chemists has yet to be fully unveiled. Because so many of these German scientists were seen as assets to the U.S. Army Chemical Corps' nerve agent programs, and were thus wanted as participants in Operation Paperclip, secret deals were made, and the many documents pertaining to these arrangements were classified. President Clinton's Interagency Working Group had access to eight million pages of declassified documents, but millions more documents remain classified. In *U.S. Intelligence and the Nazis,* the Interagency Working Group's authors write that "the truest reckoning with the official past can never be complete without the full release of government records."

Part of the problem lies in identifying where records are physically located. For example, a 2012 FOIA request to the State Department, asking for the release of all files related to Otto Ambros, was denied on the grounds that no such files exist. But it is a matter of record, owing to a May 1971 news article in the *Jewish Telegraphic Agency,* that Otto Ambros traveled to the United States twice with the State Department's assistance, despite his status as a convicted war criminal. In an interview with State Department official Fred Scott, the *Jewish Telegraphic Agency* learned, "Ambros came to the United States in 1967 after the State Department recommended to the Justice Department a waiver on his eligibility, which was granted," and that in 1969, Ambros received a second visa waiver and traveled to the United States again. In the spring of 1971 Ambros was attempting to get a third visa waiver from the State Department when the *Jewish Telegraphic Agency* reported the story. According to Fred Scott, Ambros's host for his May 1971 visit was listed as the Dow Chemical Company. After the story was published, Jewish groups held protests, and Ambros allegedly canceled his trip. But none of this information is contained in Ambros's declassified U.S. Army files, FBI files, or CIA files. Otto Ambros was a convicted Nazi war criminal. In accordance with the Nazi War Crimes Disclosure Act, all files about him should have been released and declassified. But records that cannot be located cannot be declassified. Where are the Otto Ambros records hidden? And what secrets might be guarded therein?

Names and dates continue to come to light, and researchers, journalists, and historians continue to uncover new facts. Fate and circumstance also inevitably play a part. In 2010 a cache of almost three hundred documents was found in the attic of a house being renovated in the Polish town of O'swie‚cim, near Auschwitz. The documents include information about several Nazi doctors and Farben chemists who worked at the death camp. "The sensational value of this discovery is the fact that these original documents, bearing the names of the main murderers from Auschwitz, were found so many years after the war," says Adam Cyra, a historian at the Auschwitz museum.

Otto Ambros lived until 1990, to the age of ninety-two. After his death, the chemical conglomerate BASF, on whose board of directors he had served, lauded him as "an expressive entrepreneurial figure of great charisma." Is the old German proverb really true? *Jedem das Seine.* Does everyone get what he deserves?

Still, as of 2013, the Space Medicine Association in America continues to bestow its prestigious Hubertus Strughold Award to a scientist or specialist for outstanding contribution to aviation medicine. It has done so every year since 1963. On December 1, 2012, the *Wall Street Journal* ran a page-one story about Dr. Strughold, presenting revelatory new information about his criminal activities during the war. German historian Prof. Hans-Walter Schmuhl had been researching another subject when he came across evidence that showed that Dr. Strughold had allowed epileptic children to be experimented on inside the high-altitude chamber at Strughold's Aviation Medical Research Institute of the Reich Air Ministry in Berlin. Rabbits had been put to the test first and had died. Next, Reich medical researchers wanted to see what would happen to young children with epilepsy subjected to those same conditions. Strughold authorized the potentially lethal tests on the children. "The head of the Institute is responsible," says Schmuhl; "using this expensive equipment, the head of an institute had to have been informed about the use." When the German Society for Air and Space Medicine learned about Schmuhl's discovery, they eliminated their prestigious Strughold Award, which had been given annually in Germany since the mid-1970s.

In America, the *Wall Street Journal* article renewed debate as to why the Space Medicine Association had not yet eliminated its Hubertus Strughold Award. Dr. Mark Campbell, a former president of the Space Medicine Association, insists the award will not go away. Campbell blames the Internet for maligning what he sees as Dr. Strughold's good character. "I was a member of a committee investigating Dr. Strughold to see if his name should be removed from the Space Medicine Association Strughold Award," says Campbell. "I was amazed to find that the facts that were uncovered were so different from the claims being made on the Internet." But most of Dr. Campbell's colleagues disagree. "Why defend him?" asks Dr. Stephen Véronneau, a research medical officer at the FAA's Civil Aerospace Medical Institute in Oklahoma City, and a member of the Space Medicine Association. "I can't find another example in the world of [an institution] honoring Dr. Strughold except my own association."

The National Space Club Florida Committee, one of three committees of the National Space Club in Washington, D.C., gives out a similarly prestigious space-related award called the Dr. Kurt H. Debus Award. This annual award is named in honor of Operation Paperclip's Kurt Debus, who became the first director of the Kennedy Space Center. Kurt Debus is the scientist who, during the war, was an enthusiastic member of the SS, wore the SS uniform to work, and turned a colleague over to the Gestapo for making anti-Nazi remarks and failing to give Debus the Nazi salute. Under Operation Paperclip, Kurt Debus worked on missiles for the army and for NASA for a total of twenty-eight years—many of which he spent alongside Arthur Rudolph and Wernher von Braun.

Kurt Debus retired in 1974. In 2013, after the Strughold award debate resurfaced, I interviewed Steve Griffin, the National Space Club chairman, to determine why the organization continues to give out an award that is named after someone who was once an avowed and active Nazi.

"Simple as it is, Kurt Debus is an honored American," Griffin says. I read to Griffin information from Kurt Debus's OMGUS security report. "It is a simple matter," Griffin told me. "Kurt Debus was the first director of the Kennedy Space Center."

Unlike the Strughold award, which was created in 1963 when Strughold's foreign scientist case file was still classified, the Dr. Kurt H. Debus Award was first bestowed in 1990, after Debus's OMGUS security report had been declassified and the revelation that he was an active member of Heinrich Himmler's SS had been revealed. "It is not my purview to decide if we have an award or what it is called," Griffin says. But Griffin conceded that he has been on the board of the National Space Club since the Debus award's inception, so technically it is, and always has been, within his purview. Like so many of those involved in Operation Paperclip decades ago, Griffin looks past Debus's former commitment to Nazi Party ideology. He only sees the scientist.

"What do you say when people ask you about Kurt

Debus's Nazi past?" I asked. "Not a single person has asked me this question in [twenty-three] years," Griffin said.

To report this book, I filed dozens of Freedom of Information Act (FOIA) requests, some of which were honored, many of which were denied, and most of which are still pending. I came across oblique references, circa 1945, regarding a supposed list of Nazi doctors whom the Office of U.S. Chief of Counsel, U.S. Army, sought for involvement in "mercy killings," or medical murder crimes. FOIA requests for the list turned up nothing. Then, at the Harvard Medical Library, I found a collection of papers that once belonged to Colonel Robert J. Benford, the first commander of Operation Paperclip's aviation medicine research program at the Army Air Forces Aero Medical Center in Heidelberg. Benford, with Dr. Strughold, oversaw the work of the fifty-eight Nazi doctors at the center; both Benford and Strughold worked under Colonel Harry Armstrong. In Benford's papers I came across a file labeled "List of Personnel Involved in Medical Research and Mercy Killings," but the access to the file was "Restricted Until 2015." The Harvard Medical Library informed me that the Department of Defense had classified the list and that only the DoD had the authority to declassify it.

Harvard filed a FOIA request on my behalf, and the "mercy killings" list was declassified and released to me.

Included on this list, which had been in Colonel Benford's possession, were seven Nazi doctors hired under Operation Paperclip: Theodor Benzinger, Kurt Blome, Konrad Schäfer, Walter Schreiber, Hermann Becker-Freyseng, Siegfried Ruff, and Oskar Schröder. The fact became instantly clear: U.S. Army intelligence knew all along that these doctors were implicated in murder yet chose to classify the list and hire the doctors for Operation Paperclip. Blome, Schäfer, Becker-Freyseng, Ruff, and Schröder were all tried at Nuremberg. Schreiber's public outing in 1951 and his subsequent banishment from America are now on record. Dr. Theodor Benzinger seems to have slipped away from accountability.

From his *New York Times* obituary in 1999, the world learned that Dr. Theodor Benzinger, 94, invented the ear thermometer, a nominal contribution to the medical world. As for the military world, Benzinger's research work for the navy was destroyed or remains classified as of 2013. Wernher von Braun, Arthur Rudolph, Kurt Debus, and Hubertus Strughold led the American effort to get man to the moon. The question remains, despite a man's contribution to a nation or a people, how do we interpret a fundamental wrong? Is the American government at fault equally for fostering myths about its Paperclip scientists—for encouraging them to whitewash their past so that their scientific acumen could be exploited for U.S. weapons-related work? When, for a nation, should the end justify the means? These are questions that can only be answered separately, by individuals. But as facts emerge and history is clarified, the answers become more suitably informed.

In addition to the ear thermometer, Theodor Benzinger left the world with the Planck-Benzinger equation, fine-tuning the second law of thermodynamics, which states that nothing lasts. Benzinger's lifelong scientific pursuit was studying entropy—the idea that chaos rules the world and, like ice melting in a warm room, order leads to disorder.

I prefer Gerhard Maschkowski's take on what matters and what lasts. Maschkowski was the Jewish teenager fortuitously spared the gas chamber at Auschwitz because he was of use to IG Farben as a slave laborer at their Buna factory. I was interviewing Maschkowski one spring afternoon in 2012 when I asked him the question, "What matters, what lasts?" He chuckled and smiled. He pushed back the sleeve on his shirt and showed me his blue-ink Auschwitz tattoo. "This lasts," he said. "But it is also a record of [the] truth."

ACKNOWLEDGMENTS

The idea for *Operation Paperclip* first took hold while I was reading documents about two Nazi aircraft designers, Walter Horten and Reimar Horten, both of whom play a role in my previous book, *Area 51: An Uncensored History of America's Top Secret Military Base.* In this research, I occasionally came across the name Siegfried Knemeyer, a senior adviser to the Horten brothers but, more notably, a man with quite a title from World War II: technical adviser to Reichsmarschall Hermann Göring. Not knowing much about Operation Paperclip back then, I was very surprised when I learned that just a few years after the war's end, Siegfried Knemeyer was living in America working for the U.S. Air Force, and that he would eventually be awarded the U.S. Department of Defense Distinguished Civilian Service Award — the highest civilian award given by the DoD. In 2010, I located Siegfried Knemeyer's grandson, Dirk Knemeyer, and asked if he would meet with me for an interview. He agreed.

Dirk Knemeyer is an American father, husband, entrepreneur, and businessman. By his own admission, he spends a good deal of his time thinking philosophically about concepts of "identity," "science," "time," and "life." When I first met him, Dirk Knemeyer shared with me that he had never before spoken publicly about his grandfather's wartime activities, and that most members of his family go out of their way to maintain silence about Siegfried Knemeyer's high ranking in the Nazi Party. "For the most part the family's position is: Do not discuss Siegfried's [past] with anyone, particularly not a journalist," Dirk Knemeyer says. "But what do you do when you have documents in your attic, as I do, praising your grandfather and his excellent work, and signed by people like Hermann Göring and Albert Speer?"

I find no easy answer to this question. The conundrum therein is one of the notions that made me want to learn much more about Operation Paperclip.

"Siegfried's life is complicated," Dirk Knemeyer says of his grandfather. "I am someone [who is] more interested in learning the truth about the past than denying it. Besides, some things are unwise to ignore."

I asked Knemeyer if he would share with me some of his grandfather's documents that were stored in boxes in his own attic. He said he would think about it. Eventually he agreed. Thank you, Dirk Knemeyer.

I wish to thank Dr. Götz Blome, Gabriella

Hoffmann, Paul-Hermann Schieber, and Rolf Benzinger for their time and their transparency. The indomitable John Dolibois, whom I admire: Thank you for taking the time with me on so many different points. Dr. Jens Westemeier assisted me with all things German in this book and located some very difficult to find documents in numerous German archives; thank you, Jens. I thank the author and historian Clarence Lasby for sharing his insights with me. Lasby first began his book *Project Paperclip: German Scientists and the Cold War* (1971) in the 1960s, as an extension of his college thesis—decades before the truth about the German scientists' Nazi past was revealed under the Freedom of Information Act. Lasby's access to documents was seminal. As we have seen, many of these documents have since disappeared from various collections—been destroyed or lost.

In addition to interviews and oral history recordings, the foundations of this book are from military and civilian archives in the United States and Germany. Countless individuals were helpful, the following notably so: Lynn O. Gamma of the Air Force Historical Research Agency, Maxwell Air Force Base; Michael Jenack, INSCOM Freedom of Information/Privacy Act Office; Richard L. Baker and Clifton P. Hyatt, U.S. Army Military History Institute; Werner Renz, Fritz Bauer Institute; Leon Kieres, Instytut Pamieci Narodowej, Poland; Dorothee Becker, Wollheim

Commission, Goethe University; Joerg Kulbe and Regine Heubaum, Mittelbau-Dora Concentration Camp Memorial; Peter Gohle, Bundesarchiv Ludwigsburg; Dr. Matthias Röschner, Deutsches Museum Archive; Christina Wooten, U.S. Air Force; Lanessa Hill, United States Army Garrison, Fort Detrick; Michael Fauser, U.S. Holocaust Memorial Museum; Bert Ulrich and Allard Beutel, NASA.

At the National Archives and Records Administration, I would like to thank David Fort and Amy Schmidt. At Harvard Medical School, at the Center for the History of Medicine at the Francis A. Countway Library of Medicine, Jessica B. Murphy graciously helped me to petition the Harvard Medical School privacy board, which in turn petitioned the Department of Defense to declassify files under the Freedom of Information Act — with success. Many thanks, Jessica. At the Historical and Special Collections at the Harvard Law School Library, I thank Lesley Schoenfeld for her help with The Alexander Papers; Margaret Peachy and David Ackerman for copying historical film footage of Dr. Leopold Alexander and Telford Taylor at the Nuremberg doctors' trial; Adonna Thompson, Duke University Medical Center Archives; David K. Frasier, The Lilly Library Manuscript Collection, Indiana University; Lynda Corey Claassen, Mandeville Special Collections Library, UC–San Diego; John Armstrong, Special Collections and Archives, Wright State University Libraries; Carol A. Leadenham; Hoover Institution Archives; Loma Karklins, the Caltech Archives; Anne Coleman, Archives and Special Collections, M. Louis Salmon Library, University of Alabama in Huntsville; Dr. Martin Johnson, The Thalidomide Trust, England; Jen Stepp at *Stars and Stripes;* Brett Exton, Island Farm; Nick Greene, the *Village Voice;* author Danny Parker, who helped me locate documents at the National Archives and also a very obscure document at the Landeskirchliches Archiv in Stuttgart; Julia Kiefaber, who translated many German trial transcripts and wartime Nazi Party documents; Larry Valero, with the Intelligence and National Security Studies Program at the University of Texas at El Paso; John Greenewald, founder and curator of the Black Vault.

In Germany, Manfred Kopp graciously drove me around Oberursel, taking me to the old Camp King facilities and safe houses that were once used for classified programs, including Operations Bluebird, Artichoke, and MKUltra. Together with Maria Shipley, we journeyed to Schloss Kransberg, formerly the Dustbin Interrogation Center. Jens Hermann was most helpful in showing us around the castle and its grounds. The journalist and author Egmont Koch generously shared his findings on Camp King and the CIA's Artichoke program with me. Thanks to John

Dimel for lending me a rare unpublished copy of *The History of Camp King,* and to investigative journalist Eric Longabardi for sharing his reporting on U.S. Army chemical weapons tests with me. I thank Michael Neufeld for answering questions about Wernher von Braun, Walter Dornberger, and Arthur Rudolph. Neufeld is curator of the Department of Space History at the Smithsonian National Air and Space Museum and the author of several books and papers on German rocket scientists, including monographs for NASA, all of which helped me tremendously.

I thank Albert Knoll, director of the archive and library at the Dachau concentration camp, for making the documents and photographs of medical murder experiments that were carried out at Dachau during the war available for my review, and also for showing me blueprints and maps of Experimental Cell Block Five. Riot police commissioner Mathias Korn and police historian Anna Naab took me around the expansive former SS training area grounds, which are adjacent to the Dachau concentration camp, including areas not open to the public. Thanks to them, I was able to go into the buildings used by the U.S. military to prosecute the Dachau war crimes trials, and to see where Georg Rickhey was tried and acquitted. Dr. Harald Eichinger, the warden of Landsberg Prison, gave me a comprehensive tour of that famous facility (still in use today), where Hitler wrote *Mein Kampf* and where the convicted Nuremberg war criminals were briefly imprisoned—until they were either hanged or granted clemency by U.S. High Commissioner John J. McCloy.

An author is nothing without a team. Thank you, John Parsley, Jim Hornfischer, Steve Younger, Nicole Dewey, Liz Garriga, Heather Fain, Amanda Brown, Malin von Euler-Hogan, Janet Byrne, Mike Noon, Ben Wiseman, and Eric Rayman. Thank you, Alice and Tom Soininen, Kathleen and Geoffrey Silver, Rio and Frank Morse, and Marion Wroldsen. And my fellow writers from group: Kirston Mann, Sabrina Weill, Michelle Fiordaliso, Nicole Lucas Haimes, and Annette Murphy.

The interviews I did with Gerhard Maschkowski I shall never forget. Thank you, Gerhard.

The only thing that makes me happier than finishing a book is the daily joy I get from Kevin, Finley, and Jett. You guys are my best friends.

Principal Characters

These brief character descriptions include information pertaining to the narrative of Operation Paperclip. Military titles refer to the highest rank achieved by individuals within the timeframe of this story. The term "Dr." is used to identify medical doctors.

William J. Aalmans: Post-war investigator with the U.S. War Crimes Division, he arrived on scene at the V-2 tunnel complex in Nordhausen shortly after it was liberated. Found the telephone list that implicated Georg Rickhey and Arthur Rudolph in slave labor and served on the prosecution staff during the Dora-Nordhausen war crimes trial.

Dr. Leopold Alexander: Boston psychiatrist and neurologist sent to post-war Germany to investigate medical crimes. He later served as expert consultant during the Nuremberg doctors' trial and co-authored the Nuremberg Code.

Otto Ambros: IG Farben chemist, codiscover of sarin gas and Buna synthetic rubber, he was awarded one million reichsmarks by Hitler as a scientific achievement award. Served the Reich as chief of the Committee-C for chemical warfare, manager of IG Farben's slave labor factory at Auschwitz and manager of the Dyhernfurth poison gas facility. He was tried and convicted at Nuremberg, and after an early release he worked for the U.S. chemical corporation W. R. Grace, the U.S. Department of Energy, and other European government and private sector business concerns.

Major General Dr. Harry G. Armstrong: Set up the post-war U.S. Army Air Forces Aero Medical Center in Heidelberg, employing fifty-eight Nazi doctors, thirty-four of whom followed him to the U.S. Air Force School of Aviation Medicine in Texas, where he served as commandant. He was the second surgeon general of the U.S. Air Force.

Colonel Burton Andrus: U.S. Army commandant at Central Continental Prisoner of War Enclosure Number 32, code-named "Ashcan," the interrogation facility where the major Nazi war criminals were interned after the war. Later, as governor of the Nuremberg Prison, he oversaw the hangings of the convicted high command Nazis.

Herbert Axster: Nazi lawyer, accountant and chief of staff on V-weapons under General Dornberger. Axster and his wife Ilse were among the few individuals in Operation Paperclip to be outed by the pubic as ardent Nazis. Forced to leave army employ, he opened a law firm in Milwaukee, Wisconsin.

Harold Batchelor: U.S. Army expert in weaponized bubonic plague and codesigner of the Eight Ball aerosol chamber at Camp Detrick. As a member of the Special Operations Division, he and Frank Olson conducted covert field tests across America using a pathogen that simulated how bioweapons would disperse. Consulted with Dr. Blome in Heidelberg after the Nuremberg Doctors' Trial.

Werner Baumbach: Wartime general of the bombers for the Luftwaffe. Chosen by Albert Speer and Heinrich Himmler to pilot each man's escape from Germany. Originally part of Operation Paperclip, he instead went to South America, where he died in a plane crash.

Dr. Hermann Becker-Freyseng: Nazi aviation physiologist under Dr. Strughold who oversaw medical murder experiments on prisoners at the Dachau concentration camp. He was tried at Nuremberg, convicted, and sentenced to twenty years. Granted clemency, he was released in 1952.

Dr. Wilhelm Beiglböck: Nazi aviation physiologist who oversaw salt water experiments at Dachau, he removed a piece of prisoner Karl

Höllenrainer's liver without anesthesia, one of the many crimes for which he was tried and convicted at Nuremberg. Sentenced to fifteen years, Beiglböck was granted clemency in 1951 and returned to Germany to work at a hospital.

Colonel Peter Beasley: Officer with the U.S. Strategic Bombing Survey sent to post-war Germany to locate engineers with knowledge about how underground weapons facilities were engineered. Recruited Georg Rickhey for Operation Paperclip.

Colonel Dr. Robert J. Benford: Commanding officer at the U.S. Army Air Forces Aero Medical Center, he oversaw the research efforts of fifty-eight Nazi doctors working in Heidelberg.

Dr. Theodor Benzinger: Wartime department chief of the Experimental Station of the Air Force Research Center in Rechlin and chief of medical work in the research department of the Technical Division of the Reich Air Ministry. An ardent Nazi and member of the SA (Storm Troopers) with rank of medical sergeant major, he was hired to work at the U.S. Army Air Forces Aero Medical Center in Heidelberg, then arrested, imprisoned at Nuremberg, listed as a defendant in the doctors' trial, and mysteriously released. Under Operation Paperclip he worked for the Naval Medical Research Institute in Bethesda, Maryland.

Dr. Kurt Blome: Deputy surgeon general of the Reich, deputy chief of Reich's Physicians' League, and member of the Reich Research Council, he served as chief of the Reich's bioweapons facilities in Nesselstedt, Poland, and in Geraberg, Germany. An "Old Fighter" Nazi Party member, he wore the Golden Party Badge and was a lieutenant general in the SA (Storm Troopers). The JIOA tried but failed to bring him to America; he worked for the U.S. Army at Camp King, in Oberursel, Germany.

William J. Cromartie: U.S. Army bacteriological warfare expert and officer with Operation Alsos.

Kurt Debus: V-weapons flight test director and member of the SS, he turned a colleague over to the Gestapo for making anti-Nazi remarks.

671

Under Operation Paperclip he served as part of the von Braun rocket team at Fort Bliss, Texas and became the first director of NASA's John F. Kennedy Space Center, in Florida.

John Dolibois: U.S. Army officer with military intelligence who interrogated many of the Nazi Party high command interned at Ashcan, including Hermann Göring and Albert Speer. He was trained by General William Donovan in the art of interrogation.

Major General William J. Donovan: Founding director of the Office of Strategic Services, the forerunner to the CIA. After the German surrender Donovan kept an office at Camp King, in Oberursel, Germany, where he oversaw Nazis writing reports for the U.S. Army.

Major General Walter Dornberger: Wartime German general in charge of V-weapons development and the technical staff officer in the Nordhausen slave labor tunnels. He was arrested by the British for war crimes, interned in England, and later released into U.S. custody. Under Operation Paperclip he worked for the U.S. Air Force and then Bell Aircraft Corporation. Until the late 1950s he acted as a missile and space-based weapons consultant to the Joint Chiefs, carrying a Top Secret clearance and visiting the Pentagon frequently.

Captain R. E. F. Edelsten: British officer assigned to the joint British-U.S. interrogation center at Castle Kransberg, code-named "Dustbin."

Donald W. Falconer: U.S. biological weapons explosives expert who consulted with Dr. Blome shortly after Blome's acquittal at Nuremberg. Colleague of Frank Olson at Camp Detrick.

Karl Otto Fleischer: Allegedly the V-weapons business manager in Nordhausen, he played an important role in revealing to Major Staver the whereabouts of a key document stash after the German surrender. Under Operation Paperclip he worked at Fort Bliss, Texas, until it was learned he was not a scientist but worked in food services for the Reich's missile program.

Dr. Karl Gebhardt: Himmler's personal physician and chief surgeon of the staff of the Reich Physician SS and police, he was in charge of sulfa experiments at Ravensbrück, administrated by Major General Dr. Walter Schreiber. He was convicted at the doctors' trial at Nuremberg, sentenced to death and hanged in the courtyard at Landsberg Prison.

Lieutenant General Reinhard Gehlen: Hitler's senior intelligence officer on the eastern front, he was hired by the U.S. Army to run the Gehlen Organization, a group that gathered intelligence on Soviet-bloc spies at Camp King. The organization was taken over by the CIA in 1949 and the majority of details kept classified until 2001.

Hermann Göring: Reichsmarschall, commander-in-chief of the Luftwaffe, and the long-serving designated successor to Hitler. As head of the Reich Research Council he was in charge of the coordination of all German research and streamlined Nazi science to weapons-related programs, making him the dictator of science.

Sidney Gottlieb: Director of the CIA's Technical Services Staff, he oversaw the MKUltra program and, with his deputy Robert Lashbrook, covertly drugged bacteriologist Frank Olson with LSD during an Agency weekend retreat.

Samuel Goudsmit: American particle physicist and wartime scientific director of Operation Alsos. Born in Holland, Goudsmit spoke Dutch and German. He discovered documents in Nazi scientist Eugen Haagen's Strasbourg apartment that revealed Reich doctors were conducting deadly human experiments on concentration camp prisoners.

John C. Green: Executive secretary for the Office of Publication Board, a division of the Commerce Department. He was instrumental in getting secretary of commerce Henry Wallaces to lobby President Truman to endorse Operation Paperclip.

L. Wilson Greene: Technical director of the Chemical and Radiological Laboratories at Edgewood, his secret monograph, entitled "Psychochemical Warfare: A New Concept of War," was the genesis for the

CIA's MKUltra program. Colleagues with Paperclip chemist Fritz Hoffmann.

Dr. Karl Gross: Biological weapons researcher with the Hygiene Institute of the Waffen-SS, he was assigned by Himmler to work with Dr. Blome at Posen, Poland, and Geraberg, Germany.

Dr. Eugen Haagen: Virologist and key developer in the Nazi bioweapons program, notably vaccine research. Inside Haagen's Strasbourg apartment Alsos agents discovered the first evidence that Nazi doctors were experimenting on humans in concentration camps. Before the war Haagen codeveloped the yellow fever vaccine at the Rockefeller Institute in New York.

Alexander G. Hardy: Nuremberg prosecutor during the doctors' trial. In 1951, Hardy and Dr. Alexander wrote to President Truman portraying U.S. Air Force Paperclip contract employee Dr. Walter Schreiber as a war criminal, sadist, and liar, leading to Schreiber's expulsion from the U.S.

Major James P. Hamill: Wartime officer with the U.S. Army Ordnance Corps, he was in charge of the Paperclip group at Fort Bliss, Texas.

Heinrich Himmler: Reichsführer-SS, chief of police, and Reich commissar for the consolidation of the ethnic German nation, he championed medical experiments on human beings in concentration camps and oversaw the "sale" of more than half a million slave laborers to military and industrial concerns. Himmler's all-inclusive powers gave him unparalleled responsibility for Nazi terror and atrocities.

Karl Höllenrainer: German Roma, or gypsy, arrested during the war for marrying a German woman in violation of the Nuremberg Laws. He was sent to Auschwitz, then Buchenwald, and finally Dachau, where he became a rare survivor of medical murder experiments in Experimental Cell Block Five. During Nuremberg trial testimony, he tried to stab defendant Wilhelm Beiglböck.

Dr. Ernst Holzlöhner: Senior doctor at the University of Berlin who conducted freezing experiments at Dachau with Dr. Sigmund Rascher. He committed suicide in May 1945.

Friedrich "Fritz" Hoffmann: Wartime organic chemist at the chemical warfare laboratories at the University of Würzburg, and for the Luftwaffe. Under Operation Paperclip, he worked at Edgewood in the classified research and development division, the Technical Command, synthesizing tabun gas and later VX. For the CIA, he traveled the world in search of exotic poisons.

Dieter Huzel: Wartime personal aide to Wernher von Braun, he oversaw the V-weapons document stash with Bernhard Tessmann. Under Operation Paperclip, he worked as part of the von Braun team at Fort Bliss, Texas.

Colonel Gordon D. Ingraham: Commander of Camp King from 1949 to 1951, he supervised Paperclip contract employees Dr. Walter Schreiber and Dr. Kurt Blome.

Janina Iwanska: Rare survivor of the Ravensbrück concentration camp medical murder experiments. She came to the U.S. for medical treatment, in 1951, and unexpectedly provided critical testimony to the FBI regarding the alleged war crimes of Operation Paperclip's Dr. Walter Schreiber.

John Risen Jones Jr.: American soldier with the 104th Infantry Division and one of the first men to enter the slave tunnels at Nordhausen. His iconic photographs documented the horror that had befallen thousands of V-weapon laborers.

Dr. Heinrich Kliewe: Reich chief of counterintelligence for bacterial warfare concerns, he was interned and interrogated at Dustbin with Dr. Blome. Testified as a witness in the Nuremberg doctors' trial.

Colonel Siegfried Knemeyer: Luftwaffe spy pilot, engineer, and wartime chief of Luftwaffe technical developments under Hermann Göring. Considered one of the Reich's top ten pilots, he was asked by armaments minister Albert Speer to pilot Speer's escape to Greenland. Under Operation Paperclip, he worked at Wright Field from 1947 until 1977.

Major General Hugh Knerr: Post-war commanding general at Air Technical Service Command, Wright Field. An early advocate of Operation Paperclip, he sent a memo to the War Department encouraging them to overlook German scientists' Nazi pasts. "Pride and face saving have no place in national insurance," he said.

Karl Krauch: Chairman of IG Farben board of directors and Göring's Plenipotentiary for Special Questions of Chemical Production. Courted for Operation Paperclip while incarcerated at Nuremberg, he was convicted alongside colleague Otto Ambros.

Richard Kuhn: Nobel Prize–winning organic chemist who developed soman nerve agent for the Reich and was known to begin his classes at the Kaiser Wilhelm Institute with "Sieg Heil." Under Operation Paperclip he worked for the U.S. Army Air Forces Aero Medical Center in Heidelberg, and also privately for General Loucks's Heidelberg working group on sarin production.

Brigadier General Charles E. Loucks: Long-serving U.S. Army chemical warfare officer, he oversaw Operation Paperclip scientists working at Edgewood. He was transferred to Heidelberg, Germany, in June 1948, served as chief of intelligence collection for Chemical Warfare Plans, European Command, and created the Heidelberg working group on sarin production with Hitler's former chemical weapons experts, including Schieber, Schrader, Kuhn, and von Klenck. Initiated the first U.S. Army interest in the incapacitating agent Lysergic Acid Diethylamide, or LSD.

Dr. Ulrich Luft: Luftwaffe respiratory specialist and Dr. Strughold's deputy director at the Aviation Medical Research Institute in Berlin. Under Operation Paperclip he worked for the U.S. Army Air Forces Aero Medical Center, in Heidelberg, and for the U.S. Air Force School of Aviation Medicine, in Texas.

Gerhard Maschkowski: Auschwitz death camp prisoner No. 117028, he survived the Buna-Monowitz labor-concentration camp, also called IG Auschwitz, and was nineteen years old at liberation.

John J. McCloy: Lawyer, banker, politician, and presidential advisor, as chairman of the State-War-Navy Coordinating Committee, he played a crucial role in the earliest days of the Nazi scientist program. As U.S. high commissioner in Germany, he championed Accelerated Paperclip, also known as Project 63, and granted clemency to many Nazi war criminals convicted at Nuremberg.

Charles McPherson: Member of the Special Projects Team for Accelerated Paperclip, McPherson recruited Dr. Kurt Blome to work on biological weapons research.

Hermann Nehlsen: Sixty-three-year-old German aircraft engineer working under Operation Paperclip at Wright Field, he turned in Georg Rickhey for war crimes.

Carl Nordstrom: Chief of the Scientific Research Division under U.S. High Commissioner McCloy, he oversaw the Special Projects Team in their recruitment of Nazi scientists under the Accelerated Paperclip program.

Frank Olson: Bacteriologist for the Special Operations Division at Fort Detrick and CIA operative involved in the controversial interrogation programs at Camp King, including Operations Bluebird and Artichoke. He was covertly drugged with LSD by CIA colleagues, and later fell, or was pushed, out of a New York City hotel room to his death.

Werner Osenberg: Wartime high-ranking member of the Gestapo and an engineer, he ran the Planning Office inside Goring's Reich Research Council and was the eponymous creator of the Osenberg List, a "Who's Who" record of over 15,000 Reich scientists, engineers, and doctors.

Colonel Boris Pash: Commanding officer of the Alsos Mission and later an employee of the CIA.

Albert Patin: Reich businessman whose wartime factories mass-produced aircraft instruments using a 6,000-person workforce that included slave laborers supplied by Himmler. Under Operation Paperclip he worked for the U.S. Army at Wright Field.

Colonel William R. Philp: First commander of Camp King in Oberursel, Germany. He was one of the first post-war U.S. Army officers to hire Nazi military intelligence officers to analyze information from Soviet prisoners, an action that metastasized into the Gehlen Organization.

Lieutenant General Donald L. Putt: Accomplished test pilot and engineer. He was one of the first wartime officers to arrive at Hermann Göring's secret aeronautical research center at Völkenrode, where he recruited dozens of scientists for Operation Paperclip, later supervising them at Wright Field.

Dr. Sigmund Rascher: SS doctor at Dachau who conducted medical murder experiments at Experimental Cell Block Five. His correspondence with Himmler, including a gruesome collection of photographs, was used in the Nuremberg doctors' trial. He was allegedly murdered on Himmler's orders shortly before war's end and became the scapegoat of many Luftwaffe doctors.

Georg Rickhey: General manager of the Mittelwerk slave labor facility in Nordhausen. Under Operation Paperclip he worked for the U.S. Strategic Bombing Survey and the U.S. Army Air Forces at Wright Field. Exposed by a fellow Paperclip scientist for alleged war crimes, he was returned to Germany as a defendant in the Dora-Nordhausen labor-concentration camp trial and was acquitted.

Walther Riedel: Engineer with the V-weapons design bureau. Under Operation Paperclip he worked for the U.S. Army at Fort Bliss, Texas, before returning to Germany.

Howard Percy "H. P." Robertson: Physicist, collaborator of Albert Einstein, and ordnance expert who served as an officer with Operation

Alsos. He served as General Eisenhower's post-war chief of the Scientific Intelligence Advisory Section and chief of Field Information Agency, Technical (FIAT). He was vocally opposed to hiring Nazi scientists.

Arthur Rudolph: Operations director at Mittelwerk slave labor facility in Nordhausen, he specialized in V-weapons assembly and oversaw slave laborer allocation. Under Operation Paperclip he worked for the U.S. Army at Fort Bliss, Texas, and became project manager for the Saturn V. In 1980 he was investigated by the Department of Justice and left America in 1984 to avoid prosecution.

Dr. Siegfried Ruff: Director of the Aero Medical Division of the German Experimental Station for Aviation Medicine in Berlin, close colleague and coauthor of Dr. Strughold. As supervisor to Dr. Sigmund Rascher he administrated the medical experiments in Experimental Cell Block Five at Dachau. Under Operation Paperclip he worked for the U.S. Army Air Forces Aero Medical Center, in Heidelberg, before being tried at Nuremberg and acquitted.

Emil Salmon: Nazi aircraft engineer implicated in the burning down of a synagogue in Ludwigshafen, Germany. He was hired under Operation Paperclip to build engine test stands for the U.S. Army Air Forces, who found his expertise "difficult, if not impossible, to duplicate."

Dr. Konrad Schäfer: Luftwaffe physiologist and chemist, he developed the Schäfer Process of desalination in pilot sea emergencies, which became part of the medical murder experiments at Dachau. Under Operation Paperclip he worked for the U.S. Army Air Forces Aero Medical Center, in Heidelberg, before being tried in the Nuremberg doctors' trial and acquitted. His second Paperclip contract was for the U.S. Air Force School of Aviation Medicine, in Texas.

SS-Brigadeführer Walter Schieber: Chief of the Armaments Supply Office in the Speer Ministry, he acted as liaison to the industrial production of tabun and sarin gas. One of Hitler's Old Fighters and holder of the Golden Party Badge, First Class, he served on Reichsführer-SS Himmler's personal staff. Under Operation Paperclip he worked for the

U.S. Army, in Heidelberg, for General Loucks's Heidelberg working group on sarin production, and for the CIA.

Heinz Schlicke: Electronic warfare expert and director of the Reich's Naval Test Fields at Kiel, he was on board the German submarine U-234, headed for Japan with a weapons cache, when captured. Under Operation Paperclip he worked for the U.S. Navy.

Hermann Schmitz: CEO of IG Farben and director of the Deutsches Reichsbank, it was inside a secret wall safe in Schmitz's Heidelberg home that Major Edmund Tilley discovered the photo album connecting IG Farben to IG Auschwitz.

Gerhard Schrader: IG Farben chemist who discovered tabun nerve agent for the Reich. He repeatedly turned down Operation Paperclip contract offers but worked privately for General Loucks with the Heidelberg working group on sarin production.

Major General Dr. Walter Schreiber: Surgeon general of the Third Reich, wartime chief of medical services, Supreme Command of the German Army, member of the Reich Research Council, and chief of protection against gas and bacteriological warfare. Captured by the Soviets during the Battle for Berlin, he was a surprise witness for the Russian prosecution team at Nuremberg. Under Operation Paperclip he worked for the U.S. Army, at Camp King, Oberursel, and for the U.S. Air Force School of Aviation Medicine, in Texas.

Dr. Oskar Schröder: Chief of staff of the Luftwaffe Medical Corps, he ordered and oversaw medical murder experiments at Dachau. Under Operation Paperclip he worked for the U.S. Army Air Forces Aero Medical Center, in Heidelberg, before being tried and convicted at Nuremberg. Sentenced to life in prison, he was granted clemency and released 1954.

Robert Servatius: Nazi defense counsel during the Nuremberg doctor's trial, he located a 1945 *Life* magazine story describing how U.S. military doctors experimented on U.S. prisoners during the war and read the article word-for-word in the courtroom, seriously damaging the prosecution's case.

Major Eugene Smith: U.S. Army officer assigned to investigate Georg Rickhey on war crimes accusations. He took statements from Arthur Rudolph at Fort Bliss, Texas, which later proved instrumental for the Department of Justice.

Albert Speer: Reich minister of armaments and war production, he was responsible for all warfare-related science and technology for the Reich, starting in 1942, and oversaw the placement of millions of people into labor-concentration camps. Convicted at Nuremberg, he served twenty years in Spandau Prison. After his release he wrote several memoirs, always maintaining he knew nothing of the Holocaust.

Vivien Spitz: Youngest court reporter at the Nuremberg doctors' trial, she authored *Doctors from Hell* at the age of eighty, after being outraged by a speech given by a Holocaust denier.

Major Robert B. Staver: Officer with the research and intelligence branch of U.S. Army Ordnance, he oversaw Special Mission V-2 and the capture of 100 V-weapons and related documents from Nordhausen. His actions led to the first group of rocket scientists coming to America under Operation Paperclip. He left the Army in December 1945.

Dr. Hubertus Strughold: Wartime director of the Aviation Medical Research Institute of the Reich Air Ministry in Berlin for ten years of Hitler's twelve-year rule. Despite being listed on the Central Registry of War Criminals and Security Suspects, CROWCASS, he was recruited by Harry Armstrong to codirect the Top Secret medical research program in post-war Germany. Under Operation Paperclip he worked for the U.S. Army Air Forces Aero Medical Center, in Heidelberg, and the U.S. Air Force School of Aviation Medicine, in Texas. Referred to as the Father of Space Medicine, he remains one of the most controversial figures in the history of Operation Paperclip.

Colonel Philip R. Tarr: Wartime chief officer of the Intelligence Division of the U.S. Chemical Warfare Service, Europe. Stationed at Dustbin interrogation center, he defied orders to arrest Otto Ambros on war

crimes charges and drove him to Heidelberg to meet with Chemical
Corps intelligence officers and a civilian with Dow Chemical.

Major Edmund Tilley: British officer in charge of post-war interrogations of Hitler's chemical weapons experts interned at Dustbin. Fluent
in German, he was responsible for locating evidence used to convict
Otto Ambros and other Farben chemists at Nuremberg.

General Telford Taylor: Chief prosecutor at Nuremberg. During the
doctors' trial, he described Hitler's doctors as having become proficient
in the "macabre science" of killing. He was a vocal critic of U.S. High
Commissioner McCloy's decision to pardon convicted Nazis and over-
turn ten of the tribunal's death sentences.

Bernhard Tessmann: Weapons facilities designer at Nordhausen. With
Huzel, and at the direction of von Braun, he hid the V-weapons documents in the Dörnten mine. Under Operation Paperclip he worked for
the U.S. Army at Fort Bliss, Texas.

Erich Traub: Virologist, microbiologist, and doctor of veterinary medicine, he served as the wartime deputy director of the National Research
Institute on the island of Riems, Germany. Sent by Himmler to Turkey
in search of rinderpest, he sought to weaponize the virus for the Reich.
Under Operation Paperclip he worked for the U.S. Army, the U.S. Navy,
and the U.S. Department of Agriculture. He asked to be repatriated to
Germany, in 1953.

Wernher von Braun: Technical director of V-weapons development for
the German Army and head of the Mittelbau-Dora Planning Office, a
division within the SS. Under Operation Paperclip he worked for the
U.S. Army, Fort Bliss, Texas, and became director of the Marshall Space
Flight Center and chief architect of the Saturn V launch vehicle, which
propelled Americans to the moon.

Magnus von Braun: Gyroscope engineer at Nordhausen and the
younger brother of Wernher von Braun. Under Operation Paperclip he
worked for the U.S. Army, at Fort Bliss, Texas.

Robert Ritter von Greim: WWI flying ace who mentored Dr. Strughold in pilot physiology studies between world wars. At war's end, he was appointed by Hitler to serve as the last chief of the Luftwaffe and committed suicide in May 1945.

Jürgen von Klenck: SS officer, IG Farben chemist, and deputy chief of special Committee-C for chemical warfare under Ambros, he inadvertently led Major Tilley to a document cache that led to the arrest and conviction of many Nazi colleagues. Under Operation Paperclip he worked for General Loucks's Heidelberg working group on sarin production.

Herbert Wagner: Chief armaments design engineer at Henschel Aircraft Company and inventor of the HS-293 missile. He was the first Nazi scientist to arrive in the U.S. under Operation Paperclip and worked for U.S. Naval Technical Intelligence.

Notes

Prologue

p. ix "I'm mad on technology": Hitler, *Table Talk,* 308.

p. x "These men are enemies": Maxwell AFB History Office document. Memo, Patterson to Secy, General Staff, May 28, 1945, Subj: "German Scientists."

p. xii the JIC warned the Joint Chiefs of Staff: JIC estimate, JIC 250 (Winter 1945), said that the Soviets would avoid war with the United States until 1952. JIC estimate, JCS 1696 (Summer 1946), said that war with the Soviets would be "total war."

p. xii the surgeon general: RG 319 Walter Schreiber, Schuster File, August 18, 1949; HLSL Item No. 286. The Harvard Law School Library owns approximately one million pages of documents relating to the trial of the major war criminals before the International Military Tribunal (IMT) and to the subsequent twelve trials of other accused war criminals before the United States Nuremberg Military Tribunals (NMT). As part of Harvard's Nuremberg Trial Project, many of these documents are now available online. When such a digital document exists, I list it hereafter by its HLSL item number.

p. xii sarin gas: Sarin, tabun, soman, and VX are all organophosphorus nerve agents—liquids, not gases (under most

685

combat conditions); some chemical weapons experts use the term "nerve agent," not "nerve gas."

Chapter One: The War and the Weapons

p. 3 difficult time: Pash, *The Alsos Mission*, 147–51.

p. 3 Air battles raged overhead: Goudsmit, 68.

p. 3 soldiers guarded: Ibid., 73.

p. 4 documents in the cabinets: Pash, *The Alsos Mission*, 154.

p. 4 Bill Cromartie and Fred Wardenberg: Ibid., 157.

p. 5 Reich Research Council: HLSL Item No. 2076; *Deichmann*, 4, 90–93. Originally established in 1937, the Reich Research Council was designed to focus Germany's individual science branches on a war economy — to be in sync with Hitler's Four-Year Plan, of which Hermann Göring was plenipotentiary (minister). The original president of the Reich Research Council was General Karl Becker, an engineer. On June 9, 1942, by Führer decree, Hermann Göring was made chairman of the Reich Research Council, with all research to focus on warfare.

p. 6 members of the Nazi Party: Goudsmit, 73.

p. 6 set up camp in Professor Haagen's apartment: RG 165 "Code Name: Alsos Mission, Scientific Intelligence Mission, M.I.S., G-2, W.D.," photographic history compiled by Colonel Boris T. Pash, 10.

p. 6 K-rations on the dining room table: Ibid., 11.

p. 6 "usual gossip": Goudsmit, 69.

p. 7 "Heil Hitler, Prof. Dr. E. Haagen": Goudsmit, 74–75; Pash, *The Alsos Mission*, 150.

p. 7 The document: Goudsmit, 74.

p. 8 troubling reality: Goudsmit, 79; Pash, *The Alsos Mission*, 149–52.

p. 9 The sword and the shield: HLSL Item No. 1614; HLSL Item No. 1615.

p. 9 readied for a celebration: Neufeld, *Von Braun*, 187–88. There are conflicting reports about where this ceremony may have taken place (Schlossberg?), as discussed with Michael Neufeld on April 3, 2013.

p. 9 china place setting: Photographs, Dornberger files, Deutsches Museum, Munich.

p. 10 *Christian Science Monitor* reported: April 28, 1945, http://www.nasm.si.edu/events/spaceage/vengeance.htm.

p. 10 "hitherto unknown, unique weapons": Evans, 674. The quote is from Adolf Hitler, *Reden und Proklamationen, 1932–1945* (Speeches and Proclamations, 1932–1945).

p. 11 rocket programs: The German army called the V-2 the A-4; Neufeld, *The Rocket and the Reich*, 1.

p. 11 "Around the castle in the dark forest": Neufeld, *Von Braun*, 188.

p. 13 wunderkind-physicist: Von Braun was a poor student. His excellence in physics took off after 1929, when he read *Die Rakete zu den Planetenräumen* (By Rocket into Interplanetary Space), by Hermann Oberth.

p. 13 Knight's Cross of the War Service Cross: Neufeld, *Von Braun*, 187. Neufeld explains that this award is "a very high noncombatant decoration equivalent to the Knight's Crosses awarded for valor" and illustrates von Braun's high place in the Nazi hierarchy in the final months.

p. 13 receiving the honor: Photographs, Dornberger files, Deutsches Museum, Munich.

p. 13 Albert Speer was now responsible: Evans, 324.

p. 13 "Total productivity in armaments": Ibid., 327.

p. 13 science and technology programs: Cornwell, 313.

p. 14 "sounded thoroughly utopian": Speer, *Reich*, 366.

p. 14 "excitement of the evening": Dornberger files, Deutsches Museum, Munich; Neufeld, *Von Braun*, 188.

p. 15 flutes of champagne: Photographs, Dornberger files, Deutsches Museum, Munich.

p. 15 highest death toll: See www.V2Rocket.com. This website contains an eyewitness account of the incident by Charles Ostyn under "The Rex Cinema."

p. 16 "It seemed likely that": McGovern, 62.

p. 16 Trichel was putting together: NASA History Office, 194.

p. 16 The British had the lead: McGovern, 101.

p. 17 the Middle Works: It is sometimes also referred to as the Central Works.

p. 18 move rocket production underground: Himmler wanted to take charge of the entire V-weapons program, displacing Albert Speer, but this proved impossible considering the fact that Hitler saw Speer as the miracle worker of armaments production (Evans, 320–25).

p. 18 Hans Kammler: Neufeld, *The Rocket and the Reich*, 201.

p. 18 *Jedem das Seine:* Derives from the Latin phrase *Suum cuique,* "to each his own" or "to each what he deserves." The phrase is also attributed to Cicero, who wrote, "Iustitia suum cuique distribuit." In Nazi Germany these slogans were transformed into propaganda phrases and were hung over the main entrance gate to many concentration camps. *Jedem das Seine* was over Buchenwald. *Arbeit macht frei,* "Work makes you free," was over Auschwitz.

p. 19 high school graduate: RG 330 Arthur Rudolph, JIOA Form No. 2., Basic Personnel Record. The extent of Rudolph's higher education, listed on his personnel record, was "toolmaking and machine construction" — courses he'd taken at vocational school.

p. 19 Rudolph worked with the SS construction staff: Neufeld, *The Rocket and the Reich,* 206.

p. 19 The prisoners worked: Aalmans, 10. I use the monograph Aalmans wrote for the U.S. War Crimes commission as my primary source for Dora-Nordhausen facts unless otherwise noted. The statistics vary dramatically in secondary works.

p. 19 read one report: Aalmans, 10.

p. 20 workers suffered and died: Ibid., 11.

p. 20 dead were replaceable: The Mittelbau-Dora Memorial has many details online: http://www.buchenwald.de/en/338/.

p. 20 approximately half of the sixty thousand: This is Aalmans's figure. The memorial puts the death rate at two out of three.

p. 20 innocuous-sounding division: Allen, 2. The German term is Wirtschaftsverwaltungshauptamt.

p. 20 This office was overseen by Heinrich Himmler: Himmler received his authority directly from Adolf Hitler. The SS functioned as an extrajudicial organization, carrying out the most atrocious aspects of the war, including the concentration camp system, mass exterminations, and the slave labor network.

p. 20 required partnerships: Allen, 173.

p. 21 Speer's buildings: Evans, 327.

p. 21 "The work must proceed": Aalmans, title page of the war crimes monograph.

p. 21 first six months: Evans, 665.

p. 21 "American standards," wrote Speer: Neufeld, *The Rocket and the Reich,* 212.

p. 22 Himmler reminded the Führer: Evans, 663–65.

p. 22 thirty subcamps: As explained at the memorial site, "Dora and other subcamps originally belonging to the Buchenwald complex were amalgamated to form the independent Mittelbau-Dora Concentration Camp, in which Dora now served as the main camp. Once Mittelbau had become an independent concentration camp, it gained further subcamps through the relocation of new armament projects to the Southern Harz Mountains."

p. 22 "Mittelwerk General Manager": RG 330 Georg Rickhey, JIOA Form No. 2, Basic Personnel Record. His position as such earned him a salary of 54,000 reichsmarks a year. The average German worker during this period earned 3,100 reichsmarks a year.

p. 22 Rickhey was in charge: Bundesarchiv Ludwigsburg, Georg Rickhey B-162/964; RG 330 Georg Rickhey "Statement of Work History" and "Condensed statements of my education and my activities" (March 4, 1948).

p. 23 typical rate of four marks: Neufeld, *The Rocket and the Reich,* 186. Along with slaves, writes Neufeld, the SS Business Administration Main Office supplied security guards, food, and clothing "in a manner that led to a heavy death toll from starvation, disease and overwork."

p. 23 two and three reichsmarks per man: Aalmans, 12.

p. 23 were all present: Neufeld, *The Rocket and the Reich,* 228.

p. 24 "During my last visit to the Mittelwerk": Ibid., 229.

p. 25 "Howling and exploding bombs": Speer, *Reich,* 418.

p. 26 poison gas air locks: Author tour of Schloss Kransberg estate and interview with Jens Hermann, caretaker of the castle, August 1, 2012.

p. 26 series of small cement bunkers: Photographs, Bavarian State Archive, Munich; interview with Hanns-Claudius Scharff, September 27, 2012.

p. 26 Hitler had been directing: Photographs of Hitler at Adlerhorst/Eagle's Nest, Bavarian State Archive, Munich.

p. 27 Speer later recalled: Speer, *Reich*, 419–21.

p. 28 grand Führerbunker in Berlin: RG 330 Georg Rickhey, "Statement of Work History"; Bundesarchiv Ludwigsburg, Georg Rickhey B-162/964.

p. 28 Hitler moving back: On January 15, Hitler boarded his armored train for the nineteen-hour trip to Berlin.

p. 28 captured or destroyed: Speer, *Reich*, 422.

p. 29 ferocious and fanatical Nazi Party loyalist: Kershaw, *The End*, 50.

p. 29 "I am a deserter": Photographs on display at the Dachau concentration camp memorial site, permanent exhibition.

p. 29 no one had any idea: Speer, *Reich*, 422.

p. 29 an otherwise empty hotel: Ibid., 422.

p. 29 "In my room hung an etching": Ibid.

p. 30 "The war is lost": Shirer, *Rise and Fall*, 1097. Shirer also notes that "from the afternoon of January 27, the Russians had crossed the Oder 100 miles from Berlin. Hitler's HQ had moved to the Chancellery in Berlin where he would remain until the end."

Chapter Two: Destruction

p. 32 Levi recalled: Thomson, 184.

p. 32 Auschwitz III: U.S. Army aerial photograph, "Auschwitz-Birkenau Complex, O'swie¸cim, Poland 26 June 1944." Enlarged from the original negative and captioned by the CIA: "SS War Industries, IG Farben etc. Auschwitz III Buna." National Archives; Schmaltz, Buna Monowitz monograph.

p. 33 first corporate concentration camp: Drummer and Zwilling, 80.

p. 33 had been murdered for minor infractions: Interview with Gerhard Maschkowski, June 2, 2012.

p. 33 nine thousand emaciated, starving inmates: Ibid.; see also oral history interview with Gerhard Maschkowski, June 29, 2007, Archive of the Fritz Bauer Institute, Norbert Wollheim Memorial, Frankfurt, Germany.

p. 34 "Dogs on leather leashes": Interview with Gerhard Maschkowski, June 2, 2012; Primo Levi describes this similarly.

p. 34 infectious diseases ward: Thomson, 186; the German word is Infektionsabteilung.

p. 34 "He had a high temperature": Thomson, 186.

p. 34 Primo Levi explained: In a letter to his friend Jean Samuel (Thomson, 186–87).

p. 34 plant manager of Buna-Werk IV: Author visit to Fritz Bauer Institute in Frankfurt, which is located inside the former IG Farben building. For a full listing of the Fritz Bauer collection of documents on Ambros, which include many from the NMT, see the Wollheim Memorial: http://www.wollheim-memorial.de/en/igwerksleitung.

p. 35 Ambros had been awarded: RG 319 Otto Ambros, SHAEF file card, WD44714/36, "Microfilm Project MP-B-102." The card

reads: "Name: Ambros, Career: Member of I.G. Board, Chairman of
C-Committee (i.e. Committee for Poison Gas) in Speer Ministry,
Chief of CW Production, Chief of Buna Production. In June 1944
Hitler [m]ade a donation of one million marks to him. [*sic*] sotensibly
[*sic*] as a reward to him and his collaborators for an important
discovery in Buna making Germany independent of natural rubber."

p. 35 Ambros left the concentration camp: Walther Dürrfeld, the
engineer in charge of the Buna plant, was the second man. Walther
Dürrfeld, affidavit, February 18, 1947, NI-4184. Archive of the Fritz
Bauer Institute, Subsequent Nuremberg Trials, Case VI, 73–77.

p. 35 to destroy evidence: RG 330 Jürgen von Klenck, D-64032,
July 25, 1952; RG 319 Otto Ambros, September 28, 1944.

p. 36 evacuate Dyhernfurth: "Elimination of German Resources for
War," 1282; Tucker, 70.

p. 37 kill an individual in minutes: The figure varies. Many
WWII-era documents state that liquid tabun can kill a man in
thirty seconds from exposure. The Centers for Disease Control and
Prevention (CDC) puts the figure at "within 1 to 10 minutes."

p. 37 Exposure meant: CDC Fact Sheet, "Chemical Emergencies,
Facts about Tabun."

p. 37 where accidents had happened: Tucker, 48.

p. 37 last moments of an ant: Harris and Paxman, xiv.

p. 38 Sachsenheimer's commando force: Groehler, 11.

p. 39 first cleared of pine trees: Gross-Rosen Museum, Rogoznica,
Poland. The concentration camp was established in August 1940 as a
subcamp of Sachsenhausen. Prisoners were assigned slave labor in a
local granite quarry, which belonged to the SS.

p. 39 largest corporation in Europe: Dwork and Van Pelt, 198.

p. 39 chairman of Committee-C: RG 319 Otto Ambros, SHAEF file card, WD44714/36.

p. 39 prestigious title: Ibid.; Stasi records (BStU), Dr. Otto Ambros file, MfS HA IX/11 PA 5.380. In German, the medals are Kriegsverdienstkreuz and Ritterkreuz des Kriegsverdienstkreuzes.

p. 40 official car and airplane retrofitted: Stasi records (BStU), Dr. Walter Schieber file, MfS HA IX/11 AS 253/68.

p. 40 Their reliability: Groehler, 246, 268–69.

p. 40 Dr. Walter Schieber: Nuremberg Trial Testimony, February 10, 1947. Transcript, 2788.

p. 41 similar to deep-sea diving suits: Tucker, 51.

p. 41 Witnesses in nearby villages: Groehler, 9–10; Tucker, 70–71.

p. 43 void of people and tabun: Tucker, 71–72.

p. 43 Sixteenth and Eighteenth Chemical Brigades: RG 218 Joint Chiefs of Staff, Joint Intelligence Group Report, "Intelligence on Soviet Capabilities for Chemical and Bacteriological Warfare," 15–16.

p. 43 The factory was dismantled: RG 319 Otto Ambros, "Report on Chemical Warfare Based on the Interrogation and Written Reports of Jürgen E. von Klenck, Also Comments by Speer and Dr. E. Mohrhardt," 25, 34–35.

p. 43 Chemical Works No. 91: Tucker, 402.

p. 43 Red Army's stockpile: Tucker, 107.

p. 44 Stalin, rather unexpectedly argued: From the *Telegraph*, January 1, 2006, "Churchill Wanted Hitler sent to the electric chair," based on the notebooks of Sir Norman Brook, the former deputy cabinet secretary to Churchill, who kept a shorthand account of the proceedings.

p. 44 "If the war is lost," Hitler famously told Speer: Shirer, *Rise and Fall*, 1104.

p. 44 "What will remain": Kershaw, *The End*, 290. This famous passage is translated from German in a variety of ways and has been cited many different ways.

p. 45 He told Speer: Sereny, *Albert Speer*, 475.

p. 45 Demolitions on Reich Territory: Ibid., 475–76.

p. 46 death marches from Auschwitz: Aalmans, 10–11.

p. 46 given a promotion: Mittelbau was the parent corporation to Mittelwerk and operated additional weapons factories in the area.

p. 46 Jewish factory owner: The businessman had fled to South Africa to avoid being deported to a concentration camp.

p. 46 launch ramps for the V-2: Neufeld, *Von Braun*, 194.

p. 46 "dug by workers": Ibid., 195.

p. 46 he commandeered: Ibid., 194.

p. 47 Von Braun had to have known: *United States v. Kurt Andrae et al.*, trial records.

p. 47 To send a message: Neufeld, *Von Braun*, 165.

p. 47 one war crimes report: Aalmans, 11.

p. 48 get their crane back: "Interview with Rudolph by Major Eugene Smith of the U.S. Army Air Force, June 2, 1947," (DOJ monograph); Feigin, 333.

p. 48 benefits: Neufeld, *Von Braun*, 193–94.

p. 48 facilities designer Bernhard Tessmann: McGovern, 108. Tessmann had been chief designer of the test facilities at Peenemünde.

p. 48 transported von Braun: Neufeld, *Von Braun,* 195, citing a letter that von Braun wrote to his parents after the war.

p. 48 personal aide, Dieter Huzel: McGovern, 108.

p. 49 last V-1s were fired: Neufeld, *The Rocket and the Reich,* 263.

p. 50 taken to the Alps in a private car: McGovern, 111; the rocket engineers were under guard by the SD (Sicherheitsdienst).

p. 50 wearing blindfolds: Huzel, 151–61.

p. 51 Nazi documents hidden: Huzel, 159.

p. 51 decided to make an exception: This is from Huzel's memoir. There are different versions of the story, concerning whether or not the two men were supposed to tell Fleischer.

Chapter Three: The Hunters and the Hunted

p. 53 their next scientific intelligence mission: Pash, *The Alsos Mission,* 160–62. Pash had done several reconnaissance-type missions, including in Italy, without Goudsmit, before Normandy.

p. 54 apartment belonged to IG Farben: Pash, *The Alsos Mission,* 161.

p. 54 largest task force to date: Pash, *The Alsos Mission,* 171; Goudsmit, 77.

p. 54 Montgomery's famous words: Lasby, 23.

p. 55 it truly meant plunder: Associated Press, March 24, 1945.

p. 55 Representing the United States: Gimbel, 3–17.

p. 56 Black Lists: Ibid., 3, 9.

p. 56 Frontline requests: Ibid., 8; SHAEF sent CIOS requests through the London secretariat for coordination.

p. 56 bright red *T* painted: RG 165 "Code Name: Alsos Mission, Scientific Intelligence Mission, M.I.S., G-2, W.D.," photographic history compiled by Colonel Boris T. Pash, 20.

p. 56 dispatched his team to Europe: McGovern, 101–2. Major Staver was already in London.

p. 57 "30 Assault Unit": Bower, 75–76.

p. 57 CHAOS for CIOS: Lasby, 66.

p. 57 one of the greatest competitions: Ibid., 18–19.

p. 58 "not in good shape": Pash, *The Alsos Mission*, 172.

p. 58 "A single salvo": Ibid., 173.

p. 58 the factory been heavily damaged: RG 165 "Code Name: Alsos Mission, Scientific Intelligence Mission, M.I.S., G-2, W.D.," photographic history compiled by Colonel Boris T. Pash, 18.

p. 58 destroyed or removed: Pash, *The Alsos Mission*, 174–75.

p. 59 Lieutenant Colonel Philip R. Tarr: In much of the existing secondary literature, Philip R. Tarr is misidentified as Paul R. Tarr; Kleber and Birdsell, 40, 45, 454.

p. 59 CIOS team: Kleber and Birdsell, 40, 45, 79.

p. 60 raw materials were coded: Tucker, 49.

p. 61 always said the same thing: DuBois, 37–38.

p. 61 "lied vigorously": Tilley and Whitten, "Scientific Personnel I.G. Farbenindustrie A.G., Ludwigshafen," 45.

p. 61 Alsos had been trailing: Pash, *The Alsos Mission*, 174.

p. 61 flushing documents down the toilet: McGovern, 104.

p. 62 assigned by Göring: Goudsmit, 187–89; NMT Case 1, Document No-897, "List of Plenipotentiaries."

p. 62 "Leading men of science": HLSL Item No. 2076.

p. 63 led to the release: Goudsmit, 188.

p. 63 weapons-related research programs: Goudsmit, 80–85; Osenberg also worked with the Gestapo to learn and record the habits of the individual men, their weaknesses and their strengths — if they had a drinking problem, a mistress, or were homosexual.

p. 63 captured Osenberg: Goudsmit, 93, 197.

p. 63 index of cards: Dowden, "Examination of Dr. Ing. W. Osenberg and Documents," 4, Appendix II.

p. 64 appalled by the hubris: Goudsmit, 198. In German the sign read: "z.Zt.Paris."

p. 64 "[O]ne cannot trust you": Ibid., 200.

p. 65 85 percent of the city destroyed: O'Donnell, 31.

p. 65 "as Berlin awaited": Beevor, 173.

p. 65 "The deadly Jewish-Bolshevik": Quoted in Evans, 683, 685.

p. 66 four thousand Nazi bureaucrats: Beevor notes that ironically, in April 1945, there were more than double the number of those assigned work here.

p. 66 adored Knemeyer: Knemeyer, 23.

p. 67 lost more than twenty thousand airplanes: Evans, 683.

p. 67 didn't mean much: Shirer, *Rise and Fall*, 1099.

p. 67 Speer and Knemeyer had agreed: Knemeyer, 23, 34; Letter from Albert Speer to Sigurd Knemeyer, February 7, 1981.

p. 68 escape to Greenland: Knemeyer, 31; Speer, *Inside,* 494. "It was a plan hatched in a combination of panic and romanticism," Speer claimed after the war, telling the British historian Hugh Trevor-Roper, "Greenland is simply wonderful in summer and May" (Trevor-Roper, 178).

p. 68 made the first high-altitude sortie: Knemeyer, 39.

p. 69 Baumbach and Knemeyer began gathering: Ibid., 31; Baumbach, 230.

p. 69 Speer's command: Knemeyer, 31.

Chapter Four: Liberation

p. 71 Jones's: Collection at Huntsville, AL, Company L, 414th Regiment, 104th Infantry Division.

p. 72 "It was a fabric of moans": Bower, 109.

p. 72 seven war crimes investigators: Dora-Nordhausen memorial. Aalmans's photographs have become iconic images of the liberation of Nordhausen.

p. 72 "Four people were dying every hour": Bower, 119.

p. 73 "The building and sites": RG 319 Kurt Blome, "Alsos Mission Report by Cpt. William J. Cromartie and Major J. M. Barnes," July 30, 1945.

p. 74 Dr. Gross's possessions: Ibid., 9.

p. 74 "Russian contributions on plague": Ibid., 10.

p. 75 the Alsos scientists noted in their report: Ibid., 11.

p. 76 Hermann Göring Aeronautical Research Center at Völkenrode: Samuel, 147. This institution (Luftfahrtforschungsanstalt) is referred to a number of different ways; I use the U.S. Air Force term.

p. 77 Operation Lusty: It stood for LUftwaffe Secret TechnologY (Operation LUSTY fact sheet, National Museum of the U.S. Air Force, World War II gallery).

p. 77 When Putt arrived: Samuel, 105–6.

p. 77 a legendary test pilot: USAF biography, Lieutenant General Donald Leander Putt, U.S. Air Force website; Thomas, 219–40.

p. 77 "He displayed the ability": Thomas, 224, quoting Lieutenant General L. C. Craigie.

p. 78 payload was eventually revealed: Thomas, 226.

p. 79 still unknown to American fliers: Samuel, 150.

p. 79 When Putt learned: McGovern, chapter 15; History of AAF Participation in Project Paperclip; Samuel, 4–5.

p. 79 "the most superb instruments": Samuel, 143.

p. 79 Why not also fly: McGovern, chapter 15; Samuel 4–5.

p. 79 "If we are not too proud": Samuel, 4, cites Colonel Putt's "Technical Intelligence Speech" given to the Dayton Country Club, May 7, 1945.

p. 80 Knerr and Putt believed they could: Lasby, 76.

p. 80 "Pride and face saving": History of AAF Participation in Project Paperclip, 0920.

p. 82 dispatched to Raubkammer: CIOS Report 31: "Chemical Warfare Installations in the Münsterlager Area, Including Raubkammer."

p. 83 single yellow ring: Tucker, 85; Photographs, Loucks papers (mustard, sarin, and tabun munitions).

p. 84 extracting the liquid substance: CIOS Report 31: "Chemical Warfare Installations."

p. 84 rabbit's skin: Ibid.

p. 85 menacing new breed of chemical weapons: Tucker, 85–86; Groehler, 323–24; Bower, 94–95.

p. 85 set up in a subway tunnel: RG 319 Walter Schreiber, "Capture and Preliminary Interrogation," December 17, 1948.

p. 86 not necessarily the truth: RG 319 Walter Schreiber, "Agent Report," October 27, 1948.

p. 86 Goebbels's Happy Birthday broadcast: http://archive.org/details/Hitler_Speeches.

p. 86 Hitler said a few words: Beevor, 251.

p. 87 final assault on Berlin: Beevor, 255. Beevor notes that the Red Army was still outside Berlin proper, and the artillery fire actually hit the Berlin suburbs in the northeast.

p. 87 Soviet operation to capture Berlin: Beevor, 147.

p. 87 1.8 million shells: Ibid., 262.

p. 87 Knemeyer and Baumbach would flee Berlin: Knemeyer, 34; Baumbach, 233.

p. 88 Baumbach later explained: Baumbach, 233.

p. 88 through an intermediary: Pogue, 476.

p. 89 escape with Speer: Baumbach, 230.

p. 89 manor of Dobbin: Knemeyer, 34, 35–36; Baumbach, 234–37.

p. 90 "In the very near future": Baumbach, 236. Baumbach says he told Himmler, "I was examining a map of the world yesterday to see where we could fly to. I have planes and flying boats ready to fly to any point at the globe. The aircraft are manned by trustworthy crews. I have given instructions that nothing is to take off without a verbal order from myself."

p. 90 where Knemeyer waited: Knemeyer, 35–36.

p. 91 was still on hold: Speer, *Inside,* 494. Knemeyer, Baumbach, and Speer met up in the interim in Hamburg. In his memoirs, Speer claims that Baumbach pressed him to escape, but Speer says he "rejected this idea" because he had other things to do first. Speer also says the plan was Baumbach's idea but then contradicts himself by explaining how he had wanted to go to Greenland since seeing the Udet film.

p. 91 "overwhelming desire": Ibid., 476.

p. 91 Driving alone in his private car: Ibid., 477.

p. 93 landing amid rubble piles: Ibid., 478.

p. 94 Speer's final meeting with Hitler: Ibid., 485. Speer does not specify a date on his last visit to the FHQ but writes that "six days later" Hitler wrote his political testament, which was recorded on April 29, 1945, at 4:00 a.m.

p. 94 The post was surrounded: Author tour of Dachau concentration camp memorial site; maps and old photographs in the archive and library.

p. 95 several thousand corpses: NMT Evidence Code PS-2428; War Crimes Investigation Team No. 6823.

p. 95 "concentrated" in a group: Lifton, 153.

p. 95 "legally independent administrative units": Martin Broszat, "The Concentration Camps, 1933–1945," in *Anatomy of the SS State,* edited by Helmut Krausnick, Martin Broszat, and Hans-Adolf Jacobsen (London: Collins, 1968), 429–30.

p. 96 cold and gloomy: Smith, 79.

p. 96 "One of my men weeps": Ibid., 91.

p. 96 "I cannot believe this is possible": Ibid., 92.

p. 96 "On one of these walks": Ibid., 178.

p. 97 freestanding barracks: Author tour of Dachau concentration camp memorial site, archive and library, postwar, witness drawings.

p. 97 Experimental Cell Block Five: Technical Report No. 331-45, "German Aviation Medical Research at the Dachau Concentration Camp," 6; Alexander, "Exposure to Cold," 17. The facility was also called "Block 5, Experimental Station."

p. 97 Reich's elite medical doctors: HLSL Item No. 995; HLSL Item No. 80; HLSL Item No. 4.

Chapter Five: The Captured and Their Interrogators

p. 98 planned their final assault: To the Soviets, the Reichstag symbolized the "fascist beast." To fly the Soviet flag from the top of the Reichstag on May 1, a national holiday in the Soviet Union, would be a Red Army propaganda coup.

p. 98 Sometime around 3:30: Beevor, 359.

p. 100 given to him by Albert Speer: Knemeyer, 37.

p. 100 Their resort, Haus Ingeburg: Neufeld, *Von Braun*, 198.

p. 100 "There I was living royally": Lang, "A Romantic Urge," 75.

p. 101 the radio announcer declared: McGovern, 141.

p. 101 make a deal with the Americans: Neufeld, *Von Braun*, 199.

p. 101 Dornberger was overheard saying: McGovern, 142, citing his own interviews with Dornberger and von Braun about their time at Oberjoch.

p. 101 German and Austrian intelligence sources: von Braun, *Space Man*, 12–13.

p. 102 When Private Schneikert spotted: Burrows, 116.

p. 102 CIOS Black List: McGovern, 123. "Staver had placed von Braun's name at the top of his Black List."

p. 104 "biggest liar": Neufeld, *Von Braun,* 199–201.

p. 104 Twenty-five agents: DuBois, 40; Jeffreys, 354.

p. 105 wealthiest banker in all of Germany: DuBois, 40; Wollheim Memorial IG Farben, biographies of key executives of IG Farben, Hermann Schmitz (1881–1960).

p. 105 "the legend of Schmitz": DuBois, 40.

p. 106 "a dumpy Frau": Ibid., 40.

p. 106 "Doctor of Laws Schmitz": Ibid., 41.

p. 107 justify arresting: Jeffreys, 356.

p. 107 Major Tilley could: Affidavit of Major Edmund Tilley, November 21, 1945; DuBois, 6–20.

p. 107 buried in Schmitz's office wall: Tilley, "Report on the Finding of Evidence of Hermann Schmitz's Connection with and Knowledge of the Auschwitz Concentration Camp."

p. 108 cartoonish drawing: DuBois, 43–44.

p. 108 Tilley did not yet know: Beevor, 46. Three months earlier, on January 27, 1945, Soviet reconnaissance troops with the 107th Rifle Division discovered the Auschwitz concentration camp. Red Army photographers took pictures of the atrocities they found, all of which were sent back to Moscow. A report about the liberation of Auschwitz was published on February 9, 1945, in *Stalin's Banner (Stalinskoe Znamya)*, the Red Army newspaper, but as for news for the outside world, that was being withheld by the propaganda ministry in Moscow until Germany surrendered.

p. 109 Stalin was waiting: *Pravda,* the Soviet newspaper, published a small report on January 28, 1945, and on February 1 printed a thirty-line article about the liberation of Auschwitz, stating only the size of the camp, the number of inmates left, and the nutritional condition of the prisoners left behind.

p. 109 Tilley had no idea: DuBois, 44; Tilley, "Report on the Finding of Evidence of Hermann Schmitz's Connection with and Knowledge of the Auschwitz Concentration Camp."

p. 110 at the Nuremberg trials: DuBois, 4–10.

p. 110 "a plain chemist": Ibid., 4.

p. 113 triggered an alert: RG 319 Kurt Blome, Arrest Report, May 17, 1945.

p. 114 searching for Dr. Blome: RG 330 Kurt Blome, Preliminary Interrogation Report (PIR) No. 1; RG 319 Kurt Blome, File No. 100-665, War Crimes Office, Judge Advocate General's Office.

p. 114 a dedicated and proud Nazi: RG 330 Kurt Blome, PIR No. 1.

p. 114 His book: Kurt Blome, *Arzt im Kampf,* 138–39; chapter 7, "Kampf für Adolf Hitler und der Sieg" (Battle for Adolf Hitler and the Victory). Blome also wrote *Der Krebs* (The Cancer), and conducted extensive cancer research for the Reich before the war.

p. 114 trying to piece together: RG 330 Kurt Blome, Extract from 12 Army Group Interrogation Centre, June 22, 1945; Interrogation of Doctor Kurt Blome, July 1, 1945.

p. 114 "special treatment" (*Sonderbehandlung*): Bundesarchiv Ludwigsburg, Dr. Kurt Blome file, B162/28667.

p. 115 hierarchy of the Reich Hygiene Committee: Conti hanged himself in his cell in October 1945, thereby becoming a convenient scapegoat for Blome and others during the doctors' trial.

p. 115 "[I] can not approve": RG 330 Kurt Blome, PIR No. 1.

p. 115 "mass sterilization, gassing of Jews": Ibid.

p. 116 Major Gill pressed Blome for information: all the information in the Gill-Blome interview comes from RG 330 Kurt Blome, Interrogation of Doctor Kurt Blome, July 1, 1945.

p. 118 Blome denied having any idea: Ibid.; Goudsmit, 73.

p. 119 Beasley asked around: Beasley, 76.

p. 119 "I made daily visits to the jails": Ibid.

p. 120 "a nervous little man": From Norman Beasley, "The Capture of the German Rocket Secrets," *American Legion,* October 1963. The piece was later compiled by Diane L. Hamm in *Military Intelligence: Its Heroes and Legends* (U.S. Army Intelligence and Security Command History Office, October 1987; rev. ed. Honolulu, HI: University Press of the Pacific, 2001), 73–83.

p. 120 "[W]e accept you": Beasley, 80.

p. 121 He had promised Rickhey: RG 330 Georg Rickhey, Letter to Mr. Peter Beasley, April 6, 1948.

p. 121 daily six-mile drive: Sereny, 547.

p. 122 he had been shaving: Ibid., 555.

p. 122 soldiers with antitank guns: Ibid.

p. 122 arrested and taken away: Schmidt, *Justice,* 130; also in the castle, and dually listed on Speer's arrest report, was Speer's good friend and Hitler's personal doctor, Karl Brandt.

p. 122 American officials had known: Sereny, 547–49.

p. 122 in discussions with American officials: Ibid., 556. Sereny notes that in *Spandau,* Speer made no mention of his arrest or his first interrogations.

p. 123 Nitze recalled years later: Ibid., 549. Speer's secretary, Annemarie Kempf, recalled May 12, 1945, as the date that the American officials arrived at the castle.

p. 124 "He said he couldn't comment": Ibid., 551.

p. 126 the prisoner announced: Longerich, 1; *After the Battle* 14 (August 15, 1976): 35.

p. 126 "Police Chief of Nazi Europe": The *Time* cover image was drawn by Boris Artzybasheff.

p. 126 papers that identified him: Longerich, 2.

p. 127 blue-tipped object: Ibid., 3.

p. 127 jerked his head back: Ibid.

p. 127 "[T]his evil thing": Quoted in Winston G. Ramsey, "Himmler's Suicide," in *After the Battle* 14 (August 15, 1976), 35. The officer assisting Dr. Wells was Major Norman Whittaker.

p. 127 Of the 18.2 million: Kershaw, 379.

p. 127 50 million people: A definitive figure is impossible to pin down. This one comes from the Eleanor Roosevelt papers project, George Washington University, Washington, D.C.—available online: http://www.gwu.edu/~erpapers/.

p. 128 "The question who is a Nazi": Ziemke, 380.

Chapter Six: Harnessing the Chariot of Destruction

p. 129 "The scale on which science": Bower, 102. Farren was director of Farnborough, a research facility of the Royal Aircraft Establishment.

p. 131 "German science must be curbed": Lasby 32, 63; O'Mara, "Long-Range Policy on German Scientific and Technological Research, AIF."

p. 132 a directive known as JCS 1076: RG 164, Records of the War Department General and Special Staffs; Valero, 3.

p. 132 scientists were now being held: Secondary sources vary on this number. I use Lasby's figures; telephone interview with Clarence Lasby, March 23, 2013.

p. 132 General Eisenhower sought clarification: RG 331 AC/SG-2 to Chief, MIS, May 16, 1945, Subj: "Long-Range Policy on German Scientific and Technological Research"; Lasby, 75–76.

p. 133 a "matter of urgency": Quoted in Lasby, 63–64.

p. 134 Wolfe flew to SHAEF headquarters: General Wolfe also served on a classified study regarding secret German weapons that was being conducted by the Foreign Economic Administration (FEA). In this capacity, General Wolfe was to determine if the Nazis had shipped wonder weapons out of Germany before the end of the war, with the purpose of later selling them on the black market — similar to stolen artwork and Nazi gold. The results were inconclusive, but it gave General Wolfe a broader vision of the kinds of secret weapons projects the Nazis had been working on, and their value.

p. 134 hardly a good time: Lasby, 65, 106.

p. 134 "Besieged by the countless demands": Lasby, 65.

p. 135 RG 169. Memo, Major K. B. Wolfe to War-Navy Ad Hoc Interdepartmental Committee to Handle FEA Projects, May 14, 1945.

p. 135 "V-2 had exploded overhead": McGovern, 99. Cites his own interview with Major Staver.

p. 136 learning everything he could: Ibid., 99–106.

p. 136 U.S. Army calculations: History of AAF Participation in Project Paperclip, "Letter and Report from Staver to Colonel S. B. Ritchie," May 23, 1946.

p. 138 Ask Rees, Fleischer said: McGovern, 163.

p. 138 interviewed by Staver: History of AAF Participation in Project Paperclip, "Letter and Report from Staver to Colonel S. B. Ritchie," May 23, 1946.

p. 139 rough treatment: RG 330 Walther Riedel, JIOA Form No. 4, Security Report by Sponsoring Agency.

p. 139 "short trips around the moon": Ibid.: McGovern, 166–67.

p. 139 Riedel insisted: RG 330 Walther Riedel, JIOA Form No. 2, Basic Personnel Record. The five Nazi organizations were: NSDAP, NSKK (Motor Corps) NSV (Welfare), DAF (Labor Front), NSBDT (technicians), RLB (air raid defense).

p. 140 This territory included: McGovern, 151. Even though the land had been conquered by American forces, it was agreed, per Yalta, that the Soviets would control it. The Russians lost seventeen million people fighting the Nazis and wanted reparations to represent losses. This piece of land was 400 miles long and 120 miles wide in some places.

p. 140 Dr. H. P. Robertson told Major Staver: Papers of Dr. Howard P. Robertson, California Institute of Technology, Pasadena, California; Telephone interview with Clarence Lasby, March 23, 2013; Lasby, 108.

p. 141 "hostile to the Allied cause": Papers of Dr. Howard P. Robertson, California Institute of Technology, Pasadena, California.

p. 141 Dr. Robertson was a mathematical physicist: Ibid.

p. 141 in defiance of the Nazi Party: Albert Einstein biography online at Nobel Prize.org.

p. 143 "Von Ploetz said": McGovern, 168.

p. 144 Neither Huzel nor Tessmann had shared: In his book, Huzel says that he told von Braun about Fleischer. Every other account says that he did not.

p. 145 "100% Nazi," a "dangerous type": RG 330 Arthur Rudolph, JIOA Form No. 4, Security Report by Sponsoring Agency.

p. 145 Denazification was an Allied strategy...through tribunals: Thomas Adam, ed., *Germany and the Americas: Culture, Politics, and History, O–Z* (Santa Barbara: ABC-CLIO, 2005), 275.

p. 146 *The Green Archer:* Franklin, 98–99.

p. 146 Staver was making headway: McGovern, 163–76.

p. 147 Inn of the Three Lime Trees: McGovern, 169.

p. 148 "in almost inaudible": Ibid.

p. 149 hitch a ride: Ibid.; the colleague was Major William Bromley.

p. 151 Urgently request reply: History of AAF Participation in Project Paperclip, "Letter and Report from Staver to Colonel S. B. Ritchie," May 23, 1946.

p. 151 Major Staver returned: McGovern, 175.

p. 152 " 'Go, go — the Russians are coming' ": Franklin, 98–99.

p. 152 Over one thousand Germans: McGovern, 183.

p. 154 "I wanted to blow up": Ibid., 185. The agreement required captured facilities to be "held intact and in good condition at the disposal of Allied representatives for such purposes as they may prescribe."

p. 155 Institute Rabe: Chertok, 345–48.

p. 155 set up in a two-story schoolhouse: McGovern, 195.

p. 156 Dr. Herbert Wagner and four: RG 330 Herbert A. Wagner, "Certificate from Sponsoring Department German (or Austrian) Scientist or Important Technician." Wagner had been kept in the same facility as von Braun and Dornberger in the Bavarian Alps.

p. 156 nemesis of the U.S. Navy: "Paperclip, Part I," *Office of Naval Intelligence Review*, February 1949, 22–23.

p. 157 surrendered itself: On May 8, Lieutenant Commander Johann Heinrich Fehler received a message that had been broadcast through the Japanese cipher ordering the U-234 to return to Bergen, Norway, or to continue on to Japan. Fehler chose to continue on to Japan. A few hours later, his chief radioman picked up news from Reuters that the Japanese government had severed all ties with Germany and was arresting any German citizens who were in Japan. On May 15, the USS *Sutton* intercepted the U-234 and turned her over to U.S. Coast Guardsmen, who escorted the vessel to Portsmouth, New Hampshire.

p. 157 "said to contain": William M. Blair, "Big U-Boat Arrives with High General," *New York Times*, May 19, 1945.

p. 157 Additionally, there were drawings and plans: William Broad, "Captured Cargo, Captivating Mystery," *New York Times*, December 31, 1995.

p. 158 one of the most qualified: Scalia, 144–45.

p. 158 intelligence report: RG 330 Herbert Wagner, "Certificate from Sponsoring Department German (or Austrian) Scientist or Important Technician."

p. 159 Sturmabteilung: RG 65 Herbert Wagner, Federal Bureau of Investigation, File No. 106-131, June 1, 1948; Hunt, 7.

p. 159 laws of the occupying forces: Bower, 120.

p. 159 "an opportunist": RG 65 Herbert Wagner, Federal Bureau of Investigation, File No. 106-131, June 1, 1948.

p. 160 giving classified lectures: Scalia, 42–43.

p. 160 "These men are enemies": Maxwell AFB History office document. Memo, Patterson to Secy, General Staff, May 28, 1945, Subj: "German Scientists."

p. 161 Patterson suggested: History of AAF Participation in Project Paperclip, 9028; SWNCC was created in December 1944 to address political-military issues in U.S.-occupied Germany.

Chapter Seven: Hitler's Doctors

p. 162 Nazi Party–sponsored medical journal: Weindling, 23; Bundesarchiv Ludwigsburg, Dr. Hubertus Strughold file, No. 828/73.

p. 163 Eighth Air Force: Bullard and Glasgow, 47–48. Also known by its famous moniker, the Mighty Eighth. By the time Armstrong took over as its chief surgeon in 1944, the Mighty Eighth had 190,000 personnel, 300 of whom were flight surgeons. They were the command that conducted round-the-clock bombing against Nazi-occupied Europe.

p. 163 twenty-two-pound armored jacket: USAF Biography of Major General (Dr.) Malcolm C. Grow. It also protected soldiers against a .45-caliber round fired at point-blank range.

p. 163 hatched during a meeting: Bullard and Glasgow, 54.

p. 164 new research laboratory: Benford, "Report from Heidelberg," 9. The actual quote is "bring back from Germany everything of aero medical interest to the Army Air Forces and all information of importance to medical science in general."

p. 164 list of 115 individuals: Weindling, 80.

p. 165 "became quite good friends": Bullard and Glasgow, 52.

p. 165 Aviation Medical Research Institute: Benford, "Report from Heidelberg," 6. The institute's name, Luftfahrtmedizinische Forschungsinstitut, has been translated from the German in many different ways. The name as stated in the original report: Aviation Medical Research Institute of the Reich Air Ministry in Berlin.

p. 166 specialist named Ulrich Luft: Bullard and Glasgow, 55–56.

p. 166 Luft told Harry Armstrong: Mackowski, 111.

p. 168 first documentary evidence: *Life*, "Life Behind the Picture: The Liberation of Buchenwald," May 7, 1945.

p. 169 (UNWCC): Complete History of the United Nations War Crimes Commission and the Development of the Laws of War. Compiled by the United Nations War Crimes Commission. Published for the UNWCC by His Majesty's Stationery Office (London, 1948).

p. 170 become a central player: Alexander Papers, Harvard Law School Library, Series 1, Nuremberg Materials, 1939–1947. Subseries A, Trial Documents, 1942–1947.

p. 170 saw the liberated: Alexander Papers, Harvard Law School Library, Box 2, Letter to Mrs. Alexander in Newtonville, Mass, n.d.

p. 171 brought doctors and nurses: Smith, 215; an estimated 75 percent of the former prisoners were still at Dachau.

p. 171 medical crimes were suspected: Weindling, 77.

p. 171 Fate and circumstance: Alexander Papers, Harvard Law School Library, Series II, Box 7, Subseries A, Personal Life, 1883–1985.

p. 172 pull toward medicine: Ibid.; Schmidt, *Justice*, 24.

p. 172 first woman awarded a PhD: That is, since the institution opened its doors in 1685; his mother received her PhD in 1903; Schmidt, *Justice*, 23.

p. 172 intellectual-bohemian splendor: Alexander Papers, Harvard Law School Library, Series II, Box 7, Subseries A, Personal Life, 1883–1985.

p. 173 would become his mentor: Weindling, 74; Schmidt, *Justice*, 30. In his notes, Schmidt states that Kleist has only recently become the focus of debate (301n).

p. 173 Gustav Alexander was killed: Alexander Papers, Harvard Law School Library, Series II, Box 7, Subseries A, Personal Life, 1883–1985.

p. 174 Schmidt, *Justice,* 40; on page 48, Schmidt writes, "According to a government census from July 1933, the German Reich had a total of 51527 doctors (4395 female doctors), of whom 5557 (10.9 per cent) were identified as Jewish. The figure was probably higher because hundreds of Jews had already emigrated by the time of the census. The Reich Representation of the German Jews estimated a total of 9000 'non-Aryan' doctors; about 17 per cent of all German doctors were therefore considered to be Jews. Compared to the number of Jews in the German population as a whole, which amounted to little more than 1 per cent in 1933, the Jews had acquired disproportionate representation in the medical and legal professions, in banking and the arts."

p. 174 "succumbed to the swastika": Schmidt, *Justice,* 44. Prof. Kleist warned Dr. Alexander not to come back to Germany.

p. 174 "Our very existences": Schmidt, *Justice,* 55.

p. 174 In China, Dr. Alexander: Alexander Papers, Harvard Law School Library, Series II, Box 7, Subseries A, Personal Life, 1883–1985.

p. 175 "Have no false hopes": Schmidt, *Justice,* 55.

p. 176 "The ship traveled up": Alexander Papers, Harvard Law School Library, Subseries B, 14-3, Letters, Theo Alexander Worcester, January 21, 1934.

p. 176 to thrive in America: Alexander Papers, Harvard Law School Library, Series II, Box 7, Subseries A, Personal Life, 1883–1985.

p. 177 joined the fight: Ibid. In 1938, Dr. Alexander had signed up to fight the Nazis on the front lines as part of the Army Reserve Medical Corps, but he was rejected because he was overweight. After war was declared, he signed a waiver that allowed him to join.

p. 177 who might be guilty: Alexander Papers, Harvard Law School Library, Series 1, Nuremberg Materials, 1939–1947, Subseries A, Trial Documents, 1942–1947.

p. 178 Institute for Aviation Medicine: Eckart, 111. The institute he was in charge of in Munich had previously been named the Physiological Institute in Munich, Department of Aviation Physiology.

p. 178 In their first interview Weltz: Alexander, "Exposure to Cold," 3–12.

p. 179 "unfreeze a man": Ibid., 5–7, 67.

p. 179 "startling and useful discovery": Ibid., 4.

p. 179 groundbreaking research: Ibid., 11.

p. 179 had solved an age-old riddle: Ibid., 8. The exact translation of Weltz's goal was: "Is it possible to revive a man who is apparently dead from chilling…and after what interval of time can this still be done?"

p. 180 this very technique: Ibid.

p. 180 "adult pigs": Ibid., 9.

p. 180 "Weltz explicitly stated": Ibid., 11–12.

p. 182 dirty wooden tubs: Ibid., 9.

p. 182 "were being concealed": Ibid., 8.

p. 182 "German science presents": Alexander Papers, Harvard Law School Library, Series II, Box 2, Letter to Mrs. Alexander in Newtonville, Mass, n.d.

p. 183 *Untermenschen:* Herausgegeben vom Reichsführer-SS und SS-Hauptamt. Berlin 1942 (Berlin Do 56/685), per Dr. Jens Westemeier.

p. 184 Law for the Prevention of Genetically Diseased Offspring: Correspondence with John Dolibois, 2012–2013; United States

Holocaust Memorial Museum (USHMM), *Holocaust Encyclopedia,* The Biological State: Nazi Racial Hygiene, 1933–1939. The *Holocaust Encyclopedia* is an online resource of the USHMM that provides text, historical photographs, maps, audio clips, and other artifacts related to the Holocaust, in fourteen languages.

p. 184 requested that Kleist be fired: Schmidt, *Justice,* 95.

p. 185 "It sometimes seems as if": Alexander Papers, Harvard Law School Library, letter dated December 31, 1946. Dr. Alexander cited *The Tales of Hoffmann* by example. There, the evil Dr. Coppélius, posing as the Sandman, enters the bedrooms of sleeping children at night and cuts out their eyes. "Some new evidence has come in where two doctors in Berlin, one a man and the other a woman, collected eyes of different colour," Alexander wrote in another letter to his wife, Phyllis. "It seems that the concentration camps were combed for people who had slightly differently coloured eyes. That means people whose one eye had a slightly different colour than the other. Who ever was unlucky enough to possess such a pair of slightly unequal eyes had them cut out and was killed, the eyes being sent to Berlin."

p. 186 "On my way to Göttingen": Alexander, CIOS Report 24: "Exposure to Cold," 13.

p. 186 To his mind, the experiments: Ibid., 8, 13.

p. 186 Dr. Alexander told Dr. Strughold: Ibid., 13.

p. 186 The conference: Bundesarchiv Ludwigsburg, Dr. Hubertus Strughold file, 828/73.

p. 187 Rascher: Alexander, "Exposure to Cold," 13–68; Technical Report No. 331-45, "German Aviation Medical Research at the Dachau Concentration Camp," 77–92.

p. 188 Did Strughold approve: Alexander, "Exposure to Cold," 13, 14.

p. 189 received further extraordinary, related news: Ibid., 17–18.

p. 190 "a car showed up": Telephone interview with Hugh Iltis, January 24, 2012.

p. 190 most important collection: Ibid.

p. 190 broke the original seals: Weindling, 76.

p. 190 "The idea to start the experiments": Alexander, "Exposure to Cold," 20.

p. 190 Dr. Ruff had been in charge: Ibid., 17; HLSL Item No. 28; HLSL Item No. 995.

p. 191 photographs: Author viewed copies at Dachau concentration camp memorial site archive and library.

p. 192 "Strughold at least must have": Alexander, "Exposure to Cold," 17.

p. 192 "were still being covered up by" him: Ibid., 17, 65–66.

p. 193 "too soft": Ibid., 41.

p. 193 prisoner-witnesses who offered testimony: Ibid., 42–44.

p. 193 Alexander first returned to Dachau: Ibid., 43. The men worked with Häusermann in the Dachau disinfectant plant, where cadavers were taken after they were autopsied.

p. 194 "shipwreck experiments": Alexander, "Exposure to Cold," 10, Appendix I.

p. 194 Father Michalowski described: HLSL Item No. 2585.

p. 194 help of Hugh Iltis: Telephone interview with Hugh Iltis, January 24, 2012.

p. 194 "Auschwitz is in every way": Alexander, "Exposure to Cold," 33.

p. 195 private screening at the Air Ministry: Ibid., 66.

p. 195 named Dr. Theodor Benzinger: Ibid., 66; HLSL Item No. 1320.

p. 195 Winfield wrote: Winfield, "Preliminary Report," 4–5; Strughold promised Winfield that his work was "largely academic," which Winfield conceded was "somewhat contradictory," given the work Strughold published, particularly high-altitude studies, which required that research be conducted in the field.

p. 196 "[T]he heartbeat may continue": Lovelace, 65.

p. 197 high-altitude specialist who ran: Benford, "Report from Heidelberg," 6.

p. 197 "studies in reversible and irreversible deaths": Lovelace, 67.

p. 197 efforts to make salt water drinkable: HLSL Item No. 83.

p. 198 "Strughold was not always quite honest": Weindling, 78–80.

p. 198 military intelligence objected: Ibid., 82.

p. 199 handpicked fifty-eight: Benford, "Report from Heidelberg," 1–3. Benford specifies who started when, circa October–November 1945, but indicates that deals were made in the summer/fall before.

Chapter Eight: Black, White, and Gray

p. 200 debate over the Nazi scientist program: History of AAF Participation in Project Paperclip, 9030; Bower, chapter 7.

p. 201 To "open our arms": History of AAF Participation in Project Paperclip, 9028.

p. 201 Patterson sent a memorandum: McGovern, 192–93.

p. 202 chairman of SWNCC: Bird, 192.

p. 202 helping to develop the war crimes program: Taylor, *Nuremberg Trials*, 22, 36.

p. 202 backbone of a healthy economy: Bird, 229–30.

p. 202 In McCloy's eye: This presentation is informed by Bird's biography of McCloy. It is also interesting to note that McGovern, who authored the book on the V-2 in the 1960s and interviewed many of the U.S. Army players, went on to work for McCloy in the high commissioner's office.

p. 204 wished to see action overseas: Correspondence with John Dolibois, 2012–2013.

p. 205 "the war was very real": Ibid.; Dolibois, 64–65.

p. 205 "We had reports": Dolibois, 71.

p. 206 He recalled Mondorf's "beautiful park": Correspondence with John Dolibois, 2012–2013.

p. 207 the Palace Hotel: Ziemke, chapter 5, gives an overview of the interrogation centers.

p. 208 five stories tall: Correspondence with John Dolibois, 2012–2013.

p. 209 "solution to the Jewish question": USHMM, *Holocaust Encyclopedia*, SS and the Holocaust.

p. 209 "At once I understood": Correspondence with John Dolibois, 2012–2013; Dolibois, 85.

p. 209 "He had been told": Dolibois, 85.

p. 210 fingernails had been varnished: Andrus, 29.

p. 210 "Yes," he said: Dolibois, 86; at Ashcan, Dolibois used the cover name Lieutenant John Gillen.

p. 211 noted Ashcan's commandant: Andrus, 27.

p. 211 committed suicide: Henkel and Taubrich, 40–47.

p. 211 "In a second circle": These are the descriptions of these men that Dolibois uses in his book.

p. 212 masturbating in the bathtub: Dolibois, 128.

p. 212 "Often, I was taken": Correspondence with John Dolibois, 2012–2013; Dolibois, 89.

Chapter Nine: Hitler's Chemists

p. 213 analyzing its properties: Tucker, 90.

p. 214 "by air under highest priority": Ibid.

p. 214 using chemicals to kill people: FDR speech, June 9, 1943; Harris and Paxman, chapter 5.

p. 214 "In 1945, in the aftermath": Tucker, 103.

p. 214 tons of tabun: Hilmas, chapter 2; the British had inherited the facility but willingly shared its spoils, which included three thousand bombs and five thousand artillery shells.

p. 215 "We should do everything": Bower, 121.

p. 216 He told his former colleagues: Goudsmit, 81.

p. 216 " 'Siegheil' like a true Nazi": Ibid.

p. 217 The Allies were also reorganizing: Gimbel, 60. Some of FIAT's officers saw the name of their agency as alluding to *fiat voluntas tua,* Latin for "thy will be done"; Ziemke, 272.

p. 218 Dustbin was self-contained: Author tour of Schloss Kransberg estate, August 1, 2012.

p. 218 Speer took walks: Interview with Jens Hermann, August 1, 2012.

p. 219 Dr. Schrader had been working: BIOS Report No. 542, "Interrogation of Certain German Personalities Connected with Chemical Warfare"; Harris and Paxman, 55.

p. 219 IG Farben wanted to develop: "Elimination of German Resources for War," 1156–58.

p. 221 discoveries with potential military application: Ibid., 1276–78.

p. 221 "Everyone was astounded": BIOS Report No. 542, "Interrogation of Certain German Personalities Connected with Chemical Warfare," 24–35.

p. 221 English word "taboo": Tilley interview with Karl Krauch; "Elimination of German Resources for War," 1278–80.

p. 222 Dr. Schrader was told: BIOS Secret, Final Report 714, "The Development of New Insecticides and Chemical Warfare Agents," by Gerhard Schrader, 22–23; CIOS Report 31,"Chemical Warfare Installations."

p. 222 report to Göring: Tucker, 35.

p. 222 "psychological havoc on civilian populations": Tucker, 36.

p. 223 most of what Ambros did: "Elimination of German Resources for War," 1256, 1278, 1409.

p. 223 "Who is Mr. Ambros?": Ibid., 1261.

p. 224 Tilley learned quite a bit more: Ibid., 1280–95.

p. 224 "Judging from conversations": Ibid., 1278.

p. 225 "plain chemist": Dubois, 5.

p. 226 "Case #21877. Dr. Otto Ambros": RG 238 Otto Ambros: FIAT EP, June 14, 1946.

p. 226 impossible to comprehend: Ibid.

p. 227 American interrogation center: Kleber and Birdsell, 40, 45, 73, 454.

p. 227 sniffing at the air: DuBois, 5.

p. 227 Few men were as important to IG Farben: Stasi records, Dr. Otto Ambros file, BStU MfS HA IX/11 PA 5.380, "Report on the IG Farben Ludwigshafen and leading persons of IG Farben."

p. 227 butadiene: Dwork, 199. Synthetic rubber was called Buna after its components butadiene and sodium.

p. 228 masterminding this undertaking: Otto Ambros, affidavit, April 29, 1947, NI-9542. Archive of the Fritz Bauer Institute, Subsequent Nuremberg Trials, Case VI, PDB 75 (e), 1–18.

p. 228 "Greetings from Auschwitz": Dwork, 17.

p. 228 production of synthetic rubber required: Ibid., 197.

p. 229 "settle the question regarding": DuBois 168–75; Dwork, 201.

p. 229 official company report: DuBois, 172.

p. 229 "It is therefore necessary": Otto Ambros, Letter to Ter Meer and Struss, April 12, 1941, NI-11118. Archive of the Fritz Bauer Institute, Subsequent Nuremberg Trials, Case VI.

p. 230 SS officers hosted a dinner party: Ibid.

p. 230 Farben would pay: Drummer and Zwilling, 80.

p. 230 "Our new friendship with the SS": DuBois, 172; the Farben plant would be called Buna-Werke, Auschwitz 3. Prisoners selected for work at Buna-Werke would be awakened for roll call at 4:00 a.m., marched four miles to the rubber plant to work for ten to twelve hours, and then marched home. As for the workers who died during a shift, it was up to the laborers to carry the bodies back to the main concentration camp for cremation; Jeffreys, 302–4.

p. 231 "You said yesterday": "Elimination of German Resources for War," 1262.

p. 232 "known to all the IG directors in Auschwitz": Ibid.

p. 233 "30 or 40 drawings": RG 319 Otto Ambros, File CML-SP-la, February 9, 1949.

p. 233 trusted Ambros: Ibid.

p. 233 concentration camp workers: RG 319 Otto Ambros, Case 21877, Microfilm Project MP-B-102; FIAT EP 254-82, June 14, 1946; File XE021877, April 14, 1958.

p. 234 "The CIC personnel": RG 319 Otto Ambros, File CML-SP-la, February 9, 1949.

p. 234 "network of spies": RG 319 Otto Ambros, FIAT E 254-82, September 13, 1945.

p. 234 guesthouse that IG Farben maintained: Ibid.

p. 235 found him residing: Ibid.

p. 236 something was amiss: Tucker, 95.

p. 236 given a job as plant manager: RG 319 Otto Ambros, FIAT E 254-82, September 18, 1945.

p. 236 Wilson saw the situation in much darker terms: RG 319 Otto Ambros, P. M. Wilson letter, September 4, 1945.

p. 238 "I would look forward": Tucker, 95–96. For the Hirschkind quote to Ambros, Tucker cites a letter written by Hirschkind dated July 21, 1967, and located at the Carroll County, Maryland, library.

p. 238 "It is believed": RG 319 Otto Ambros, FIAT E 254-82, September 13, 1945.

p. 240 initial interview at Dustbin: RG 330 Kurt Blome, Alsos report, July 30, 1945 (23 pages).

p. 240 Blome had been observed: Ibid.

p. 240 "Kliewe claims": RG 330 Kurt Blome, Alsos report, 2, 14, 16; RG 65 Heinrich Kliewe, No. 655815-1, July 12, 1945.

p. 240 what he told them: RG 330 Kurt Blome, Alsos report, 1–23. Alsos agents interviewed Blome first in Dustbin, then in Heidelberg, then back in Dustbin. On October 1, 1945, he was moved to Oberursel, per letter from FIAT director of intelligence.

p. 241 specific plans: Ibid., 4.

p. 241 Himmler ordered him: Ibid., 6, 17.

p. 242 "History gives us examples": Ibid., 17–18.

p. 243 they settled on Nesselstedt: Ibid., 6.

p. 243 "infected rats on to U-boats": Ibid., 11. Blome says the original idea came from a Dr. Strassburger, assistant to Dr. Kisskalt, of the Hygiene Institute.

p. 244 Himmler assigned: Ibid.; see also Erhard Geissler et al., *Conversion of Former Btw Facilities* (Berlin, Germany: Kluwer/Spring Verlag, 1998), 60.

p. 244 He told his interrogators: RG 330 Kurt Blome, Alsos report, 6.

p. 246 There, Dr. Traub acquired: Ibid., 13.

p. 248 alarmed interrogators: Ibid., 18.

p. 248 The Russians had the laboratory: Ibid., 18.

Chapter Ten: Hired or Hanged

p. 249 "Our task was to prepare": Taylor, 49.

p. 250 "guilt for war atrocities": Andrus, 57.

p. 250 "atrocity films taken at Buchenwald": Ibid., 54.

p. 251 Funk started to cry: Andrus, 55–56.

p. 252 "They could sit out in the garden": Ibid., 57.

p. 255 fifty-two Ashcan internees were going: Correspondence with John Dolibois, 2012–2013; Dolibois, 133.

p. 256 His prisoner list included: Dolibois, 133.

p. 256 chatter among the Nazis in his backseat: Ibid.

p. 257 Hanns Scharff, kept a diary: Telephone interview with Hanns-Claudius Scharff, September 27, 2012, in California; Toliver, 16–18.

p. 258 "What is that horrible smell?": Correspondence with John Dolibois, 2012–2013; Dolibois, 134.

p. 260 airport at Luxembourg City: Correspondence with John Dolibois, 2012–2013; Dolibois, 135.

p. 261 even Heinrich Himmler was frightened: Last Days of Ernst Kaltenbrunner, CIA Center for the Study of Intelligence, September 22, 1993.

p. 262 middle-aged man: Dolibois, 137.

p. 263 approved — on paper — a Nazi scientist program: History of AAF Participation in Project Paperclip, 0938–0942.

p. 263 "formulate general policies": History of AAF Participation in Project Paperclip, 0941.

p. 265 Operation Backfire: McGovern, 200–204.

p. 266 "lustily singing, *Wir Fahren gegen England*": Franklin, 99–100.

p. 266 "The British pulled a sneaky on us": McGovern, 202.

p. 266 "interrogated for a week": History of AAF Participation in Project Paperclip, 0994.

p. 267 issued a brown jumpsuit: McGovern, 203.

p. 267 "the most hated man": correspondence with Brett Exton, October 14, 2012. The source was a guard at Island Farm, Sergeant Ron Williams. More at: http://www.islandfarm.fsnet.co.uk.

p. 267 five midlevel: They were Erich Neubert, Theodore Poppel, August Schulze, William Jungert, and Walter Schwidetzky. These men signed several of the first contracts for Project Paperclip's precursor, Operation Overcast.

p. 267 crass anti-American jokes: Neufeld, *Von Braun,* 212.

p. 267 Sipser overheard von Braun: Ibid., 502n.

p. 268 von Braun later told *New Yorker* magazine writer Daniel Lang: Lang, "A Romantic Urge," 89–90.

p. 268 handpicked by Colonel Putt: History of AAF Participation in Project Paperclip, 1016.

p. 268 the Germans boarded a C-54 military transport: Samuel, 185.

p. 268 "Quickly the plane moved": Ibid., 374.

p. 269 "They were all seasick as can be": David Boeri, "Looking Out: Nazis on the Harbor," NPR, August 19, 2010.

p. 270 ideal place for a secret military program: History of AAF Participation in Project Paperclip, 0997–0998.

p. 270 an insidious unease: Ibid; intelligence officer Henry Kolm had been working with many of these prisoners already—he was part of a Top Secret German prisoner of war interrogation program at Fort Hunt, Virginia, code-named P.O. Box 1142.

p. 270 Operation Overcast hotel: History of AAF Participation in Project Paperclip, 0997–0998.

p. 270 the "capitalists' game": Samuel, 380.

p. 271 Major Hamill was required: McGovern, 207–8.

p. 272 "Well it turned out": Neufeld, *Von Braun*, 215. Cites a speech given by Hamill on October 19, 1961.

p. 272 Major Hamill later recalled: Ibid.; McGovern, 208.

p. 273 "He is known to have spoken to Hitler": RG 319 Otto Ambros, FIAT file, November 7, 1945; FIAT EP 254-82, June 14, 1946.

p. 273 "he is the key man": Ibid.

p. 273 "Ambros claimed to be unable": Ibid.

p. 274 "Saw Ambros at LU [Ludwigshafen]": RG 319 Otto Ambros, Report by Captain Edelsten, August 28, 1945.

p. 274 Whenever the U.S. Army showed up: Ibid.

p. 275 "private intelligence center": Ibid.

p. 276 "Sorry that I could not": RG 238 Otto Ambros, 201 File, letter from Ambros dated August 27, 1945.

p. 276 a sizable Dustbin dossier: RG 330 Jürgen von Klenck, D-64032, July 25, 1952.

p. 277 Horn confirmed: RG 330 Jürgen von Klenck, summary report by Walter Spike, April 23, 1952.

p. 277 Otto Ambros's right-hand man: RG 330 Jürgen von Klenck, summary report by Lawrence R. Feindt, July 25, 1952.

p. 279 "buried at the lonely farm": RG 319 Otto Ambros, FIAT file, November 7, 1945.

p. 279 of nerve agent contracts: RG 319 Otto Ambros, IG Farbenindustrie AG Ludwigshafen, September 28, 1944, signed "Ambros."

p. 279 "full details": RG 319 Otto Ambros, FIAT file, November 7, 1945.

p. 279 "I (SPEER)": RG 319 Otto Ambros, SHAEF report, Microfilm MP-B-102; RG 319 Otto Ambros, FIAT file, November 7, 1945.

p. 280 "CW plant at AUSCHWITZ": RG 319 Otto Ambros, FIAT file, November 7, 1945.

p. 280 Tilley's intelligence report: Ibid.

p. 281 discovered in a Gendorf safe: Ibid.; RG 330 Jürgen von Klenck, summary report by Lawrence R. Feindt, July 25, 1952.

p. 281 an alternative theory: Ibid.

p. 281 issued a warrant: RG 319 Otto Ambros, FIAT file, November 7, 1945.

Chapter Eleven: The Ticking Clock

p. 285 removed from the Military Intelligence Division: History of AAF Participation in Project Paperclip, 1032; Lasby, 107, 128; telephone interview with Clarence Lasby, March 23, 2013.

p. 286 "The JIC structure": Telephone interview with Larry Valero, May 24, 2013.

p. 286 "The most important JIC estimates": Ibid.; Valero, 1.

p. 286 would postpone "open conflict": Valero, 4–5.

p. 287 most likely been captured by the Soviets: Ibid., 5-6.

p. 288 On its governing body: RG 330 Defense Secretary, Joint Intelligence Objectives Agency. http://www.archives.gov/iwg/declassified-records/rg-330-defense-secretary/.

p. 289 Klaus had hands-on experience: NARA, Holocaust-Era Assets, Overview. Klaus is an unsung hero in this story; most of the State Department files in his name remain classified. Of note, in 1946, a Foreign Economic Administration historian wrote: "There is, so far as this writer knows, no record which names one man as the originator of the Safe Haven project idea. Internal evidence from the records, however, supports the testimony of many participants in the project's work that Samuel Klaus must be credited with formulating the concepts upon which the program was based." Klaus was also credited with helping to set up the Berlin Document Center, a central repository for Reich documents, many of which would be used during the Nuremberg trials. The center was used by G-2 intelligence to source German scientists' past history with the NSDAP, as 90 percent of its member documents survived the war.

p. 290 Klaus's sentiments were shared: Hunt 121–23; telephone interview with Clarence Lasby, March 23, 2013; Bower, 164–65. "The military had been won over by phony propaganda that the Germans were the greatest scientists," the State Department's Herbert Cummings told journalist Tom Bower decades after the war.

p. 290 unabashedly vocal: Telephone interview with Clarence Lasby, March 23, 2013.

p. 290 "less than a dozen": Bower, 188.

p. 290 Green came up with an idea: Truman Library, Howland H. Sargeant Papers, 1940–1943; Technical Industrial Intelligence Committee — General, 1944–1950.

p. 291 different kind of restitution: Library of Congress, Technical Reports and Standards, PB Historical Collection, http://www.loc.gov/rr/scitech/trs/trspb.html. The Publication Board soon changed its name to the equally bland Office of Technical Services, which continued running a public relations campaign.

p. 292 sixty million jobs: Marlow, "60 Million Jobs. Late Henry Wallace's Dream Comes True," Associated Press, April 6, 1966.

p. 292 Subject headings included: Office of Technical Services, "Classified List of OTS Printed Reports," John C. Green, Director, Library of Congress, October 1947.

p. 292 The man in charge of both lists: Lasby, 129–31.

p. 292 "Specialized knowledge": Ibid.

p. 293 first group of six: History of AAF Participation in Project Paperclip, 0990; Lasby, 119–21. Note that there is some discrepancy as to who the six were, and the lists (made later) differ in both primary and secondary sources. Luftwaffe test pilot Karl Bauer had already come to the United States and been repatriated, which adds to the confusion.

p. 293 they began compiling: History of AAF Participation in Project Paperclip, 1705. Also contains a map of Wright Field and surrounding area.

p. 293 original scientists at Wright Field: History of AAF Participation in Project Paperclip, 0986, 0988.

p. 294 the Germans did not pay U.S. taxes: Ibid., 1720.

p. 294 At the Hilltop: Ibid., 0962.

p. 294 introductory pamphlet: Ibid., 0992, 0988; Lasby, 120.

p. 295 "The mere mention": History of AAF Participation in Project Paperclip, 1055.

p. 295 having handpicked: Samuel, 383.

p. 296 A War Department memo: War Department General Staff Memo, MIL 920, September 26, 1946 (FOIA).

p. 296 specialists were offended: Lasby, 123.

p. 296 like "caged animals": History of AAF Participation in Project Paperclip, 0989.

p. 297 "Intangibles of a scientist's daily life": Ibid., 0989–0990.

p. 298 hideous monster Medusa: Author tour of Courtroom 600, Palace of Justice, Nuremberg; Museen der Stadt Nürnberg audio guide; Henkel and Täubrich, 30.

p. 298 twenty-one present: Three defendants were missing: leader of the German Labor Front Robert Ley had committed suicide in his Nuremberg jail cell; arms magnate Gustav Krupp was eighty-five years old and deemed too frail to stand trial; Martin Bormann had disappeared while attempting to flee from Berlin and was believed dead.

p. 298 news about Nuremberg: Taylor, 279. Two hundred fifty journalists from twenty countries crowded into the upper balcony area of Courtroom 600; eighty of them were from America. The only nation that seemed relatively uninterested in the trial of the major war criminals was Germany. "Considering the shock, the horror and the destruction Germany was confronted with, getting them interested proved to be very difficult," said prosecutor Telford Taylor after the trial.

p. 298 swastikas painted on their tails: Wright Field Air Fair footage at http://www.youtube.com/watch?v=jrXTHMtX5Nc.

p. 299 make use of cutting-edge science: *The Code of Federal Regulations of the United States of America,* Volume 12, Title 15, p. 2306.

p. 299 questions for Putt: History of AAF Participation in Project Paperclip, 1020.

p. 300 Putt wrote in a memo: Ibid., 1027–29; Lasby, 128, 306n.

p. 300 Air Material Command: Note nomenclature change, from USAF.gov: Redesignated Army Air Forces Technical Service Command on August 31, 1944; Air Technical Service Command on July 1, 1945; Air Materiel Command on March 9, 1946.

p. 300 "letters of interest": History of AAF Participation in Project Paperclip, 1021–1029.

p. 301 continued to voice objections: Lasby, 129–32.

p. 301 temporary military program: History of AAF Participation in Project Paperclip, 0938.

p. 302 Wallace urged the president: Office of Technical Service, Letter from Wallace to Truman, December 4, 1945; Lasby, 133.

p. 302 once they found out about it: On October 1, 1945, the War Department Bureau of Public Relations issued a two-paragraph press release about the program with instructions to bury the story. No major news organizations reported on it. The headline of the press release was "Outstanding German Scientists Being Brought to U.S."

p. 302 air of democracy: History of AAF Participation in Project Paperclip, 1044–1050 (includes Wallace letter to Honorable Robert Patterson, Secretary of War, November 9, 1945); Lasby, 33–35.

p. 303 likened Hitler to Satan: Henry A. Wallace, U.S. vice president, in an address to the Free World Association, New York, August 5, 1942.

p. 303 a major news scoop: Weindling, 83.

p. 303 their meeting in Saint-Germain: Bullard and Glasgow, 54.

p. 304 war work in Heidelberg under army supervision: History of AAF Participation in Project Paperclip, 1208.

p. 305 "follow-on plan": Benford, "Report from Heidelberg," 9, notes.

p. 305 Control Council Law 25: Gimbel, 175. Peaceful research was allowed.

p. 305 Armstrong: USAF biography. At the time, his official title was surgeon of the Air Division of OMGUS, with headquarters in Berlin.

p. 306 Nickles who inspired Armstrong: Bullard and Glasgow, 2.

p. 306 When Nickles "hinted": Ibid., 18–20.

p. 307 closed his practice in Minneapolis: Dempsey, 1.

p. 307 arrived with his family at Wright Field: Bullard and Glasgow, 3, 6.

p. 308 He envisioned a future: Benford, *Doctors in the Sky*, 29.

p. 308 spotted a trapdoor: Bullard and Glasgow, 21.

p. 308 unusual-looking chamber: Ibid.; photographs from Wright Field archives.

p. 308 wrote a letter to the engineering division : Bullard and Glasgow, 6–8; Mackowski, 20.

p. 309 death at high altitude was caused by: Bullard and Glasgow, 18–20.

p. 310 he dissected the rabbit: Ibid., 19.

p. 310 Armstrong's discovery: Dempsey, 5, 63, 116.

p. 311 Halley's comet: Kokinda, 4. Halley's comet lasted fourteen days.

p. 311 "When I looked": Ibid., 6.

p. 311 pursued auto-experimentation: Mackowski, 43; his PhD thesis, which earned him a medical diploma in 1922, was called "The Distribution of Pain Spots on the Skin."

p. 312 Adolf Hitler needed a pilot: Mackowski, 46.

p. 313 Strughold packed his bags: Hasdorff, 10, 20.

p. 314 until one of them blacked out: Thomas, 27–28; Hasdorff, 3.

p. 315 apes and humans: Thomas, 32; Mackowski, 51.

p. 315 a haven for risk takers: Thomas, 37–38.

p. 315 officials from the Nazi Party: Ibid.; photographs, Bundesarchiv Ludwigsburg Collection. Note: In "Biologists under Hitler," Ute Deichmann explains that Nazi Party membership was never a requirement for doctors or professors. To be appointed to a university teaching position (*Habilitation* in German) did not require NSDAP membership. Only 45 percent of doctors joined the Nazi Party. Of scientists between the ages of thirty-one and forty, 63 percent became members (Mackowski, 65).

p. 316 "Our studies are all very risky": In his interview with Thomas, Strughold uses the story as a means to illustrate how he had to regularly outfox the Nazi Party in order not to succumb to their pressure to join. He says that he suggested the Nazi Party officials try out the low-pressure chamber test themselves. "That did it," Strughold told Thomas. "The older one said to the younger one, 'Herr Oberregierungsrat, we must go in five minutes. We can not stay.'"

p. 317 the pilot physiology challenges grew: Heinz Beauvais, "Performance and Characteristics of German Airplanes in Relation to Aviation Medicine," *German Aviation Medicine in World War II*, Volume I, 55–68; Eckart, *Man, Medicine, and the State*, 117.

p. 317 Dr. Theodor Benzinger: RG 330 Theodor Benzinger; Benford, "Report from Heidelberg," 6. Benzinger was also head of medical work in the research department of the Technical Division of the Reich Air Ministry.

p. 317 put each man in charge: Ibid.

p. 318 a committed Nazi: Bundesarchiv Militärarchiv Freiburg, Benzinger file, Pers 6/138768; RG 330 Theodor Benzinger, JIOA Form No. 2.

p. 318 In service of this idea: Benford, "Report from Heidelberg," 6.

p. 319 In addition to researching aviation medicine: RG 330 Theodor Benzinger, JIOA Form No. 2, June 1947; War Department, Intelligence Division, Basic Personnel Record, n.d.

p. 319 Ruff was an avowed and dedicated: HLSL Item No. 28, 995.

p. 319 Dr. Ruff who oversaw: Alexander, "Exposure to Cold," 17, 39; HLSL Item No. 28; HLSL Item No. 995.

p. 320 coauthored several papers: Weindling, 23. Weindling says the U.S. Air Corps circulated 250–300 copies to individual flight surgeons and air force bases across the country.

p. 320 coauthored a book: Weindling, 372n.

p. 320 His wartime research work: HLSL item 1878.

p. 321 A contest was proposed: HLSL Item No. 80; HLSL Item No. 83, NMT Trials Document No. 02626002.

p. 323 Becker-Freyseng was held in great esteem: HLSL Item No. 229.

p. 323 a self-experiment he did in a chamber: Ibid.,; HLSL Item No. 83.

p. 323 symptoms of paralysis: HLSL No. 2626.

p. 323 continued their work: Benford, "Report from Heidelberg," 3–28.

p. 324 military pose: Ibid., 4.

p. 324 "This property": Ibid., 36; Exhibit 5.

p. 324 There was equipment here: Ibid., 1, 10–16. A photograph shows the Freising chamber being installed in a corner of the institute. The caption reads, "[T]he low pressure chamber…was moved to Heidelberg from the Munich Institute of Aviation Medicine at Freising."

p. 326 number of German doctors believed: Weindling, 1, 162.

p. 326 classified list: Office of U.S. Chief of Counsel, APO 124-A. U.S. Army. List of Personnel Involved in Medical Research and Mercy Killings, n.d. (FOIA).

Chapter Twelve: Total War of Apocalyptic Proportions

p. 327 end of January 1946: Lasby, 185. Lasby writes that the secretary of war revealed that there were 130 scientists in the country and that approximately 140 more would arrive in the near future. A group of rocket men arrived January 15, and it took them a little more than a month to get to Fort Bliss.

p. 327 men resided in a two-story barracks: McGovern, 210.

p. 327 "romantic Karl May affair": Neufeld, *Von Braun,* 222; Neufeld says that work on the sci-fi novel began in 1947.

p. 328 "Frankly we were disappointed": Quoted in Lasby, 116; V-2 Firing Tables summarizing all flights at White Sands—www.wsmr.army.mil.

p. 328 The actual rocket firings: McGovern, 211.

p. 328 one of the fins fell off: Neufeld, *Von Braun,* 220.

p. 329 marry his first cousin: Ibid., 228.

p. 329 "The conditions of employment": Huzel, 217.

p. 329 swimming pool: Franklin, 102.

p. 330 "half a dozen discredited SS Generals": RG 65 Magnus von Braun, September 25, 1948; RG 330 JIOA list: Neufeld, *Von Braun,* 508n.

p. 330 job of club manager: Bower, 200.

p. 331 some work opportunities: History of AAF Participation in Project Paperclip, 1030–1050.

p. 331 the groups' complaints: Ibid., 0989; Bower, 158.

p. 332 Patin's industrial vision: RG 319 Albert Patin, "Statements Made by Paperclip Specialist Albert Patin, 18 October 1948."

p. 333 Patin acknowledged that his wartime access: Ibid.

p. 333 "improve the morale": History of AAF Participation in Project Paperclip, 0989–990; Bower, 158–60.

p. 334 Brigadier General Samford's office: History of AAF Participation in Project Paperclip, 1008, 1055.

p. 334 created a perfect storm: Bower, 161.

p. 336 Patterson, now secretary of war, shifted: Maxwell AFB History office document, Memo, Patterson to Secy, General Staff, May 28, 1945, Subj: "German Scientists"; Lasby, 71, 303n.

p. 337 left to their own devices: Bower, 165–68.

p. 338 new program protocols: History of AAF Participation in Project Paperclip, 1190–1192; Bower, 168.

p. 339 program would be called Operation Paperclip: History of AAF Participation in Project Paperclip, 1190-1192.

p. 340 McNarney wrote to JIOA: RG 319, JIOA, General Correspondence 1946–1952.

p. 340 "These [men] cannot now": Bower, 176.

p. 340 America's "national interest": Ibid.; Lasby, 174–75.

p. 341 legendary Long Telegram: Thompson, 59. The telegram was sent on February 22, 1946. Keenan's official title was "The Charge in the Soviet Union."

p. 341 regarding Soviet-American relations: "American Relations with the Soviet Union," September 24, 1946; Report by Clark Clifford, Subject File; Conway Files; Truman Papers.

p. 342 one thousand German scientists: The word "Austrian" was also added to the mission statement, even though there were only twelve Austrians on the list.

p. 342 With presidential approval official: History of AAF Participation in Project Paperclip, 1190–1192; telephone interview with Clarence Lasby, March 23, 2013. On June 3, 1963, Lasby interviewed Truman and asked the former president about his classified decision to hire Hitler's former scientists en masse. Truman told Lasby that because of America's precarious relationship with Russia at the time, "this had to be done and was done." These former Nazis, Truman insisted, "should always have an American 'boss.'"

p. 344 "in the capacity of a doctor": RG 238 Kurt Blome, 201 Prisoner file, May 9, 1946.

p. 344 unwanted spotlight: Deichmann, 282–89.

p. 345 like Posen, only bigger: Covert, 21. Of the 2,273 personnel, 1,770 were military.

p. 345 199 other germ bomb projects: Regis, 79.

p. 345 Top Secret program: Covert, 15; Regis, 93.

p. 346 made and sold vaccines: "Key Facts About Merck," Associated Press, November 3, 2005.

p. 348 he held the title: RG 330 Walter Schreiber, File RT-758-48, December 17, 1948.

p. 348 Schreiber held the position: RG 330 Walter Schreiber, "Memorandum to President Truman from Boston physicians," February 1952.

p. 348 moved around various interrogation facilities: RG 330 Walter Schreiber, "The Case of Walter Schreiber," February 17, 1952.

p. 349 Dr. Schreiber took the stand: HLSL Item No. 286.

p. 350 Wehrmacht's medical chain of command: RG 330 Walter Schreiber, Report: Interrogation of General Schreiber, December 16, 1948.

p. 352 invited Holzlöhner: RG 330 Walter Schreiber, Dr. Alexander and Hardy's letter to the Physicians Forum, February 1952; NMT-1 Document No. 922.

p. 357 Dr. Blome's plague research: HLSL Item No. 286.

p. 358 employed by the army: RG 238 Kurt Blome, 201 Prisoner file, May 9, 1946.

p. 358 the prison complex: Henkel and Taubrich, 32.

p. 359 "confidential change of status report": RG 238, Kurt Blome 201, Prisoner file, No. 31G5173069.

p. 359 fifty-eight German physicians: Benford "Report from Heidelberg," 1–5, photograph Exhibit 16; the army also requisitioned the Helmholtz Institute and set up Strughold as director there, with a salary of 28,000 marks a year. The Helmholtz Institute had been home to Philipp Bouhler, head of the Action T-4 euthanasia program.

p. 359 They all reported: Benford, "Report from Heidelberg," 37.

p. 359 regularly visited the facility: Ibid., 20.

p. 360 compiled into a two-volume monograph: Ibid., 1–2.

p. 360 five arrest warrants: RG 238 Theodor Benzinger, 201 File.

p. 361 twenty-three defendants: Weindling, 6; sixteen Nazi doctors, four non-Party physicians, and three SS administrators.

p. 361 "beyond the pale": New York Times, "Germans on Trial in 'Science' Crimes," December 10, 1946.

p. 361 university skeleton collection of the Untermenschen: RG 319 August Hirt, OSS Biographical Report, n.d. A student of Hirt's described him as missing part of his jaw, which made him speak with a "hissing sound."

p. 362 "the dregs of": New York Times, "Germans on Trial in 'Science' Crimes," December 10, 1946.

p. 362 once been internationally esteemed: Interview with Dr. Götz Blome, August 3, 2012, in Germany; Report on the Third International Congress for Medical Postgraduate Study, 1937 and 1938.

p. 362 listed the individual names: "Nazi 'Doctors' to Be Tried Next," *Stars and Stripes,* October 12, 1946.

p. 362 October 1942 conference: Bundesarchiv Ludwigsburg, Dr. Hubertus Strughold file, 828/73.

p. 363 film screening: HLSL Item No. 1320; NMT-I Document No. 224.

p. 363 that was not a crime: interviews with Rolf Benzinger, February 19, 2013, and April 10, 2013.

p. 363 "After the showing of the film": NMT-1 Document No. 224, August 21, 1946.

p. 363 Benzinger insisted: Wright Library Papers. "Sworn Statement of Theodor Hannes Benzinger," U.S. Department of Justice, Office of Special Investigations, November 22, 1983; telephone interviews with Rolf Benzinger, February 19, 2013, and April 10, 2013.

p. 364 a month in the Nuremberg jail: RG 238, Theodor Benzinger, 201 File.

p. 364 Wright Field circulated his report: Weindling, 192. Years later, in an interview with journalist Linda Hunt, Theodor Benzinger blamed his arrest and incarceration at Nuremberg on Dr. Strughold. Benzinger said he was "set up" by Strughold — that Strughold did so as a means of deflecting his own participation in war crimes. Strughold "had to put the blame on someone [else] because he was so vulnerable," Benzinger told Hunt. "He was wedged in amongst all the criminals and his way out was to finger me."

p. 364 "interrogations were sloppy": Weindling, 193.

p. 365 thirty-four of the doctors remaining: Bullard and Glasgow, 64–66.

p. 365 what sentences would be imposed: Steinbach, 84.

p. 366 "I wanted the condemned men": Andrus, 186.

p. 367 alternating its hiding place: Ibid., 184. Taylor, in his book, suggests that Göring might have had help from an American guard from Texas in hiding the poison vial.

p. 368 river: Andrus, 198. Taylor also refers to this in his book, but adds that he could not verify it as fact.

p. 368 "I hanged those ten Nazis": Quoted in *Time*, October 28, 1946, p. 34.

Chapter 13: Science at Any Price

p. 369 "State would accept as final": Bower, 180.

p. 370 Russian army's newspaper: *New York Times*, "U.S., Britain Hold German Experts, Berlin Communist Papers Charge," October 27, 1946.

p. 371 "apply for citizenship": "U.S. To Offer Citizenship to German Scientists," Associated Press, November 24, 1946.

p. 371 sanitized version of its program: History of AAF Participation in Project Paperclip, 0874–0875.

p. 372 "I wish we had more of them": "Nazi Brains Help Us," *Life*, December 9, 1946; *Newsweek*, December 9, 1946, pp. 68–69; Herbert Shaw, "Wright Field Reveals 'Project Paperclip,'" *Dayton Daily News*, December 4, 1946.

p. 373 "This Command is cognizant": RG 330 Emil Salmon, JIOA Form No. 3.

p. 373 provided photographs: *Time* and *Life* photographs are now Getty Images and were taken by Thomas D. McAvoy.

p. 373 news stories about the scientists: History of AAF Participation in Project Paperclip, 0867–0871; Lasby, 186; Hunt, 36.

p. 374 "We object not because they are citizens": Bower, 189.

p. 376 With its more than thirty rooms: O'Donnell, 24.

p. 377 Raven Rock Mountain Complex, or Site R: Interviews with Dr. Leonard Kreisler, 2012–2013, in Nevada. Kreisler served as the post doctor at Fort Detrick and Site R in the mid- to late 1950s.

p. 377 Reich's Demag motorcar company: Bundesarchiv Ludwigsburg, Georg Rickhey file, B162/25299.

p. 377 oversaw the construction: RG 330 Georg Rickhey, "Condensed statements of my education and my activities," March 4, 1948; "Transcript of Conference of May 6, 1944 in the office of Director General Rickhey." This document is from the DOJ, Office of Special Investigations.

p. 378 Rickhey and Patin's black market business: RG 319 Georg Rickhey, Summary, "Georg Rickhey's mail from wife, sister, brother," October 7, 1946.

p. 378 liked to gamble: Ibid.

p. 379 Nehlsen decided he had had enough: RG 330 Hermann Nehlsen, October 17, 1947.

p. 380 Putt had a gentleman's agreement: RG 330 Georg Rickhey, May 19, 1947; Hunt, 38.

p. 380 Wright Field mail censors: RG 330 Hermann Nehlsen, October 17, 1947; Hunt, 37–42.

p. 381 "One of the group who acted": RG 330 Georg Rickhey. Memorandum for the Director of Intelligence, December 19, 1946.

p. 381 The Pentagon assigned: Ibid. Colonel Lewis also requested "a more comprehensive investigation of scientists requested for shipment into the Unites States."

p. 381 five-year contract: RG 330 Georg Rickhey, Document No. 1258.

p. 382 Putt suggested: RG 319 Georg Rickhey, December 19, 1946.

p. 382 Nehlsen swore: Ibid.

p. 382 Voss also testified: Werner Voss changed his story twice. Smith interrogation of Werner Voss, in *United States of America v. Arthur Kurt Andrae et al.*, Preliminary Investigation file, roll 1.

p. 383 At Fort Bliss, in the evenings: Huzel, 215: According to Dieter Huzel, these were "simple, average people who were happy to have escaped the more serious consequences of the war, and who, with nothing else to do, devoted themselves with enthusiasm to their tasks.... As a result, the food was excellent."

p. 386 his findings: Major Eugene Smith to Air Provost Marshal, "Investigation Regarding Activities of Dr. Georg Rickhey, Former Director-General of the Underground Mittelwerk Factory Near Nordhausen, Germany," June 10, 1947, in *United States of America v. Arthur Kurt Andrae et al.*, Preliminary Investigation file, roll 1; Hunt, 62–69.

p. 386 three thousand SS officers: Aalmans, postscript.

p. 386 "I just saw a tiny headline": Bower, 201. Author's note: Coming across this quote in Bower's book was one of my favorite moments researching the book.

p. 387 "making arrangements": RG 330 Georg Rickhey, Office of the Deputy Director of Intelligence, Hq., European Command, n.d.

p. 387 escorting Rickhey: RG 319 Georg Rickhey, May 19, 1947.

p. 388 at least twenty thousand laborers: *United States of America v. Arthur Kurt Andrae et al.;* a copy of the trial data, charges, finding, and sentences can be found online at the John F. Kennedy Presidential Library and Museum.

p. 388 former SS barracks: Author's private tour of former SS barracks.

p. 389 "unprecedented move": Hunt, 77.

p. 389 Cox's call: RG 330 Georg Rickhey, Office Memorandum, March 1948.

p. 390 "He is absorbed": RG 330 Siegfried Knemeyer, Report on Siegfried Knemeyer, Siegfried, n.d.

p. 390 hated provincial life: Interview with Dirk Knemeyer, May 21, 2012, in California.

p. 391 to make significant contributions: RG 319 Siegfried Knemeyer, "Knemeyer contributions to Wright Field."

p. 391 "a genius in the creation": Knemeyer, 63.

p. 391 last-minute change: History of AAF Participation in Project Paperclip, 1447; photograph collection of Baumbach and Peron, Hans Ulrich Rudel.

p. 392 "menace of the first order": Lasby, 113; telephone interview with Clarence Lasby, March 23, 2013. Lasby interviewed and corresponded by mail with 175 Paperclip scientists in the mid-1960s.

p. 392 "Such a program must": Dornberger files, Deutsches Museum archive, Munich. The monograph is titled "Centralized vs. Decentralized Development of Guided Missiles by Walter Dornberger."

p. 393 were outraged: Minutes of the Council, FAS, New York, February 1–2, 1947; "Hiring of German Scientists," W.A.S. Bulletin, February 1947; Lasby, 201.

p. 395 sent threatening letters: History of AAF Participation in Project Paperclip, 0132.

p. 396 neighbors told army intelligence: RG 330 Herbert Axster, "The Axster Couple," March 25, 1948.

p. 396 Axster opened a law firm: Hunt; Bower, 206–7. Later the couple returned to West Germany.

p. 396 demanding an explanation: Lasby, 207; Delbert Clark, *New York Times*, "Nazis Sent to U.S. as Technicians," January 4, 1947.

Chapter 14: Strange Judgment

p. 397 "Mere punishment of the defendants": HLSL Item No. 565.

p. 398 in both languages: Papers of Dr. Leopold Alexander, Duke University Medical Center Archives, "Log Book, Journey to Nuremberg."

p. 399 writing the Nuremberg Code: Schmidt, *Justice*, 169. Schmidt writes, "On 7 December 1946, two days before the start of the trial, Alexander completed the first of two key texts on the ethics of human experimentation, which he addressed to [General] Taylor. The second memorandum was completed in April 1947. Both memoranda contributed to the debate about human experimentation inside the prosecution team, and ultimately shaped parts of the Nuremberg Code." Schmidt dedicates chapter 7 of his book to a full discussion of the Nuremberg Code.

p. 399 Ruff told the judges: HLSL Item No. 28, HLSL Item No. 995.

p. 402 dueling scars: NMT-1 photograph, Beiglböck in profile.

p. 402 clutched a dagger: interview with Vivien Spitz, January 17, 2012. In secondary accounts, it has been said that Höllenrainer punched or slapped Beiglböck. Vivien Spitz was at the trial and I report her account.

p. 402 "He was reaching": Spitz, 161.

p. 402 shock in the courtroom: Ibid.

p. 404 "My heart broke": Telephone interview with Vivien Spitz, January 17, 2012.

p. 405 how conflicted he felt: Papers of Dr. Leopold Alexander, Duke University Medical Center Archives, "Log Book, Journey to Nuremberg," n.d.

p. 405 "tremendous feeling of inner rage": Schmidt, *Justice,* 237.

p. 406 continue his testimony: *United States of America v. Karl Brandt et al.,* July 1, 1947.

p. 406 "You Gypsies stick together, don't you?": Spitz, 172.

p. 407 private letter to General Telford Taylor: Alexander Papers, Harvard Law School Library, Box 2, Letter to Brigadier General Telford Taylor, December 7, 1946.

p. 407 Dr. Alexander remembered it: Alexander Papers, Harvard Law School Library, Box 2, Letter to Mrs. Alexander, November 27, 1946.

p. 408 scores of documents: HLSL Item No. 180; HLSL Item No. 194; HLSL Item No. 279.

p. 408 Blome's defense: HLSL Item No. 276.

p. 409 left looking like hypocrites: Schmidt, *Justice,* 135.

p. 409 Seven doctors were acquitted: HLSL Item No. 184; HLSL Item No. 185; HLSL Item No. 186.

p. 410 "professional advisor" to Colonel Armstrong: RG 330 Hubertus Strughold, JIOA Form No. 3.

p. 410 "overall supervision of": Ibid.

p. 411 "the Jews had crowded the medical schools": RG 263 Central Intelligence Agency, Hubertus Strughold file, A-1-2062. This file was kept classified until the Nazi War Crimes Disclosure Act forced its declassification and release in 2006.

p. 411 "ethical principles": RG 330 Hubertus Strughold, "Sworn Statement of K. E. Schäfer," September 23, 1947.

Chapter 15: Chemical Menace

p. 416 he wrote in a State Deparment memo: Hunt, 107–9.

p. 416 security reports: RG 59 General Records of the Department of State, Doc. No. 862.542; Bower, 176, 237–39.

p. 417 wore his Nazi uniform to work: RG 330 Kurt Debus, File No. D-34033.

p. 417 revelation in his OMGUS security report: RG 330 Kurt Debus, File No. 384.201.

p. 418 transcript form: Ibid. The full transcript, in German and English, is in Debus's file.

p. 418 on November 30, 1942, "Craemer": RG 330 Kurt Debus, "Certified True Copy, F. C. Groves," n.d.

p. 419 sponsored by Heinrich Himmler: Hunt, 44.

p. 419 Klaus refusing to sign: Bower, 180–81.

p. 419 expert in tabun nerve agent synthesis: RG 330 Friederich Hoffmann, Basic Personnel Record.

p. 420 dedicated his life: Johnson, 2, 5–14; Loucks Papers (USAMHI), U.S. Army Military History Institute, Department of Defense Press Office, Biography of Charles E. Loucks.

p. 421 Mickey Mouse face: Brophy et al., 264. As technical director at Edgewood, Major Loucks was in charge of updating all soldiers' gas masks.

p. 421 incendiary bombs: Johnson, 202; Loucks Papers (USAMHI), photographs.

p. 421 four hundred battalions: Mauroni, 13.

p. 421 experts like Charles Loucks: Johnson, 59.

p. 422 Hoffmann arrived: RG 330 Friedrich Hoffmann, Security Report by Employing Agency.

p. 422 PhD in philosophy: Telephone interviews with Gabriella Hoffmann, September 27, 2012, and October 17, 2012; Hoffmann personal papers.

p. 423 risked life and limb: Hoffmann personal papers; Telephone interview with John Dippel, October 19, 2012.

p. 424 quartered inside a barracks: Hoffmann personal papers.

p. 424 "synthesizing new insecticides": RG 330 Friedrich Hoffmann, JIOA Form 3.

p. 424 code-named AI.13: Tucker, 104. Tabun was given the code name GA; sarin, GB; and soman, GD.

p. 424 under pressure to catch up: Krause and Mallory, 114–18; Loucks considered the greatest stumbling block to production to be the missing silver-lined cooking kettles that had been precision-designed to withstand the highly corrosive nature of tabun gas, items that had been at the very center of the Otto Ambros escape debacle in July 1945.

p. 424 Edgewood lagged behind: U.S. Army monograph, "Soviet Research and Development Capabilities for New Toxic Agents," Project No. A-1735, July 28, 1958, 6–7.

p. 425 "work of a high order": RG 330 Friedrich Hoffmann, JIOA Form 3.

p. 425 redouble efforts: Loucks Papers (USAMHI), "Germans Have Nerve Gas: 9GB0" (handwritten document). In addition to

serving as commanding general of the Army Chemical Center at
Edgewood Arsenal, Loucks was the deputy chief chemical officer for
research and development there.

p. 426 *United States of America v. Carl Krauch et al.:* His name was
spelled with both a *K* and a *C.*

p. 426 According to Pearson: Bower, 193.

p. 426 advocated for the use: Tucker, 35.

p. 427 briefed Eisenhower: Bower, 189, 194–95.

p. 427 "The public relations": Bower, 195.

p. 427 get rid of Samuel Klaus: Hunt, 132, 133.

p. 427 "obnoxiously difficult": Bower, 189.

p. 428 buried in scandals: Bower, 189, 194–95.

p. 428 was born: Center for Studies in Intelligence, "A Look Back,
The National Security Act of 1947," available online at www.cia.gov.

p. 429 observing the tabun tests: Office of the Historian, U.S.
Department of State, Memorandum from the Director of the Central
Intelligence Dulles to Secretary of Defense Wilson, "Subject: Research
on Psychochemicals," December 3, 1955.

p. 429 Greene was a short man: Interview with Gabriella Hoffmann,
October 17, 2012; Hoffmann personal papers (photographs). In the
1960s, the Greenes and the Hoffmanns were neighbors.

p. 430 His seminal vision: L. Wilson Greene, "Psychochemical
Warfare: A New Concept of War," Army Chemical Center, August
1949; Memorandum from Director of Central Intelligence Dulles to
Secretary of Defense Wilson, December 3, 1955, Document 244,
"Research of Psychochemicals," U.S. Department of State, Office of
the Historian.

p. 431 "hallucinogenic or psychotomimetic drugs": U.S. Army Inspector General, Report, DAIG 21-75, p. 12.

p. 431 "There can be no doubt": L. Wilson Greene, "Psychochemical Warfare: A New Concept of War," Army Chemical Center, August 1949; U.S. Army Inspector General, Report, DAIG 21-75, pp. 12–14.

p. 431 These sixty-one compounds: Ibid.

p. 431 recognized at Edgewood: Hoffmann personal papers; telphone interview with Gabriella Hoffmann, March 22, 2013; telephone interview with Dr. James Ketchum, November 7, 2012; Hunt, chapter 10.

p. 432 travel the world: Hoffmann personal papers; interview with Gabriella Hoffmann, September 27, 2012.

p. 433 the Black Maria: Covert, 39–40.

p. 434 the Eight Ball: Author tour of Fort Detrick, July 20, 2012; Covert, 40, 83, 95.

p. 435 back on the Paperclip List: RG 319 Kurt Blome, October 2, 1947, Ref No. 3047; Batchelor's colleague, Dr. Norbert Fell, had just returned from Japan where he had been working in secret with General Shiro Ishii, the dominant figure in the Japanese biological weapons program, located inside a secret facility on the Manchurian Peninsula and code-named "Water Purification Unit 731." The U.S. Chemical Warfare Corps struck a deal with General Ishii whereby in exchange for immunity, he agreed to "write a treatise on the whole subject" of his work during the war. Ishii's sixty-page report, in English, promised full disclosure "on B.W. activities directed against man [including] full details and diagrams." The report was worthless. Norman Fell returned to Detrick in June 1947. Harold Batchelor prepared for his visit to Germany. Fell later committed suicide.

p. 435 marked "Secret-Confidential": RG 319 Kurt Blome, October 2, 1947.

p. 435 everything discussed would be classified: RG 319 Kurt Blome, "Report of Interview of German Scientist, German Research on Biological Warfare," 86.

p. 436 "all the research for BW": Ibid., 86.

p. 437 Reich's outpost on the island of Riems: Ibid., 89.

p. 441 "the plague got more attention that any others": Ibid., 98.

p. 441 "Schreiber, as the head of the department": Ibid., 99.

p. 442 "everybody who knew Schreiber": Ibid., 99.

p. 443 everything they knew: RG 330 Erich Traub, JIOA Form No. 2, Basic Personnel Record.

p. 444 chose to return to Germany: RG 330 Erich Traub, Military Government of Germany, Fragebogen (questionnaire).

p. 444 lure Traub away: RG 330 Erich Traub, JIOA No. 461. In May 1948, Traub was appointed director of the institute by the Russians. Around that same time, with the aid of British intelligence, Traub began plotting his escape. On August 20, 1948, the following note was written and attached to his dossier: "He escaped the Russian zone carrying with him cultures of the Hoof and Mouth disease and has them stored in Marburg at the Behring Werke. Wants to secure employment in England, Canada, or the United States preferably." On July 1, 1949, he signed a Paperclip contract.

p. 445 shared with him by Dr. Blome: RG 319 Kurt Blome, "Report of Interview of German Scientist, German Research on Biological Warfare," 93.

Chapter Sixteen: Headless Monster

p. 446 made brigadier general: Loucks Papers (USAMHI). Loucks was made brigadier general on December 11, 1944. He was later regraded to colonel and then appointed commanding general of the Army Chemical Center in January 1951.

p. 446 working relationship with Richard Kuhn: Loucks Papers (USAMHI), speech to the Daughters of the American Revolution, Lynchburg Chapter, n.d.

p. 447 developed soman nerve agent: Tucker, 63, 89, 91–92.

p. 447 Nazi complicity: Loucks Papers (USAMHI), Letter to Mr. L. Patrick Moore, April 12, 1949.

p. 447 "I was under the impression": Loucks Papers (USAMHI), speech to the Daughters of the American Revolution, Lynchburg Chapter (n.d).

p. 448 gathering of the Swiss Society: Loucks Papers (USAMHI), "Lysergic Acid Compound" speech, n.d. Loucks wrote three different drafts of the speech. Some are typed, some handwritten, and some typed with handwritten notes.

p. 448 "Went back to the house": Loucks Papers (USAMHI), "Desk Diary 1948."

p. 448 "foreigner, dark, nationality unknown": Ibid.

p. 449 "lunch of pork chops": Ibid.

p. 449 "Lysergic Acid Diethylamide": Loucks Papers (USAMHI), "Lysergic Acid Compound."

p. 450 "enormous use as a psychiatric aid": Robert Stone, "Albert Hofmann, b. 1906: Day Tripper," *New York Times*, December 24, 2008.

p. 450 first article on LSD: Stafford, 30.

p. 450 chemists from Operation Paperclip: That these programs were originally linked to Operation Paperclip through General Charles E. Loucks was not documented before this book.

p. 451 "I can help," the caller said: Loucks Papers (USAMHI), "Desk Diary 1948."

p. 451 "We put samples [of sarin] in front of them": Loucks Papers (USAMHI); Johnson, 53–54.

p. 451 false teeth: RG 330 Walter Schieber, JIOA Form No. 2, Basic Personnel Record, n.d.

p. 452 Hitler's inner circle: Bundesarchiv Berlin-Lichterfelde, Dr. Walter Schieber file, No. 39002.

p. 452 on the personal staff of Heinrich Himmler: Ibid.

p. 452 an engineer and a chemist: RG 165 Walter Schieber, "Schuster File," Report from Dustbin, 31 pages.

p. 452 "Designs for concentration camp": Allen, 176.

p. 452 Schieber designed a "nourishment" program: Schmidt, *Karl Brandt*, 262.

p. 453 "confidential clerk of IG Farben": RG 319 Walter Schieber, Ref-No: S-3338.

p. 454 "I'm free now": Loucks Papers (USAMHI), "Desk Diary 1948"; Johnson, 53–54.

p. 454 signed a Top Secret Project Paperclip contract: RG 330 Walter Schieber, JIOA Form No. 2, Basic Personnel Record.

p. 454 "Subject is Dr. Walter Schieber": RG 330 Walter Schieber, HQ EUCOM, Frankfurt, Germany, to JIOA, January 9, 1948.

p. 455 "shipping subject via air under escort": Ibid.

p. 456 "Ship Dr. Walter Schieber to Wright Field": RG 330 Walter Schieber, War Department, Staff Message Center, Outgoing Classified Message, January 2, 1948.

p. 456 "Trusting he would be placed": RG 319 Walter Schieber, Ref No: S-2929, April 2, 1948.

p. 457 told his Paperclip handler: RG 319 Walter Schieber, March 16, 1948.

p. 458 "Schieber believes that": RG 319 Walter Schieber. Ref No: S-2929, April 2, 1948.

p. 459 "Walther Schieber started his business career": RG 330 Walter Schieber, File No. 8060622.

p. 460 "constantly profited from being a party man": Ibid.

' p. 460 "Cancel Air Force request": RG 330 Walter Schieber, April 1, 1948.

p. 460 "exception to present policy": RG 330 Walter Schieber, April 7, 1948.

p. 461 set up a meeting: Loucks personal papers; Tucker first found this story in the Loucks papers archived at the U.S. Army Military History Institute. The quotes Tucker uses differ slightly from those I found in Loucks's desk diary (which also differ slightly from references in Loucks's oral history interview), meaning that there could be a third source in Loucks's voluminous papers where he discusses his work with the German chemists.

p. 461 "Classified matters" were discussed: Loucks Papers (USAMHI), "Desk Diary 1948."

p. 461 Loucks recorded his thoughts: Ibid.

p. 462 "was more interesting": Ibid.; Johnson, 58.

p. 462 "Schieber is interesting": Ibid.

p. 464 One photograph in the album: Loucks Papers (USAMHI); photographs.

p. 464 "Driving one day in a Jeep": Johnson, 201–2.

p. 465 "one of those incidents that didn't mean anything": Ibid.

p. 465 "Could you develop the process": Loucks Papers (USAMHI), "Desk Diary 1948."

p. 465 recalled the next conversation: Johnson, 55.

p. 466 "We will pay all their expenses": Loucks Papers (USAMHI), "Desk Diary 1948."

p. 467 memorandum to the chief of the Army Chemical Corps: Ibid.

p. 467 "Hope the chief will support us": Ibid.

p. 467 industrial amounts of sarin gas: Johnson, 55–57.

p. 467 "One of the team": Loucks, "German Nerve Gas (GB)," 3 Sheets, "Written from memory but believed to be correct," signed Charles E. Loucks, April 10, 1972, Arlington, Virginia.

p. 468 "That's when we built the plant": Johnson, 56.

p. 468 code-named Gibbett-Delivery: Tucker, 123, 128. In May 1948, the Chemical Corps decided that sarin would be the standard U.S. nerve agent it would mass-produce. The plant was constructed under the code name Gibbett.

p. 468 unknown item: Loucks Papers (USAMHI), Letter, "Sehr geehrten Herr Loucks!" August 31, 1949.

p. 468 Christmas cards: Loucks Papers (USAMHI), cards dated 1949, 1952, 1955.

p. 469 "I don't like this," Loucks wrote: Loucks Papers, "Desk Diary 1950."

p. 469 Schieber had spent the night: Loucks Papers, "Desk Diary 1949."

p. 469 "long session with H.Q. Int.": Loucks Papers, "Desk Diary 1950."

p. 470 office at CIA handling the Paperclip: CIA Executive Assistant Director was Kenneth K. Addicott.

p. 470 "a photostatted copy": "CIA Memo to JIOA, Subject: Werner Osenberg Files on German Scientists," December 4, 1947 (FOIA).

p. 470 "production of intelligence": Karl H. Weber, "The Directorate of Science and Technology, Historical Series, Top Secret, The Office of Scientific Intelligence, 1949–1968, Volume II, Annexes IV, V, VI and VII," OSI-1, June 1972 (FOIA/Declassified 2008), Annex IV, p. 1.

p. 471 "Priority was accorded": Weber, 24. Weber was chairman from 1954 to 1972. The Scientific Intelligence Committee created joint intelligence subcommittees in each group: the Joint Atomic Energy Intelligence Committee was JAEIC, the Joint Biological Warfare Committee was JBWIC, the Joint Chemical Warfare Committee was JCWIC, the Joint Medical Sciences Intelligence Committee was JMSIC, and so forth. JIOA recruited German scientists and brought them to America, and the others produced and coordinated information that their work generated.

p. 471 half of the one thousand German scientists: These numbers and dates vary in secondary sources. I use JIOA records at the National Archives.

p. 472 moving a scientist from military custody: I use a summation of multiple scientists' case files, including those of Benzinger, Strughold, Traub, Schäfer, Dornberger, and Knemeyer.

p. 473 "Germany then became a new battlefield": Ruffner, ix.

p. 473 its plans for covert action: Weiner, 33, 36.

p. 474 spent time in displaced-persons camps: Ibid., 44.

p. 474 strategically located: Author tour of Camp King, which is about halfway between the Dustbin interrogation center and the former EUCOM headquarters in Frankfurt (also the former IG Farben building). Interview with Manfred Kopp, August 1, 2012, in Germany.

p. 474 the facility had two other names: General Order No. 264, Headquarters USFET, September 19, 1946, Designation of Military Installation (FOIA).

p. 474 significance of the informal name: "History of Camp King," 77. King had been assigned to the VII Corps of the First U.S. Army Corps, G-2.

p. 475 prisoners were Soviet-bloc spies: Ruffner, xlviii; "History of Camp King," 82. The partnership was effective from July 1, 1949, to May 31, 1952.

p. 475 Operation Artichoke: Marks, 31–36; Koch and Wech, 89, 97, 113. According to documents declassified by the Clinton administration, the sponsoring agency was the Joint Medical Sciences Intelligence Committee, or JMSIC, the JIOA's "intelligence production" counterpart. The Clinton panel identified the man in charge of Operation Bluebird as "Dr. Yaeger [*sic*] from the Central Intelligence Agency."

p. 476 first commanding officer: Memo CO No. 205, Headquarters USFET, August 25, 1945 (FOIA); interview with Egmont Koch, August 6, 2012, in Germany.

p. 476 intelligence reports on subjects: Correspondence with John Dolibois, 2012–2013: "I directed a study of the history of the German General Staff. Among those Donovan had me write intelligence monographs [of] were General Walter Warlimont, Field Marshal Albert Kesselring and Minister of Labor Robert Ley." The Donovan Nuremberg Trial Collection at Cornell contains 150 bound volumes of Nuremberg trial transcripts and documents from the personal archives of General William J. Donovan (1883–1959). Many of the source documents originated at Camp King.

p. 476 lacked a greater context: Silver, 2.

p. 477 analyze information from Soviet defectors: Ruffner, xv, xiv–xv. The Soviets repeatedly requested that Wessel and Baun be extradited.

p. 477 "gradual drift into operations": "Report of Interview with Brigadier General Edwin L. Sibert on the Gehlen Organization," March 26, 1970.

p. 477 make Gehlen head: Ruffner, xxi–xvii. Gehlen's official title was chief of intelligence collection, Foreign Armies East (Fremde Heere Ost). He had been a prisoner in the United States, at Fort Hunt, Virginia, since 1945.

p. 478 a village called Pullach: Ruffner, li–lviii; photographs from Central Intelligence Agency. The attachment shows photograph of the building, code-named Nikolaus Compound.

p. 478 finally realized the true nature: Ruffner, xvii–xviii.

p. 479 a million dollars a year: Bird, 353. Bird reports that in 1949, "Gehlen signed a contract with the CIA—reportedly for a sum of $5 milllion a year." Various former Agency sources suggest that sum was likely $1 million a year for five years; Breitman et al. use a figure of $500,000 a year.

p. 479 The two parties agreed: Ruffner, xlviii.

p. 479 CIA created the Office of Scientific Intelligence: Weber, IV, 23.

p. 480 "to apply special methods of interrogation": Marks, 23; CIA Memorandum for the Record, Subject "Project Artichoke," n.d., No. 75/42-75/46 (FOIA).

Chapter Seventeen: Hall of Mirrors

p. 481 most unusual press conferences: RG 65 Walter Schreiber, press onference, November 19, 1948, transcript.

p. 484 The GRU's notorious official emblem: Reuters, *Factbox:* Five facts about Russian military intelligence, April 24, 2009; GRU, abbreviation of Glavnoye Razvedyvatelnoye Upravlenie (Russian: Chief Intelligence Office).

p. 487 discussing his testimony: RG 65 Walter Schreiber, Agent Report, File VIII-III49, October 27, 1948.

p. 490 Hitler refused to allow Paulus: Shirer, *Rise and Fall,* 836–37.

p. 490 "Heroic endurance": Ibid., 837.

p. 491 "91,000 German soldiers": Ibid., 839. When Hitler learned that Field Marshal Paulus had surrendered to the Soviets, he became enraged. " 'The man [Paulus] should have shot himself just as the old commanders who threw themselves on their swords when they saw that the cause was lost.... What hurts me most, personally, is that I still promoted him to field marshal. I wanted to give him this final satisfaction. That's the last field marshal I shall appoint in this war. You mustn't count your chickens before they're hatched' " (Shirer, *Rise and Fall,* 840).

p. 491 he was living comfortably: RG 65 Walter Schreiber, File VI-878.16, October 22, 1948. Paulus, like Schreiber, testified at Nuremberg against his fellow Nazi Bonzen. "The mere presence of Paulus in Nuremberg was far more startling than anything he had to say," wrote General Telford Taylor.

p. 492 Special Agent Wallach: RG 65 Walter Schreiber, Agent Report, File VIII-III49, October 27, 1948.

p. 495 Loucks found Schreiber: RG 330 Walter Schreiber, Memo, File No. D-249361, December 15, 1949; Loucks Papers (USAMHI), "Desk Diary 1948."

p. 496 "Loucks stated subject was energetic": RG 330 Walter Schreiber, Memo, File No. D-249361, December 15, 1949.

p. 496 to serve as post physician: Ibid. Starting on November 18, 1949, "employed as physician, 7707 ECIC, Oberursel."

p. 497 name of the Soviet handler: In the English transcript of the press conference, dated November 19, 1948, Schreiber's handler's name is translated/transcribed as "Fisher."

p. 498 "I would have liked to stand up: Schmidt, *Justice*, 189.

p. 498 incarcerated for roughly one year: RG 319 Otto Ambros, 201 File; the dates Ambros was in prison, according to his official Prison File: "confined since 16 August 1948," he was "discharged 3-2-1951 [February 3, 1951]." Dates vary, including those cited at the Nuremberg Military Tribunal Museum.

p. 499 boarding school–like campus: Author tour of Landsberg with Prison Warden Dr. Harald Eichinger.

p. 499 "Politic[s] is a bitter disease": RG 319 Otto Ambros, Letter to Mrs. Prof. C. Ambros from O. Ambros, n.d.

p. 499 "my father is [being] illegally held": Landeskirchliches Archiv, Nachlass Landesbischof Wurm file (folder 307/2), correspondence file for the IG Farben Case, letter of Dieter Ambros, August 26, 1949.

p. 499 disciplinary action: RG 549 Otto Ambros, Disciplinary Report, January 18, 1951.

p. 500 requested permission: RG 549 Otto Ambros, "Special Permit," July 26, 1949; August 16, 1949.

p. 500 wrote up his annual health report: RG 549 Otto Ambros, Report of Physical Examination, August 8, 1949.

p. 501 given "victors' justice": Frei, 104–6. Meanwhile, men who had served Hitler began the creep back into German politics and industry. Three notable former Nazis were in Adenauer's cabinet: Dr. Thomas Dehler, Minister of Justice, Dr. Hans Seebohm, Minister of Transportation, and Dr. Hans Globke, a former member of Hitler's Interior Ministry and coauthor of Reich race laws.

p. 501 the "so-called prisoners of war": Bird, 329–30.

p. 501 McCloy served: Ibid., 193.

p. 502 credited in World Bank literature: Archives, World Bank, www.worldbank.org.

p. 502 that had belonged to Adolf Hitler: Bird, 316. McCloy had two trains at his disposal. The larger of the two belonged to Hitler.

p. 502 Many Germans wanted: Frei, 114.

p. 503 the largest office building in Europe: Drummer and Zwilling, 44–45.

p. 503 McCloy settled in: Ibid., 105.

p. 503 It was located: Author tour of the IG Farben building, which is now home to the Johann Wolfgang Goethe Universität in Frankfurt am Main, and the Wollheim Memorial, Fritz Bauer Institute, Frankfurt.

p. 503 something had to be done: Frei, 94, 178.

p. 504 The legal department: Bird, 330. There exists a famous and controversial story about McCloy when he was the assistant secretary of war (Bird, 214–23). Toward the end of the war, Nahum Goldmann, president of the World Jewish Congress, read in the *New York Times* that between July 7, 1944, and November 20, 1944, ten fleets with sometimes more than three hundred heavy bombers had bombed targets within thirty-five miles of Auschwitz. Goldmann went to see John McCloy at his office in the Pentagon to appeal to the assistant secretary of war, who was able to bring recommendations to European commanders. According to Goldmann, he pleaded with McCloy for Auschwitz to be bombed, saying some would die but a hundred thousand lives might be saved. McCloy said there was little he could do and instead passed off Goldmann to his British counterpart in Washington, Sir John Dill. Goldmann met with Dill, who also rejected the idea, arguing that bombs needed to be saved for important military targets. A second appeal was made to McCloy: to destroy "the execution chambers and crematories at [Auschwitz] Birkenau through direct bombing action." McCloy turned the matter over to an American lieutenant general named John Hull for quick evaluation; no commanders in Europe were consulted. Hull rejected the appeal, claiming, "The target is beyond the maximum range of medium bombardment, dive bombers and fighter bombers located in the United Kingdom, France or Italy"—a statement, Bird reminds readers, that was contrary to the facts. Goldmann pointed out that the Royal Air Force had already targeted IG Farben's Buna factory, which was four miles away from the gas chambers.

p. 504 deliberate, shameless murderers: Bird, 331.

p. 504 McCloy never responded: Ibid.

p. 504 Peck Panel: David W. Peck was at the time the presiding justice of the New York Supreme Court Appellate Division.

p. 505 McCloy's adjunct in Bonn: Schwartz, 165.

p. 505 "the widest possible clemency": Diefendorf et al., 445.

p. 505 "Due to the threat": History of AAF Participation in Project Paperclip, Office of the Sec of Defense, 7/14/50; Bower, 253.

p. 506 Dr. Nordstrom maintained a thick file: RG 330 Records of the U.S. High Commission for Germany, Project 63, 1948–1952.

p. 506 Accelerated Paperclip program: RG 330 Records of the U.S. High Commission for Germany, Files of Research Control Group, Dr. Nordstrom.

p. 506 "especially dangerous top level scientists": RG 330 Records of the U.S. High Commission for Germany, Project 63, November 22, 1950: Hunt, 203–6; Bower, 253.

p. 507 Accelerated Paperclip, or Project 63, meetings: RG 330 Records of the U.S. High Commission for Germany, Project 63, 1948–1952; the actual meeting summary notes are missing from the HICOG files—destroyed or misfiled. Only a cover letter remains, indicating who was there.

p. 507 Representatives from JIOA: RG 330 Records of the U.S. High Commission for Germany, August 2, 1951. The CIA worked under the code name "7955 Scientific Detachment," which can be identified as the CIA by its representative, Karl H. Weber.

p. 508 U.S. Army's "primary interest": RG 330 Records of the U.S. High Commission for Germany, Project 63, "Instructions for German and Austrian Nationals residing in the United States Under the Terms of Contract Agreement," 5–7.

p. 509 Otto Ambros was placed on the JIOA list: RG 319 Otto Ambros, File 291888, "Release of Inmate from Landberg Prison," RG 319 Otto Ambros, January 11, 1951.

p. 509 The Peck Panel suggested: Schwartz, 162–65.

p. 509 1 million reichsmarks: RG 319 Otto Ambros, SHAEF file card dated 10/44, WD44714/36, "Microfilm Project MP-B-102"; when contacted in 2012, Ambros's son, Dieter Ambros, declined to comment on the figure.

p. 510 panel's recommendations: Bird, 360–61.

p. 510 John J. McCloy commuted: Congressional Record—Senate, February 1951, page 1581. Drew Middleton, "7 Nazis Executed for War Murders," *New York Times,* June 7, 1951.

p. 510 one-third of the inmates tried at Nuremberg were freed: RG 549 Otto Ambros, Case Record. The "Order with Respect to Sentence of Otto Ambros" was signed by John J. McCloy and dated January 31, 1951.

p. 510 Otto Ambros traded in his red-striped denim prison uniform: Ibid.

p. 511 "Why are we freeing so many Nazis?": This is the way the quote appears in many books and papers. The way the quote appears in her column (February 28, 1951) is "the fact that we have freed so many Nazis of late must be puzzling the German People."

p. 511 "Doctors who had participated": Nachama, 379.

p. 512 Charles McPherson learned: RG 330 Kurt Blome, Special Projects Team, March 27, 1951.

p. 513 Blome said he needed some time: RG 330 Kurt Blome, Special Projects Team, June 25, 1951.

p. 513 took their boys out of school: Interview with Dr. Götz Blome, August 3, 2012, in Germany.

p. 515 Blome's secret Accelerated Paperclip: RG 330 Kurt Blome, Document No. 384.63.

p. 516 "Suspend shpmt Dr. Kurt Blome": RG 330 Kurt Blome, October 12, 1951.

p. 516 "In view of adverse publicity": RG 330 Kurt Blome, October 19, 1951.

p. 516 "Recommend Blome be shipped": RG 330 Kurt Blome, October 24, 1951.

p. 517 particularly upset because Traub: RG 330 Kurt Blome, November 27, 1951.

p. 519 have a nice house: Ibid.

p. 519 She was not interested: Interview with Dr. Götz Blome, August 3, 2012, in Germany.

p. 520 recently received a check: RG 319 Walter Schreiber, January 18, 1951.

p. 520 a home in San Antonio and a car: Ibid.

Chapter Eighteen: Downfall

p. 521 The doctors' trial had affected him: Alexander Papers, Harvard Law School Library, Series II, Box 7, Subseries A, Personal Life, 1883–1985; Papers of Dr. Leopold Alexander, Duke University Medical Center Archives, "Log Book, Journey to Nuremberg."

p. 523 continued to suffer: *New York Times*, "Cured in U.S. of Her Ills, Left as Nazi Guinea Pig," March 7, 1952.

p. 523 "I regard it as my duty": RG 330 Walter Schreiber, December 3, 1951.

p. 524 Dr. Walter Schreiber heard the telephone ring: RG 330 Walter Schreiber, "Department of the Air Force Headquarters United States Air Force Washington," December 14, 1951.

p. 524 expertise was extremely rare: RG 330 Walter Schreiber, Memo No. 24-170.

p. 525 long-winded stories: RG 319 Walter Schreiber, Basic Personnel Record for Paperclip Specialist, 30.

p. 525 identified himself as Mr. Brown: RG 330 Walter Schreiber, "Department of the Air Force Headquarters United States Air Force Washington," December 14, 1951.

p. 526 did not tell anyone: RG 330 Walter Schreiber, Memo No. 24-170.

p. 527 "He was of the opinion": Ibid.

p. 528 FBI got involved: RG 65 Walter Schreiber, File No. A8091581.

p. 529 "I have been advised": RG 330 Walter Schreiber, January 3, 1952.

p. 530 General Benson "stated": RG 330 Walter Schreiber, affidavit, January 23, 1952, p. 2.

p. 531 sent secret messages: Weindling, 15.

p. 532 FBI agents arranged to interview: RG 330 Walter Schreiber, "Statement taken from Miss Janina Iwanska in connection with an investigation of Dr. Walter Emil Wilhelm Paul Schreiber, at the Boston Office," February 27, 1952.

p. 534 "I don't know if he gave the orders": Ibid.

p. 534 Dr. Schreiber began plotting: RG 330 Walter Schreiber, letter dated March 27, 1952.

p. 536 Schreiber's wife of forty years: RG 330 Walter Schreiber, NSDAP Party document 160-75, Nazi Party No. 917,830.

p. 538 "I am fighting for justice": *Washington Post*, "Charges Denounced as 'Lies' by Schreiber," February 13, 1952.

p. 538 "I never worked in a concentration camp" RG 330 Walter Schreiber, affidavit, January 23, 1952.

p. 539 "a man should be given": RG 330 Walter Schreiber, transcript of interview between Colonel Heckemeyer, director of JIOA, and Miss Moran, *Time*, February 26, 1952.

p. 539 "Here are the facts regarding the Nazi doctor": Drew Pearson, "Air Force Hires Nazi Doctor Linked to Ghastly Experiments," Associated Press, February 14, 1952.

p. 540 Letter to President Truman: RG 330 Walter Schreiber, telegram to the president dated April 24, 1952, 9:34 p.m.

p. 541 "We are not going to make": RG 330 Walter Schreiber, transcript of interview between Colonel Heckemeyer, director of JIOA, and Miss Moran, *Time*, February 26, 1952.

p. 544 made by General Aristobulo Fidel Reyes: RG 330 Walter Schreiber, Department of the Air Force staff message No. 52571.

p. 544 air force paid for police protection: RG 330 Walter Schreiber, memorandum, Travis Air Force Base, California, April 4, 1952.

p. 544 say the family documents: Records of Schreiber's mother-in-law, Marie Conrad, and of Walter E. W. Paul Schreiber, at www.my heritage.com.

Chapter Nineteen: Truth Serum

p. 545 "the use of drugs and chemicals": CIA Memorandum for the Record, Subject Project Artichoke, January 31, 1975. This was part of the review prompted by John Marks's FOIA request and is part of the John Marks Collection at the National Security Archives. "Between 1950 and 1952, responsibility for mind-control went from the [CIA's] Office of Security to the Scientific Intelligence Unit back to Security again," writes Marks, who successfully petitioned the government in 1975 to release to him the MKUltra documents under the Freedom of Information Act.

p. 545 Blome's file: RG 330 Kurt Blome, Contract, DoD DA-91-501, December 3, 1951.

p. 545 "Bluebird was rechristened": Marks, 31.

p. 545 "modifying behavior through covert means": Ibid., 61.

p. 546 "We felt that it was our responsibility": Interview with Richard Helms, History Staff, Center for the Study of Intelligence,

Central Intelligence Agency, adapted from an interview with
Mr. Helms taped by David Frost in Washington, May 22–23, 1978.

p. 546 "to avoid duplication of effort": Marks, 61.

p. 547 "50 million doses": "Joint Hearing Before the Select
Committee on Intelligence," 91.

p. 547 SO Division: Marks, 61, 70–72; Regis, 116–19.

p. 548 secret memo to Richard Helms: Helms was Wisner's deputy
at this time. CIA's Program of Research in Behavioral Modification,"
August 3, 1977; "Joint Hearing Before the Select Committee on
Intelligence," 72–76.

p. 548 " 'Interrogation Techniques' ": CIA Memorandum for Deputy
Director (Plans), Subject Special Interrogation, February 12, 1951.

p. 548 since foreign governments: "Joint Hearing Before the Select
Committee on Intelligence," 101. This memo was read aloud in the
hearing.

p. 549 safe house called Haus Waldorf: Koch and Wech, 115–16;
the CIA called Camp King ECIC, or European Command Intelligence
Center, DOJ Klaus Barbie investigation, iii, List of Abbreviations.

p. 549 "Between 4 June 1952 and 18 June 1952": "Joint Hearing
Before the Select Committee on Intelligence," 68; Koch and Wech, 113.

p. 549 make them forget: CIA Memorandum for Director of
Central Intelligence, Subject: Artichoke, June 1952.

p. 550 Beecher was paid by the CIA: Interview with Egmont Koch,
August 6, 2012, in Germany; Koch and Wech, 102–5.

p. 550 "He had a tough time after Germany": The documentary
Codename Artichoke (2001), by Egmont Koch and Michael Wech,
minute 16:00.

p. 551 expanding its Artichoke program: the CIA assigns code cryptonyms to each of its projects. One vein of its drug-induced interrogation program at Detrick was called MKDETRIC until agents decided that that was too obvious and changed it to MKNAOMI, after Abramson's secretary, Naomi Busner.

p. 552 "the most frightening experience": Marks, 84.

p. 553 "Dr. Olson was in serious trouble": "Joint Hearing Before the Select Committee on Intelligence," 77.

p. 553 LSD tolerance experiments: Regis, 158. He was also working on a book about LSD, which would be published in 1959.

p. 554 Frank Olson's problems were all in his mind: "Joint Hearing Before the Select Committee on Intelligence," 75–78; Abramson gave Frank Olson a bottle of bourbon and the sedative Nembutal to sleep.

p. 554 "the delivery of various materials": Marks, 84.

p. 554 Olson became suspicious: "Joint Hearing Before the Select Committee on Intelligence," 75–76.

p. 555 room 1018A: Marks notes that room 1018 was actually thirteen floors up.

p. 556 empty his pockets: A curious detail from the Regis account—Lashbrook carried a slip of paper with thirty letters in a row, which Lashbrook said was the coded combination to a safe.

p. 557 from aircraft and crop dusters: home movie film footage from Frank Olson's camera, found by his son Eric, can be seen in *Codename Artichoke*, minute 11:00.

p. 557 as Senate hearings later revealed: "Joint Hearing Before the Select Committee on Intelligence," 3, 23, 68–78.

p. 558 "We used a spore," Cournoyer explained: *Codename Artichoke*, minute 16:00.

Chapter Twenty: In the Dark Shadows

p. 561 violated NATO regulations: Hunt, 182.

p. 562 fearing it would draw the ire: NARA 59 General Records of the Department of State, HICOG Cable from John McCloy to Secretary of State, February 21, 1952.

p. 562 the CIA continued to do: NARA 59 General Records of the Department of State, HICOG Memo from Karl Weber to P. G. Strong, March 18, 1952. It is worth noting that the CIA had its own photostatted copies of the Osenberg files, thanks to the JIOA, which included the names, biographical records, and addresses of nearly eighteen thousand German scientists from the Reich Research Council years—which the CIA used as a recruiting list.

p. 562 twenty-man team to Frankfurt: Dornberger files, "Calendar 1952," Deutsches Museum, Munich.

p. 562 When McCloy learned of the trip: Hunt, 182, 304n.

p. 562 A compromise was reached: Ibid., 182–83.

p. 563 target list: RG 330 U.S. High Commissioner for Germany, "Project 63 (Index 1948-52)."

p. 563 JIOA renamed Paperclip: Ibid.

p. 563 During his year-long tenure: Hunt, 2–3.

p. 563 FBI learned that Whalen: Hunt, 202–3.

p. 564 Whalen was a spy: Hunt, 2; RG 319, Records of the Army Staff, Subject: Technical Assesment Re: Former Retired LTC Willian Henry Whalen, 8 April, 1967; "Damage Assessment of Classified Documents," January 18, 1965.

p. 564 a grand jury was presented: "Indicted in Espionage," Associated Press, July 13, 1966.

p. 565 "truly revolutionary military offspring": Neufeld, *The Rocket and the Reich,* 274.

p. 565 "Developing Technology": http://www.acq.osd.mil.

p. 566 "the acquisition of unwarranted influence": York, 144.

p. 566 "Scientists and technologists had acquired": Ibid., 145.

p. 567 "Eisenhower's warnings": Ibid., 148.

p. 568 In 1950, military intelligence: RG 319 Walter Schieber, Agent Report, January 10, 1950.

p. 568 "first class business deal": Ibid.

p. 569 Special Agent Maxwell worried: Ibid.

p. 570 even worse news: RG 319 Walter Schieber, Agent Report, March 22, 1950; Agent Report, May 1, 1950. Schieber was also collecting a regular paycheck of 880 deutschmarks from the Office of the Land Commission for Baden-Württemberg in Stuttgart, in their Scientific Research Division. This meant that he was working for three governments at the same time.

p. 571 Operation Paperclip payroll until 1956: RG 263 Walter Schieber, "Official Dispatch, Chief of Mission, Frankfurt," November 18, 1963.

p. 572 "a small clique of senior congressmen": Tucker, 213.

p. 572 Sarin production took off: Ibid., 128; Keogh and Betsy, 10.

p. 572 Rocky Mountain Arsenal: Hylton, 59–70.

p. 572 Rocky Flats munitions loading plant: Hylton, 60; Tucker, 133.

p. 573 *Collier's* magazine published: Ryan, 89.

p. 574 VX by the thousands of tons: Tucker, 164–69.

p. 574 working on VX munitions: Hoffmann personal papers; Tucker, 160.

p. 575 "He was our searcher": Hunt, 159–61.

p. 575 He always flew military": Telephone interviews with Gabriella Hoffmann, October 17, 2012, and March 22, 2013.

p. 576 "non discernible microbioinoculator": "CIA poison arsenal explained," Associated Press, September 17, 1975.

p. 576 The SO Division's Agent Branch; "Joint Hearing Before the Select Committee on Intelligence," 9–12, 67–71; Regis, 151.

p. 576 favorite drink, a milkshake: Marks, 81.

p. 577 "We wish give every": Regis, 184.

p. 577 Gottlieb's plan: Housen, 1–21; Regis, 182–85.

p. 578 botulinum toxin lost its potency: Housen, 8.

p. 579 thousands of U.S. soldiers: Senate Committee on Veterans' Affairs, "Is Military Research Hazardous to Veterans' Health?" December 8, 1994.

p. 579 "the inhalation toxicity": Weeks and Yevich, 622–29.

p. 579 became even more deadly: Weeks, Weir, and Bath, 663. At the laboratory at Edgewood, Maurice Weeks conducted "inhalation exposure" experiments on rodents.

p. 579 recalls Gabriella Hoffmann: Telephone interview with Gabriella Hoffmann, October 17, 2012.

p. 580 American diplomat Sam Woods: Telephone interview with John Dippel, October 19, 2012.

p. 580 "I remember it very clearly": Telephone interview with Gabriella Hoffmann, October 17, 2012; "Agent Orange" Product Liability Litigation, 1–115.

p. 581 Hoffmann Trip Report: "Agent Orange" Product Liability Litigation, 52–53.

p. 583 Traub worked on virological research: RG 65 Erich Traub, Special Contract, War Department.

p. 583 Traub became friendly: Interview with Dr. Rolf Benzinger, February 19, 2013; the navy was conducting research on explosive decompression techniques in order to create what a Senate hearing later termed "the perfect concussion," meant to give an enemy combatant amnesia.

p. 583 agents and diseases being studied: Moon, 22.

p. 583 Congress mandated: Loucks Papers (USAMHI), "Fort Terry Historical Report: 1 January 1951–30 June 1952"; U.S. Army Chemical Corps; Robert Hurt, "Fort Terry Historical Report: 1 April–30 June 1953," July 31, 1953.

p. 584 choice for a director: RG 65 Erich Traub, File A-7183623.

p. 584 accepted a position: RG 65 Erich Traub, Special Contract. In German, the institute is called Bundesanstalt für Virusforschung.

p. 584 at his new home: RG 65 Traub, Agent Report, neighborhood check, January 28, 1963.

p. 585 Razi Institute: RG 65 Erich Traub, November 11, 1965.

p. 586 an entire group: Tucker, "Farewell to Germs: The U.S. Renunciation of Biological and Toxin Warfare, 1969, 1970," *International Security* 27, no. 1 (Summer 2002): 107–48.

p. 587 dumped in the ocean: Not until 1972 was dumping chemicals into the ocean prohibited by the Marine Protection, Research and Sanctuaries Act of 1972.

p. 587 not made to ever be dismantled: Tucker, 221–22.

p. 588 robotic separation of the chemical agent: Keogh and Betsy, 9.

p. 587 "The numbers speak volumes": U.S. Army Chemical Materials Agency, "A Success Story—Johnston Atoll Chemical Agent Disposal System," press release, September 21, 2005.

p. 588 program cost: Keogh and Betsy, 10.

Chapter Twenty-One: Limelight

p. 590 he kept track of: Dornberger files, Calendars 1949–1959, Deutsches Museum, Munich.

p. 591 beneficent science pioneer: Dornberger, *V-2: The Nazi Rocket Weapon;* McGovern, 46; In Dornberger papers housed at the Deutsches Museum, the unedited manuscript drafts of this book include re-creations of conversations Dornberger said he had with Himmler. In one, on the use of slave labor, Dornberger cites Himmler as having said to him, "[T]he power of Germany [meant] a return to the era of slavery." To this Dornberger says he wondered aloud if other nations might object, to which Himmler said, "After our victory they will not dare!"

p. 591 "It would be nice to know": Neufeld, *Creating a Memory of the German Rocket Program for the Cold War,* 78.

p. 592 There was "deafening silence": Heppenheimer, 133.

p. 592 "in the role of a double agent": RG 65 Walter Dornberger, FBI file No. 39-137.

p. 593 Fort Bliss rocket team moved: Biography of Dr. Wernher von Braun, First Center Director, July 1, 1960–Jan. 27, 1970 (available at www.nasa.gov).

p. 593 necessary first step: Neufeld, *Von Braun,* 246.

p. 594 "patriotic writing": Ibid., 271.

p. 594 "taken as a fundamental source": Neufeld, *Creating a Memory of the German Rocket Program for the Cold War,* 77.

p. 595 The first Disneyland TV broadcast: Wolper, *The Race to Space*, narrated by Mike Wallace; Brzezinski, 91.

p. 596 NASA constructed the Vertical Assembly Building: NASA facts, "Building KSC'S Launch Complex 39" (www.nasa.gov); McGovern, 251.

p. 596 designed by Bernhard Tessmann: Wolper, *The Race to Space*.

p. 596 chose not to correct the newsman: Neufeld, *Creating a Memory of the German Rocket Program for the Cold War*, 76.

p. 597 "Human skin, of course?": Neufeld, *Von Braun*, 271, 514n.

p. 598 the rocket would not hit London: Cornwell, 423.

p. 598 *The Secret of Huntsville:* Neufeld, *Creating a Memory of the German Rocket Program for the Cold War*, 84.

p. 598 The space agency's three top officials: Neufeld, *Von Braun*, 405–6.

p. 599 success in the Eastern bloc: Ibid., 408.

p. 600 "I don't know much more": Bundesarchiv Ludwigsburg, Rickhey Files, B-35/66, April 20, 1967.

p. 600 Essen-Dora Trial: Bundesarchiv Ludwigsburg, Rickhey Files, B-162/964.

p. 600 NASA lawyers suggested New Orleans: Neufeld, *Von Braun*, 428–29.

p. 601 "to broadcast the connection": Neufeld, *Von Braun*, 428; Telephone interview with Michael Neufeld, April 3, 2013.

p. 601 von Braun said no: Feigin, 330–35.

p. 601 von Braun and Dornberger corresponded: Neufeld, *Von Braun*, 429.

p. 603 "Sorry, but I can't support": Neufeld, *Von Braun*, 471.

p. 603 first ever U.S. military panel discussion on biology: Mackowski, 125–26.

p. 604 Albert was strapped into a harness: Tara Gray, "A Brief History of Animals in Space," NASA History Office, 1998, http://history.nasa.gov/printFriendly/animals.html.

p. 604 decided it was time: Hasdorff, 9.

p. 605 "we needed much larger accommodations": Bullard and Glasgow, 69.

p. 605 sell the idea to Congress: Ibid., 70–75.

p. 605 "I appointed myself Director": Ibid., 77.

p. 605 he traveled to Germany: RG 330 Hubertus Strughold, November 16, 1945.

p. 606 leaked to journalist Drew Pearson: Hunt, 54.

p. 606 at SAM, he worked: RG 330 Konrad Schäfer, March 27, 1951.

p. 606 make Mississippi River water drinkable: RG 263 Central Intelligence Agency, Hubertus Strughold file, A-1-2062.

p. 606 "The experience of this Headquarters": RG 330 Konrad Schäfer, March 27, 1951.

p. 607 that he be sent back to Germany: Ibid.

p. 607 He moved to New York City: RG 65 Konrad Schäfer, FBI File No. 105-12396. The FBI paid Schäfer at least one visit in New York, during which they "advised him of the jurisdiction of the FBI concerning espionage, sabotage and subversive activities." Schäfer promised the FBI that if the Russians were to try and recruit him, he would contact the FBI and let them know.

p. 608 in violation of the Nuremberg Code: "Memorandum to Advisory Committee on Human Radiation Experiments," April 5, 1995; Jacobsen, 302–3, 331.

p. 608 aboveground nuclear weapons tests: Jacobsen, xi–xvi, 107–23.

p. 608 Top Secret studies: USAF School of Aviation Medicine, Randolph Air Force Base, Randolph Field, Texas, "Trip Report," August 19, 1952 (FOIA).

p. 608 studies involving "flashblindness": Ibid. In addition to the flashblindness studied, SAM conducted experiments involving "Blasts, Burns, and Psychological" effects.

p. 609 "That's the thing that gave us curiosity": Welsome, 292.

p. 609 members of the SA: Ibid., 293.

p. 609 forty miles from ground zero: Welsome, 295.

p. 610 porthole-like windows: Harbert and Whittemore, 54.

p. 610 Some of the soldier volunteers: Ibid., 55.

p. 610 in declassified records as "S.H.": Ibid., 290.

p. 611 "one of the most beautiful images of a fireball": Harbert and Whittemore, 33.

p. 611 Monkey astronaut rocket tests: It would be another ten years before two monkeys, Able and Miss Baker, traveled into space and returned to earth alive. The taxidermied body of Able, strapped into his flight seat, is on display at the National Air and Space Museum.

p. 612 personally escort Farrell: Reichhardt, "First Up? Even Before NASA Was Created, Civilian and Military Labs Were in Search of Spacemen," *Air & Space,* September 2000.

p. 613 Benson singled out Strughold: Thomas, 54.

p. 613 "Johnson asked me": Ibid.

p. 613 "Nazi doctors during World War II": "Himmler the Scientist" by Julian Bach Jr. Letters to the Editor, *Saturday Review,* August 9, 1958.

p. 614 twenty-ninth meeting of the Aero Medical Association: Benford, *Doctors in the Sky: The Story of the Aero Medical Association,* 56.

p. 615 Strughold authored papers and journals: Mackowski, 215–16.

p. 616 "Only the janitor": Thomas, 49–56.

p. 616 Simon Wiesenthal: Bundesarchiv Ludwigsburg, Hubertus Strughold Files, 828/73.

p. 617 provided Wiesenthal with a copy: Ibid.

p. 617 included a list: Charles Allen Jr., "Hubertus Strughold, Nazi in U.S.," *Jewish Currents,* September 28, 1974.

p. 618 the freezing experiments in detail: Ralph Blumenthal, "Drive on Nazi Suspects a Year Later; No U.S. Legal Steps Have Been Taken," *New York Times,* November 23, 1974.

p. 619 told interviewer James C. Hasdorff: Hasdorff, 15, 42. Strughold also said he married a "pretty secretary" named Mary Webb Dalehite, in 1971, at the age of seventy-two.

p. 619 Hitler's "so-called enemy list": Hasdorff, 16. Strughold relayed to Hasdorff an almost certainly invented story involving the Nazi general and the American general George Patton. The scene was allegedly set in Garmisch-Partenkirchen, Germany, in the days or weeks after the end of the war. "This [American] General covered in full medals asked Dornberger, 'Are you the guy who was in charge of development of the V-2 rockets?'" Strughold quoted Dornberger as saying. "He [Dornberger] said, 'Ja wohl, Herr General. Yes, General.' Then [Patton] gave [Dornberger] three cigars and said, 'My Congratulations. I could not have done it [myself]!' Dornberger told me this story himself," said Strughold (Hasdorff, 19–20).

p. 619 distortion of reality: RG 330 Hubertus Strughold, JIOA Form No. 2, Basic Personnel Record.

Chapter Twenty-Two: Legacy

p. 620 large-scale public trial: Pendas, 1–3.

p. 621 "The trial was triggered": Hermann Langbein narrates this story in *Verdict on Auschwitz* (2007), directed by Rolf Bickel and Dietrich Wagner. The sequence begins at minute 7:00. The film is based on 430 hours of original audio tapes from the trial.

p. 622 filed murder charges: Bundesarchiv Koblenz, Otto Ambros Files, B 162/3221.

p. 622 leading normal lives: This is all told with stunning precision in *Verdict on Auschwitz*. The details are incredible. For example, Prosecutor Joachim Küger found Rudolph Höss's right-hand man by a twist of fate. He, Küger, was attending the Olympic Games in Rome when he noticed that an athlete with the last name of Mulka had won a medal for sailing. "Mulka is not a common name," explained Küger. This is how Küger was able to track down the Auschwitz adjunct Robert Mulka. The Olympic sailor was the son of the war criminal.

p. 625 were business colleagues: Bundesarchiv Koblenz, Dr. Otto Ambros Files, B 162.

p. 626 "I and my colleagues are the victims of the Third Reich": Ibid.

p. 627 "Former War Criminal": Ibid. The newspaper is not identified, only referred to in one of Ambros's letters.

p. 627 "The whole affair": Bundesarchiv Koblenz, Dr. Otto Ambros file, B 162; the letters are both dated April 25, 1964.

p. 628 "how I behaved during the Reich": Ibid.

p. 628 W. R. Grace: correspondence with Ambros's former colleague (from the 1950s) Michael Howard, 2012. Ambros was "always the most intelligent of those with whom he consorted," recalls Howard. "Ambros was a puppet master, the éminence grise."

p. 629 "Dr. Ambros had contacts": Memorandum for the president, from James W. Nance, the White House, Washington, D.C., March 23, 1982 (FOIA).

p. 630 Even the president: Ibid.; Letter to Congressman Tom Landon from James W. Nance, the White House, Washington, D.C., April 13, 1982 (FOIA).

p. 630 "It involved Jews": Hilberg, 1089.

p. 630 "to all the doctors and dentists": Bundesarchiv Ludwigsburg, Dr. Kurt Blome file, B 162/28667.

p. 632 *Sonderbehandlung:* Ibid.

p. 633 alone and estranged: interview with Götz Blome, August 3, 2012, in Germany.

p. 633 "I have a very strong conviction": Bird, 375.

p. 634 two sympathetic Dutch Red Cross nurses: Weindling, 309, 315.

p. 634 and a reported $350,000: Sereny 646; Van Der Vat, 328.

p. 635 "How often I thought of you": Neufeld, *Von Braun*, 472.

p. 635 Pilot Factors Program: Knemeyer personal papers; telephone interview with Dirk Knemeyer, June 20, 2012.

p. 635 "immediately taken to a crematory": Knemeyer, 70.

p. 636 "One seldom recognizes when the Devil": O'Donnell, "The Devil's Architect," SM45. The fate of Wilhelm Beiglböck is worth noting. He was the Nazi doctor who performed seawater experiments at Dachau, and he removed a piece of Karl Höllenrainer's liver without anesthesia. Convicted and sentenced to fifteen years in Landsberg Prison, he was granted clemency by U.S. High Commissioner John McCloy in 1951. Within a year, Beiglböck was back practicing medicine, at a hospital in Buxtehude, Germany, thanks to a former SS colleague, August Dietrich "Dieter" Allers, who ran the hospital. Beiglböck published medical papers and enjoyed prominence in the German medical community until 1962, when he traveled to his native Vienna to give a lecture. In Austria, there were open war crimes charges against Beiglböck, and he was arrested. Back in Germany, and likely as a result of attention from the Frankfurt Auschwitz trial, Dieter Allers was arrested around the same time. With no job to return to at the hospital, Beiglböck committed suicide on November 22, 1963. He was fifty-eight years old. He left all of his money to Die Stille Hilfe, or Silent Help, a clandestine society that Allers ran. It provided aid to fugitive SS members (Weindling, 309, 315).

Chapter Twenty-Three: What Lasts?

p. 638 "I do not reproach": Michel, 98.

p. 639 Rudolph relates: Ordway and Sharpe, 70–71, 77–79; Neufeld, *Von Braun*, 471.

p. 639 Rosenbaum later recalled: Feigin, 2.

p. 640 Arthur Rudolph's activities: Feigin, 300–342.

p. 643 One of Rudolph's leading proponents: Hunt, photographic insert opposite page 149.

p. 644 "not a single document": Neufeld, *The Rocket and the Reich*, 186–87, 227.

p. 644 "There is little doubt": Telephone interview with Michael Neufeld, April 3, 2013. Neufeld adds that in the summer of 1990, while researching *The Rocket and the Reich,* he came across a document in a German archive showing that Arthur Rudolph had been a staunch advocate of slave labor even before the creation of the Mittelwerk. On April 12, 1943, "Rudolph took a tour of the Heinkel aircraft plant and returned to Peenemünde excited about the use of slave labor," says Neufeld, notably because it offered partnership with the SS and "greater protection for secrecy."

p. 645 a bill for $239,680: Hunt's book, *Secret Agenda: The United States Government, Nazi Scientists, and Project Paperclip, 1945–1990,* was preceded by a book by British journalist Tom Bower, *The Paperclip Conspiracy: The Hunt for the Nazi Scientists,* for which Bower accessed unreported stories from both American and British archives. Both books helped me tremendously in my understanding of Operation Paperclip.

p. 646 Dr. Strughold had been listed: *New York Times*, "Portrait of Nazi Prompts Protest," October 26, 1993.

p. 647 "[t]he notion that": Breitman, Goda, Naftali, and Wolfe, 7.

p. 647 "there was no compelling reason": Ibid.

p. 648 chairman of the advisory committee: Letter to Congressman Tom Landon from James W. Nance, the White House, Washington, D.C., April 13, 1982 (FOIA).

p. 649 Grünenthal was a safe haven for many Nazis: Roger Williams and Jonathan Stone, "The Nazis and Thalidomide: The Worst Drug Scandal of all Time," *Newsweek,* September 10, 2012.

p. 649 "given with complete safety": Statistics about the release of the drug are available on the company's website: http://www.contergan.grunenthal.info.

p. 650 group of Nazi-era documents: Daniel Foggo, "Thalidomide 'Was Created by the Nazis,'" *Times* (London), February 8, 2009.

p. 650 Dr. Johnson points out: Correspondence with Dr. Martin Johnson, May 15, 2012; June 1, 2012; October 25, 2012.

p. 650 "The patents suggest that thalidomide": Andrew Levy, "Nazis Developed Thalidomide and Tested It on Concentration Camp Prisoners, Author Claims," *Daily Mail*, February 8, 2009.

p. 651 "the spring of 1943 Kuhn asked": *Nachrichten aus der Chemie,* May 2006, 514; the revelatory information is also contained in the 2007 edition of the *Complete Dictionary of Scientific Biography,* in the entry for Richard Kuhn.

p. 651 the many documents pertaining: Author's note: Pursuant to my FOIA denial and subsequent appeal, these records are awaiting review and possible release. However, the incumbent president has privilege over their release, per Executive Order 13489.

p. 652 "Ambros came to the United States in 1967": Jewish Telegraphic Agency, "Ambros, Convicted Nazi War Criminal, Abandons Plans to Visit U.S.," May 3, 1971.

p. 653 "The sensational value": *Daily Mail,* "Mengele's Food Coupons Found: Bizarre Insight into the Lives of Nazi Death Doctors Unearthed in Auschwitz House," March 23, 2010.

p. 653 "an expressive entrepreneurial figure": Quoted at the Wollheim Memorial, Fritz Bauer Institute, http://www.wollheim -memorial.de/en/home.

p. 654 page-one story: Lucette Lagnado, "Space Medicine Group Jousts over Fate of Strughold Prize," *Wall Street Journal,* December 1, 2012.

p. 654 came across evidence: Telephone interview with Hans-Walter Schmuhl. Schmuhl was in the archive of the Max-Planck Institute in or around 2008, researching Dr. Nachtsheim, when he came across the Strughold information. There is an even earlier reference about exactly this matter in Ute Deichmann, *Biologists Under Hitler* (237). I cite

Schmuhl, as it was the public release of his information that led to a policy change.

p. 654 Campbell blames the Internet: Interview and correspondence with Dr. Mark Campbell, February 4, 2013, and March 30, 2013.

p. 655 "Why defend him?": Lagnado, "Space Medicine Group Jousts."

p. 656 "Simple as it is, Kurt Debus is an honored American": Telephone interview with Steve Griffin, February 8, 2012.

p. 658 classify the list: Office of U.S. Chief of Counsel, APO 124-A. U.S. Army. List of Personnel Involved in Medical Research and Mercy Killings, n.d. (FOIA).

p. 658 obituary in 1999: Nick Ravo, "Dr. Theodor H. Benzinger, 94, Inventor of the Ear Thermometer," *New York Times*, October 30, 1999.

p. 659 lifelong scientific pursuit: Telephone interview with Rolf Benzinger, February 19, 2013.

p. 659 "This lasts," he said: Interview with Gerhard Maschkowski, June 2, 2012, California.

BIBLIOGRAPHY

*AUTHOR INTERVIEWS AND CORRESPONDENCE
CONDUCTED 2011–2013*

John Dolibois: Interrogator, prisoners of war, Central Continental Prisoner of War Enclosure Number 32, aka Ashcan. Army Intelligence, G-2

Vivien Spitz: American court reporter, Nuremberg war crimes trial, doctors' trial

Hugh Iltis: Translator of captured Nazi documents, Case No. 707-Medical Experiments, Himmler Papers

Michael Howard: British Intelligence Office No. 1, T-Force; W. R. Grace employee circa 1958

Ib Melchior: Agent, OSS and U.S. Army Counter Intelligence Corps

Gerhard Maschkowski: Prisoner at Auschwitz concentration camp, IG Auschwitz, Buna-Monowitz

Herman Shine: Prisoner at Auschwitz concentration camp, IG Auschwitz, Buna-Monowitz

Hanna Marx: Prisoner at Stutthof concentration camp

William Jeffers: Flight engineer, U.S. Army Air Forces, prisoner of war, Luftwaffe-Dulag Luft, Oberursel

Dr. Leonard Kreisler: Post doctor, Camp Detrick (later Fort Detrick), Site "R," also known as Raven Rock Mountain Complex

Dr. James Ketchum: Edgewood Army Chemical Center

Dr. Götz Blome: Son of Dr. Kurt Blome
Dirk Knemeyer: Grandson of Siegfried Knemeyer
Gabriella Hoffmann: Daughter of Dr. Friedrich "Fritz" Hoffmann
Dr. Dieter Ambros: Son of Dr. Otto Ambros
Dr. Rolf Benzinger: Son of Dr. Theodor Benzinger
Paul-Hermann Schieber: Son of Dr. Walter Schieber
Katrin Himmler: Grandniece of Heinrich Himmler
Eric Olson: Son of Dr. Frank Olson
Hanns-Claudius Scharff: Son of Hanns Scharff
Clarence Lasby: Author and historian
Werner Renz: Director, Fritz Bauer Institute
Dr. Harald Eichinger: Warden, Landsberg Prison
Albert Kroll: Director, Dachau concentration camp, archive and
library
Mathias Korn: Police commissioner, Dachau
Anna Naab: Dachau Police Department, historian
Manfred Kopp: Historian, Camp King, Oberursel
Jens Hermann: Caretaker, Castle Kransberg
Dr. Jens Westemeier: Historian, University of Potsdam
Egmont Koch: Journalist and author
Joe Houston: Civilian contributor, U.S. Army Intelligence and
Security Command
Dr. Martin Johnson: Thalidomide Trust, England
Mark Campbell: former present of the Space Medicine Association
Steve Griffin: Chairman, National Space Club
Michael J. Neufeld: Smithsonian National Air and Space Museum,
curator, Department of Space History

ARCHIVES

National Archives and Records Administration, College Park, Md.
RG 549 Records of United States Army, Europe
RG 466 Records of the U.S. High Commissioner for Germany
RG 341 Records of Headquarters U.S. Air Force
RG 331 Records of Supreme Headquarters Allied Expeditionary Force

RG 330 JIOA Foreign Scientist Case Files 1945–1958
RG 319 Records of the Investigative Records Repository Case Files
RG 319 Army Intelligence and Security Command
RG 319 Records of the Military Intelligence Division
RG 263 Records of the Central Intelligence Agency
RG 238 World War II War Crimes Records ("201 Files")
RG 226 Records of the Office of Strategic Services
RG 218 Records of the U.S. Joint Chiefs of Staff
RG 165 Records of the War Department General and Special Staffs
RG 153 Records of the Judge Advocate General (Army)
RG 65 Records of the Federal Bureau of Investigation
RG 59 General Records of the Department of State

U.S. Army Military History Institute, Carlisle, Pa.
Papers of Brigadier General Charles E. Loucks
Photograph Collection of Charles E. Loucks

Library of Congress, Washington, D.C.
Papers of the Office of Scientific Research and Development
Papers of the Veterans History Project

Harvard Law School Library, Cambridge, Ma.
Papers of Dr. Leopold Alexander

Harvard Medical School, Countway Library of Medicine, Boston, Ma.
Papers of Robert J. Benford
Papers of Henry K. Beecher

Wright State University, Special Collections and Archives, Dayton, Oh.
Papers of Theodor Benzinger

Duke University Medical Center Archives, Durham, NC.
Papers of Dr. Leopold Alexander

Hoover Institution, Stanford University, Palo Alto, Ca.
Papers of Boris Pash

Lilly Library, Indiana University, Bloomington, In.
Papers of Shirley Thomas

Pollak Library, California State University, Fullerton, Ca.
Papers of William Aalmans, Dora-Nordhausen trial

Mandeville Special Collections Library, UC San Diego, San Diego, Ca.
Papers of U.S. Army Air Forces, Aero Medical Center

Air Force Historical Research Agency, Maxwell Air Force Base, Montgomery, Al.
Air Materiel Command Historical Study No. 214 & No. 215, History of AAF Participation in Project Paperclip (Exploitation of German Scientists)

Archives of the California Institute of Technology, Pasadena, Ca.
Papers of Dr. Howard P. Robertson

Louis Salmon Library, University of Alabama, Huntsville, Al.
Papers of John Risen Jones Jr.

GERMAN ARCHIVES

Bundesarchiv Koblenz
Bundesarchiv Berlin: records of the former Berlin Document Center
Bundesarchiv Berlin-Lichterfelde
Bundesarchiv Militärarchiv Freiburg
Bundesarchiv Ludwigsburg
Bayerisches Staatsarchiv Nürnberg
Dachau concentration camp memorial site, archive and library, Dachau
Deutsches Museum archive, Munich
Wollheim Memorial, Fritz Bauer Institute, Frankfurt
The Agency of the Federal Commissioner for the Stasi records (BStU)

PERSONAL PAPERS, BOOKS, AND UNPUBLISHED MANUSCRIPTS

Siegfried Knemeyer
Dr. Kurt Blome
Friedrich Hoffmann
Paul-Hermann Schieber

BOOKS AND MONOGRAPHS

Alibek, Ken, with Stephen Handelman. *Biohazard: The Chilling True Story of the Largest Covert Biological Weapons Program in the World—Told from Inside by the Man Who Ran It.* New York: Random House, 2000.

Allen, Michael T. *The Business of Genocide: The SS, Slave Labor, and the Concentration Camps.* Chapel Hill: University of North Carolina Press, 2002.

Andrus, Col. Burton C., with Desmond Zwar. *I Was the Nuremberg Jailer.* New York: Coward-McCann, Inc., 1969.

Bar-Zohar, Michel. *The Hunt for German Scientists.* New York: Avon Books, 1970.

Baumbach, Werner. *The Life and Death of the Luftwaffe: Germany's "Lost Victories" of the Air by the Commander of Bomber Forces.* New York: Ballantine Books, 1972.

Beevor, Antony. *The Fall of Berlin 1945.* New York: Penguin Books, 2002.

Benford, Robert J., M.D. *Doctors in the Sky: The Story of the Aero Medical Association.* Springfield, IL: Charles C. Thomas Publisher, 1955.

Blome, Götz. *Bachflower Therapy: A Scientific Approach to Diagnosis and Treatment.* Rochester, VT: Healing Arts Press, 1999.

Blome, Kurt. *Arzt im Kampf.* Leipzig: Johann Ambrosius Barth, 1942.

Bower, Tom. *The Paperclip Conspiracy: The Hunt for the Nazi Scientists.* Boston: Little, Brown and Company, 1987.

Breitman, Richard, Norman J. W. Goda, Timonthy Naftali, and Robert Wolfe. *U.S. Intelligence and the Nazis.* New York: Cambridge University Press, 2005.

Burrows, William E. *This New Ocean: The Story of the First Space Age.* New York: Modern Library, 1999.

Campbell, Mark R., Stanley R. Mohler, Viktor A. Harsch, and Denise Baisden. "Hubertus Strughold: The 'Father of Space Medicine.'" *Aviation, Space, and Environmental Medicine* 78, no. 7 (July 2007): 716–19.

Chertok, Boris E. *Rockets and People.* Washington, D.C.: NASA History Office, 2005.

Clifford, Clark. *American Relations with the Soviet Union.* Washington, D.C.: Special Counsel to the President, September 24, 1947.

Cornwell, John. *Hitler's Scientists: Science, War and the Devil's Pact.* New York: Penguin, 2004.

Covert, Norman M. *Cutting Edge: A History of Fort Detrick, Maryland, 1943–1993.* Frederick, Maryland: Headquarters, U.S. Army Garrison, 1994.

Deichmann, Ute. *Biologists Under Hitler.* Translated by Thomas Dunlap. Cambridge, MA: Harvard University Press, 1996.

Diefendorf, Jeffry M., Axel Frohn, and Hermann-Josef Rupieper. *American Policy and the Reconstruction of West Germany, 1945–1955.* Cambridge: Cambridge University Press, 1994.

Dolibois, John E. *Pattern of Circles: An Ambassador's Story.* Kent, OH: Kent State University Press, 1989.

Dornberger, Walter. *V-2: The Nazi Rocket Weapon.* New York: Ballantine Books, 1954.

Drummer, Heike, and Jutta Zwilling. *Von Der Grüneburg Zum Campus Westend: Die Geschichte des IG Farben-Hauses.* Frankfurt: Präsidium der Johann Wolfgang Goethe-Universität, 2007.

DuBois Jr., Josiah E. *The Devil's Chemists: 24 Conspirators of the International Farben Cartel Who Manufacture Wars.* Boston: Beacon Press, 1952.

Dwork, Debórah, and Robert Jan van Pelt. *Auschwitz: 1270 to the Present.* New York: W. W. Norton, 1996.

Eckart, Wolfgang U., ed. *Man, Medicine, and the State: The Human Body as an Object of Government-Sponsored Medical Research in the 20th Century.* Stuttgart, Germany: Franz Steiner, 2006.

Evans, Richard J. *The Third Reich at War.* New York: Penguin, 2009.

Feigin, Judy. *The Office of Special Investigations: Striving for Accountability in the Aftermath of the Holocaust.* Washington, D.C.: Department of Justice Criminal Division, December 1996.

Ferencz, Benjamin B. *Less Than Slaves: Jewish Forced Labor and the Quest for Compensation.* Bloomington, IN: Indiana University Press, 2002.

Frankenthal, Hans. *The Unwelcome One: Returning Home from Auschwitz.* Evanston, IL: Northwestern University Press, 2002.

Franklin, Thomas. *An American in Exile: The Story of Arthur Rudolph.* Huntsville, AL: Christopher Kaylor Company, 1987.

Frei, Norbert. *Adenauer's Germany and the Nazi Past: The Politics of Amnesty and Integration.* Translated by Joel Golb. New York: Columbia University Press, 2002.

Gehlen, Reinhard. *The Service: The Memoirs of General Reinhard Gehlen.* Translated by David Irving. New York: Popular Library, 1972.

Gimbel, John. *Science, Technology, Reparations: Exploitation and Plunder in Postwar Germany.* Palo Alto, CA: Stanford University Press, 1990.

Goudsmit, Samuel A. *Alsos.* Los Angeles and San Francisco: Tomash Publishers, 1983.

Groehler, Olaf. *Der Lautlose Tod: Einsatz und Entwicklung deutscher Giftgase von 1914 bis 1945.* Hamburg: Rowohlt Taschenbuch Verlag GmbH, 1989.

Hager, Thomas. *The Alchemy of Air: A Jewish Genius, a Doomed Tycoon, and the Scientific Discovery That Fed the World but Fueled the Rise of Hitler.* New York: Three Rivers Press, 2008.

———. *The Demon Under the Microscope: From Battlefield Hospitals to Nazi Labs, One Doctor's Heroic Search for the World's First Miracle Drug.* New York: Harmony Books, 2006.

Halberstam, David. *The Fifties.* New York: Random House, 1993.

Harris, Robert, and Jeremy Paxman. *A Higher Form of Killing: The Secret History of Chemical and Biological Warfare.* New York: Random House, 2002.

Heigl, Peter. *Nuremberg Trials.* Amberg: Druckhaus Oberpfalz, 2007.

Henkel, Matthias, and Hans-Christian Taubrich. *Memorium Nuremberg Trials.* Museen det stadt Nürnberg, 2012.

Hersh, Seymour M. *Chemical & Biological Warfare: America's Hidden Arsenal.* New York: Anchor Books, 1969.

Hilberg, Raul. *The Destruction of the European Jews.* New Haven, CT: Yale University Press, 2003.

Hitler, Adolf. *Reden und Proklamationen, 1932–1945.* Edited by Max Domarus. Wauconda, IL: Bolchazy-Carducci Publishers, 2003.

———. *Hitler's Second Book: The Unpublished Sequel to Mein Kampf.* Translated by Krista Smith. Edited by Gerhard L. Weinberg. New York: Enignna Books, 2003.

———. *Table Talk, 1941–1944: His Private Conversations.* Translated by Norman Cameron and R. H. Stevens. Edited by H .R. Trevor-Roper. London: Enigma Books, 2000.

Howard, Michael. *Otherwise Occupied: Letters Home from the Ruins of Nazi Germany.* United Kingdom: Old Street Publishing Ltd., 2010.

Hunt, Linda. *Secret Agenda: The United States Government, Nazi Scientists, and Project Paperclip, 1944–1990.* New York: St. Martin's Press, 1991.

Huzel, Dieter K. *Peenemünde to Canaveral.* United States: Prentice-Hall, Inc., 1962.

Irving, David. *The Rise and Fall of the Luftwaffe: The Life of Field Marshal Ernhard Milch.* Focal Point, 2002.

Isaacson, Walter, and Evan Thomas. *The Wise Men: Six Friends and the World They Made.* New York: Simon & Schuster, 1986.

Jacobsen, Annie. *Area 51: An Uncensored History of America's Top Secret Military Base.* New York: Little, Brown and Company, 2011.

Jeffreys, Diarmuid. *Hell's Cartel: IG Farben and the Making of Hitler's War Machine.* New York: Metropolitan Books, 2008.

Karman, Theodore von. *Where We Stand: A Report Prepared for the AAF Scientific Advisory Group.* Dayton, OH: Headquarters, Air Materiel Command, Publications Branch, Intelligence T-2, May 1946.

Kershaw, Ian. *The End: The Defiance and Destruction of Hitler's Germany, 1944–1945.* New York: Penguin, 2011.

Kleber, Brooks E., and Dale Birdsell. *The Chemical Warfare Service: Chemicals in Combat.* Washington, D.C.: Center of Military History,

United States Army, 1990. The series title was *United States Army in World War II: The Technical Services.*

Koch, Egmont R., and Michael Wech. *Deckname Artischocke: Die geheimen Menschenversuche der CIA.* Munich: C. Bertelsmann Verlag, 2002.

Koritz, Thomas F. *USAF Pilot/Physician Program: History, Current Program, and Proposals for the Future.* Brooks Air Force Base, TX: USAF School of Aerospace Medicine, July 1989.

Krause, Joachim, and Charles K. Mallory. *Chemical Weapons in Soviet Military Doctrine: Military and Historical Experience, 1915–1991.* Boulder, CO: Westview Press, 1992.

Lasby, Clarence G. *Project Paperclip: German Scientists and the Cold War.* New York: Atheneum, 1971.

Lebert, Norbert, and Stephan Lebert. *My Father's Keeper: Children of Nazi Leaders—An Intimate History of Damage and Denial.* Translated by Julian Evans. Boston: Little, Brown and Company, 2001.

Lifton, Robert Jay. *The Nazi Doctors: Medical Killing and the Psychology of Genocide.* New York: Basic Books, 1986.

Lockwood, Jeffrey A. *Six-Legged Soldiers: Using Insects as Weapons of War.* New York: Oxford University Press, 2009.

Longden, Sean. *T-Force: The Race for Nazi War Secrets, 1945.* London: Constable & Robinson Ltd., 2009.

Longerich, Peter. *Heinrich Himmler.* Translated by Jeremy Noakes and Lesley Sharpe. New York: Oxford University Press, 2012.

Lord, M. G. *Astro Turf: The Private Life of Rocket Science.* New York: Walker & Company, 2006.

Mackowski, Maura Phillips. *Testing the Limits: Aviation Medicine and the Origins of Manned Space Flight.* College Station: Texas A&M University Press, 2006.

Marks, John. *The Search for the "Manchurian Candidate": The CIA and Mind Control: The Secret History of the Behavioral Sciences*. New York: W. W. Norton, 1991.

Mauroni, Albert J. *Chemical and Biological Warfare: A Reference Handbook*. Santa Barbara, CA: ABC-CLIO, 2007.

McGovern, James. *Crossbow & Overcast*. London: Hutchinson & Co., 1965.

Melchior, Ib. *Case by Case: A U.S. Army Counterintelligence Agent in World War II*. Novato, CA: Presidio Press, 1993.

———. *Order of Battle: Hitler's Werewolves*. Novato, CA: Presidio Press, 1991.

Michel, Ernest W. *Promises to Keep: One Man's Journey Against Incredible Odds*. New York: Barricade Books, 1993.

Michel, Jean, and Louis Nucera. *Dora: The Nazi Concentration Camp Where Modern Space Technology Was Born and 30,000 Prisoners Died*. New York: Holt Rinehart and Winston, 1979.

Miller, Judith, Stephen Engelberg, and William Broad. *Germs: Biological Weapons and America's Secret War*. New York: Simon & Schuster, 2001.

Mole, Robert L., and Dale M. Mole. *For God & Country*. New York: Teach Services, Inc., 1998.

Moreno, Jonathan D. *Undue Risk: Secret State Experiments on Humans*. New York: W. H. Freeman and Company, 1999.

Nachama, Andreas. *Topography of Terror—Gestapo, SS, and Reich Security Main Office on Wilhelm- and Prinz Albrecht-Strasse*. Berlin: Stiftung Topagraphie des Terrors, 2010.

Naimar, Norman M. *The Russians in Germany: A History of the Soviet Zone of Occupation, 1945–1949*. Cambridge, MA: Belknap Press, 1997.

Nelson, Donald M. *Arsenal of Democracy: The Story of American War Production*. New York: Harcourt, Brace and Company, 1946.

Neufeld, Michael J. *Creating a Memory of the German Rocket Program for the Cold War*. Washington, D.C.: NASA Special Publications, 2008.

————. *The Rocket and the Reich: Peenemunde and the Coming of the Ballistic Missile Era*. New York: The Free Press, 1995.

————. *Von Braun: Dreamer of Space, Engineer of War*. New York: Alfred A. Knopf, 2007.

Newell, Homer E., Jr. *Beyond the Atmosphere: Early Years of Space Science*. Washington, D.C.: NASA, 1980.

O'Donnell, James P. *The Bunker: The History of the Reich Chancellery Group*. Boston: Houghton Mifflin Company, 1978.

Oleynikov, Pavel V. "German Scientists in the Soviet Atomic Project." *The Nonproliferation Review* 7, no. 2 (Summer 2000): 1–30.

Ordway, Frederick I. III, and Mitchell R. Sharpe. *The Rocket Team*. New York: Thomas Y. Crowell, 1979.

Overy, Richard. *Interrogations: The Nazi Elite in Allied Hands, 1945*. New York: Penguin, 2002.

Pash, Boris T. *The Alsos Mission*. New York: Charter Books, 1969.

Pendas, Devin O. *The Frankfurt Auschwitz Trial, 1963–1965. Genocide, History, and the Limits of the Law*. New York: Cambridge University Press,

Pogue, Forrest C. *U.S. Army in World War II, European Theatre of Operations, the Supreme Command*. Washington, D.C.: Center of Military History, United States Army, 1954.

Regis, Ed. *The Biology of Doom: The History of America's Secret Germ Warfare Project*. New York: Henry Holt and Company, 1999.

Ryan, Allan A. Jr. *Quiet Neighbors: Prosecuting Nazi War Criminals in America*. San Diego: Harcourt Brace Jovanovich, 1984.

Samuel, Wolfgang W. E. *American Raiders: The Race to Capture the Luftwaffe's Secrets*. Jackson: University Press of Mississippi, 2004.

Sasuly, Richard. *IG Farben*. New York: Boni & Gaer, 1947.

Sayer, Ian, and Douglas Botting, with the *London Sunday Times*. *Nazi Gold: The Biggest Robbery in History*. New York: Grove Press, 1986.

———. *The Untold Story of the Counter Intelligence Corps, America's Secret Army*. New York: Franklin Watts, 1989.

Scalia, Joseph M. *Germany's Last Mission to Japan: The Failed Voyage of U-234*. Annapolis, MD: Naval Institute Press, 2000.

Schmaltz, Florian. *The Death Toll at the Buna/Monowitz Concentration Camp*. Frankfurt am Main: J. W. Goethe-Universität, 2010.

Schmidt, Amy, and Gudrun Loehrer. *The Mauthausen Concentration Camp Complex: World War II and Postwar Records*. Washington, D.C.: National Archives and Records Administration, 2008.

Schmidt, Ulf. *Justice at Nuremberg: Leo Alexander and the Nazi Doctors' Trial*. New York: Palgrave Macmillan, 2004.

———. *Karl Brandt: The Nazi Doctor: Medicine and Power in the Third Reich*. London: Hambledon Continuum, 2007.

Schwartz, Thomas A. *America's Germany: John J. McCloy and the Federal Republic of Germany*. Cambridge, MA: Harvard University Press, 1991.

Sereny, Gitta. *Albert Speer: His Battle with Truth*. New York: Alfred A. Knopf, 1995.

Shirer, William L. *End of a Berlin Diary.* New York: Alfred A. Knopf, 1947.

———. *The Rise and Fall of the Third Reich.* New York: Simon & Schuster, 1960.

Simpson, Christopher. *Blowback: The First Full Account of America's Recruitment of Nazis and Its Disastrous Effects on Our Domestic and Foreign Policy.* New York: Weidenfeld & Nicolson, 1998.

Smith, Bradley F. *Adolf Hitler: His Family, Childhood, and Youth.* Palo Alto, CA: Hoover Institution Press/Stanford University, 1986.

Smith, Marcus J. *Dachau: The Harrowing of Hell.* Albany, NY: State University of New York Press, 1995.

Speer, Albert. *Inside the Third Reich: Memoirs.* Translated by Richard Winston and Clara Winston. New York: Galahad Books, 1995.

———. *Spandau: The Secret Diaries.* Translated by Richard Winston and Clara Winston. New York: Macmillan, 1976.

Stafford, Peter. *Psychedelics Encyclopedia.* Berkeley, CA: Ronin Publishing, 1993.

Strughold, Hubertus. *Your Body Clock.* New York: Charles Scribner's Sons, 1971.

Swift, Shayla. *Lost Lessons: American Media Depictions of the Frankfurt Auschwitz Trials 1963–1965.* Lincoln: DigitalCommons@ University of Nebraska, April 8, 2006.

Taylor, Telford. *The Anatomy of the Nuremberg Trials: A Personal Memoir.* New York: Little, Brown and Company, 1992.

———. *Nuremberg and Vietnam: An American Tragedy.* New York: Bantam Books, 1971.

———. *Final Report to the Secretary of the Army on the Nuernberg War Crimes Trials Under Control Council Law No. 10.* Washington, D.C.: U.S. Printing Office, August 15, 1949.

Thomas, Shirley. *Men of Space, Volume 7: Profiles of the Leaders in Space Research, Development, and Exploration.* Philadelphia: Chilton Books, 1965.

Thompson, Nicholas. *The Hawk and the Dove: Paul Nitze, George Keenan, and the History of the Cold War.* New York: Henry Holt and Company, 2009.

Thomson, Ian. *Primo Levi: A Life.* New York: Metropolitan Books, 2002.

Toliver, Raymond F. *The Interrogator: The Story of Hanns Joachim Scharff—Master Interrogator of the Luftwaffe.* Atglen, PA: Schiffer Military History, 1997.

Tucker, Jonathan B. *War of Nerves: Chemical Warfare from World War I to Al-Qaeda.* New York: Pantheon, 2006.

Urban, Markus. *The Nuremberg Trials: A Short Guide.* Translated by John Jenkins. Nuremberg: Geschichte für Alle e.V.—Institut für Regionalgeschichte, 2012.

Van Der Vat, Dan. *The Good Nazi: The Life and Lies of Albert Speer.* New York: Houghton Mifflin Company, 1997.

Wagner, Jens-Christian. *Produktion Des Todes: Das Kz Mittelbau-Dora.* Göttingen: Wallstein, 2001.

Wegener, Peter P. *The Peenemünde Wind Tunnels: A Memoir.* New Haven, CT: Yale University Press, 1996.

Weichert, Klaus. *100 Jahre JVA Landsberg am Lech: Ein Chronik über 100 Jahre.* Nuremberg, Germany: 2008.

Weindling, Paul Julian. *Nazi Medicine and the Nuremberg Trials: From Medical War Crimes to Informed Consent.* New York: Palgrave Macmillan, 2004.

Weiner, Tim. *Legacy of Ashes: The History of the CIA.* New York: Anchor Books, 2008.

Weinreich, Max. *Hitler's Professors: The Part of Scholarship in Germany's Crimes Against the Jewish People.* New Haven, CT: Yale University Press, 1999. (Originally published in 1946 by the Yiddish Scientific Institute, with a new introduction by Martin Gilbert.)

Welsome, Eileen. *The Plutonium Files: America's Secret Medical Experiments in the Cold War.* New York: Dial Press, 1999.

Wheelis, Mark, Lajos Rózsa, and Malcolm Dando, eds. *Deadly Cultures: Biological Weapons Since 1945.* Cambridge, MA: Harvard University Press, 2006.

Wiesenthal, Simon. *The Sunflower: On the Possibilities and Limits of Forgiveness.* New York: Schocken Books, 1970.

Winterbotham, F. W. *The Nazi Connection: The True Story of a Top-Level British Agent in Pre-War Nazi Germany.* London: Granada, 1978.

York, Herbert F. *Arms and the Physicist.* Woodbury, NY: American Institute of Physics Press, 1995.

GOVERNMENT DOCUMENTS

Aalders, Gerard. *Operation Safehaven.* Amsterdam: Netherlands Institute for War Documentation, 1996.

Aalmans, William J. *A Booklet with a Brief History of the "Dora"-Nordhausen Labor Concentration Camps and Information on the Nordhausen War Crimes Case of the United States of America Versus Arthur Kurt Andrae et al.: Trial Commenced at Camp Dachau, Germany, August 7, 1947.* [Germany]: Prosecution Staff, 1947.

"Agent Orange" Product Liability Litigation, United States Court of Appeals for the Second Circuit, Brief for Defendants-Appellees on the Government Contractor Defense, No. 05-1760-CV: New York, n.d.

"The Air Force and the Worldwide Military Command and Control System." Montgomery, AL: USAF Historical Division, Liaison Office.

"An Air Force History of Space Activities, 1945–1959." United States Air Force Historical Division, Liaison Office, August 1964.

Alexander, Leo. "German Military Neuropsychiatry and Neurosurgery." CIOS Item 24, August 2, 1945.

————. "Miscellaneous Aviation Medical Matters." CIOS Item 24, File no. XXIX-21, n.d.

————. "Neuropathology and Neurophysiology, Including Electro-encephalography, in Wartime Germany." CIOS Item 24, File no. XXVII-1, n.d.

————. "The Treatment of Shock from Prolonged Exposure to Cold, Especially in Water." CIOS Report 24/CIOS Target No. 24 Medical, File no. XXVI-37, n.d.

Anderson, Arthur O. *Biowarfare to Biodefense: Operation Whitecoat & USAMRIID History.* U.S. Army Medical Research Institute of Infectious Diseases Office of Human Use and Ethics, 2011.

Andrews, A. H. "The Aviation Medicine Organization of the Luftwaffe." CIOS Target Number 24/36, File No. XXIII-10, May 1945.

Armed Services Technical Information Agency. AD 20374. Dayton, OH: Document Service Center, February 4, 2009.

Army Air Forces Aero Medical Center. "Report from Heidelberg: The Story of the Army Air Forces Aero Medical Center in Germany, 1945–1947."

"Aviation Medicine: From the Aeronauts to the Eve of the Astronauts." Bethesda, MD: National Library of Medicine, February 1979.

Benford, Robert J. "Report from Heidelberg: The Story of the Army Air Forces Aero Medical Center in Germany, 1945–1947." Unpublished.

Brophy, Leo P., with Wyndham D. Miles and Rexmond C. Cochrane. *The Chemical Warfare Service: From Laboratory to Field.* U.S. Center of Military History Publication 10-2. Washington, D.C.: U.S. Government Printing Office, 1959.

Central Intelligence Agency. "50 Years of Supporting Operations." Office of Technical Services, March 7, 2007.

———. Participation in Planning of Department of Defense Experiments, Advisory Committee on Human Radiation Experiments, September 1, 1994.

Central Registry of War Criminals and Security Suspects, Consolidated Wanted Lists, Part 1, Part 2, Supplementary Wanted List No. 2, CROWCASS Allied Control Authority, APO 742, U.S. Army. September 1947.

Cole, Ronald H., with Walter S. Poole, James F. Schnabel, Robert J. Watson, and Willard J. Webb. *The History of the Unified Command Plan, 1946–1993.* Washington, D.C.: Office of the Chairman of the Joint Chiefs of Staff, 1995.

Committee on Human Resources. Biological Testing Involving Human Subjects by the Department of Defense, 1977. Washington, D.C.: U.S. Government Printing Office, 1977.

Committee on Toxicology, National Research Council. "Review of Acute Human-Toxicity Estimates for Selected Chemical-Warfare Agents." Washington, D.C.: National Academies Press, 1997.

"Counter Intelligence Corps History and Mission in World War II." The Counter Intelligence Corps School. Baltimore, MD: Counter Intelligence Corps Center.

Cromartie, William J., and Carl Henze. "Medical Targets in the Strasbourg Area." CIOS Black List Item 24 Medical, File no. XIII-8, XXX-9, and XIV-3, n.d.

Dean, Jay B. *Aviator vs. the Environment: Aeromedical Research and the Physiology of High-Altitude Flight During WWII.* Dayton OH: Wright State University School of Medicine, n.d.

Dempsey, Charles A. *Fifty Years of Research on Man in Flight.* Air Force Aerospace Medical Research Lab, Wright-Patterson AFB, OH: U.S. Air Force, 1985.

Department of Commerce. Classified List of OTS Printed Reports. Washington, D.C.: Office of Technical Services Reports Division, October 1947.

Dick, Steven J., ed. *Remembering the Space Age.* Washington, D.C.: National Aeronautics and Space Administration Office of External Relations, History Division, 2008.

Dornberger, Walter. *Centralized vs. Decentralized Development of Guided Missiles.* Fort Bliss, TX: Ordnance Department U.S.A., 1948.

Dowden, H. J. "Examination of Dr. Ing. [Engineering] W. Osenberg and Documents." CIOS Target No. 28, File no. XXXI-49, June 25th–July 2nd, 1945.

"Elimination of German Resources for War." Hearings Before a Subcommittee of the Committee on Military Affairs, Volumes 10–11. Washington, D.C.: U.S. Government Printing Office, April 1946.

Feigin, Judy. The Office of Special Investigations: Striving for Accountability in the Aftermath of the Holocaust, Department of Justice, Criminal Division. New York, December 2006.

Greene, Wilson L. *Psychochemical Warfare: A New Concept of War.* Army Chemical Center, August 1949.

Harbert, John, and Gil Whittemore. "Advisory Committee on Human Radiation Experiments, Interview with Colonel John Pickering." Transcript. Albuquerque, NM, November 2, 1994.

Helms, Richard. History Staff, Center for the Study of Intelligence, Central Intelligence Agency, Vol. 44, No. 4; Adapted from an interview with Helms taped by David Frost in Washington, D.C., May 22–23, 1978.

Hilmas, Corey J., Jeffery K. Smart, and Benjamin A. Hill, "History of Chemical Warfare." In *Medical Aspects of Chemical Warfare,* edited by Martha K. Lenhart. Washington, D.C.: Borden Institute/U.S. Government Printing Office, 2008.

Housen, Roger T. "Why Did the U.S. Want to Kill Prime Minister Lumumba of the Congo?" Washington, D.C.: National War College, 2002.

Hylton, A. R. *The History of Chemical Warfare Plants in the United States.* Kansas City: Midwest Research Institute, U.S. Arms Control and Disarmament Agency, 1972.

Intelligence Throughout History: U.S. Intelligence and the German Invasion of the Soviet Union, Studies in Intelligence, CIA Story Archive, 2011.

Joint Hearing before the Select Committee on Intelligence and the Subcommittee on Health and Scientific Research of the Committee on Human Resources, United States Senate, Ninety-Fifth Congress, First Session, August 3, 1977. "Project MKUltra, the CIA's Program of Research in Behavioral Modification." Washington, D.C.: U.S. Government Printing Office, 1977.

Kent, Sherman. *Sherman Kent and the Board of National Estimates: Collected Essays.* Washington, D.C.: History Staff, Center for the Study of Intelligence, Central Intelligence Agency, 1994.

Keogh, James, and Tom Betsy. *Assessment of the Continuing Operability of Chemical Agent Disposal Facilities and Equipment.* Committee on Continuing Operability of Chemical Agent Disposal Facilities and Equipment, National Research Council. Washington D.C.: National Academies Press, March 16, 2007.

"The Last Days of Ernst Kaltenbrunner: Personal Recollections of the Capture and Show Trial of an Intelligence Chief." Washington, D.C.: Central Intelligence Agency, September 22, 1993.

Lovelace, W. R. "Research in Aviation Medicine for the German Air Force." CIOS Item Number 24 Medical, File no. XXVI-56, n.d.

Nazi War Crimes & Japanese Imperial Government Records Interagency Working Group: Final Report to the United States Congress. Washington, D.C.: National Archives, April 2007.

Office of the Chief, Chemical Corps. "Classification of Quick-Acting, Non-persistent Agent, GB, as a Substitute Standard Type." Edgewood, MD: Army Chemical Center, 1948.

O'Mara, John. "Long-Range Policy on German Scientific and Technological Research, AIF." CIOS XXXII-66, May 16, 1945.

"Paperclip, Part I." *Office of Naval Intelligence Review,* February 1949.

Pash, Boris T. "Code Name: Alsos Mission, Scientific Intelligence Mission, A Photographic History of the Alsos Mission, 11 May 1944– 15 November 1945."

Peyton, G. *Fifty Years of Aerospace Medicine.* AFSC Historical Publication Series No. 67-180, Bethesda, Maryland, 1968.

"Report to the Secretary of War by Mr. George W. Merck, Special Consultant for Biological Warfare." Washington, D.C.: January 1946.

Ruffner, Kevin C. "Forging an Intelligence Partnership: CIA and the Origins of the BND, 1945–1949." A Documentary History, Vol. 1.

CIA History Staff, Center for the Study of Intelligence, European Division, 1999 (declassified 2002 per Nazi War Crimes Disclosure Act).

Savage, Charles. *Lysergic Acid Diethyl Amide (LSD-25): A Clinical-Psychological Study.* Naval Medical Research Institute. Bethesda, MD: National Naval Medical Center, September 9, 1951.

Schreiber, Walter Paul, M.D. Affidavit of January 23, 1952, recorded by Dolly C. Buechel, Notary Public, Bexar County, Texas.

Simon, Leslie M. "Special Mission on Captured German Scientific Establishment: Artillery and Weapons, Rockets and Rocket Fuels, Guided Missiles, Aircraft Instruments and Equipment." CIOS Items 2, 4, 6, 25 & 27, File no. XXX-71, June 1, 1945.

Technical Report No. 331-45, "German Aviation Medical Research at the Dachau Concentration Camp." October 1945. U.S. Naval Technical Mission in Europe.

Tilley, Edmund. "Report on the Finding of Evidence of Hermann Schmitz's Connection with and Knowledge of the Auschwitz Concentration Camp," April 11, 1947. Archive of the Fritz Bauer Institute.

Tilley, Edmund, and J. M. Whitten. "Interrogation of German Scientific Personnel I.G. Farbenindustrie A.G., Ludwigshafen, March 25–31, 1945." CIOS Item No. 30, File no. XXV-49, April 10, 1945.

United States Army. Army Intelligence & Security Command. On the Trail of Military Intelligence History: A Guide to the Washington, D.C., Area. Washington, D.C.: U.S. Army Intelligence and Security Command History Office, n.d.

United States of America v. Karl Brandt et al., Case No. 1 (The Medical Case), Vols. 1 and 2. Washington, D.C.: U.S. Government Printing Office, 1946–June 1948.

U.S. Department of the Army. "Soviet Research and Development Capabilities for New Toxic Agents." Intelligence Staff Study, Project

No. A-1735. Washington, D.C.: Office of the Assistant Chief of Staff for Intelligence, July 28, 1958.

Weber, Karl H. "The Directorate of Science and Technology, Historical Series, Top Secret, The Office of Scientific Intelligence, 1949–1968, Volume II, Annexes IV, V, VI and VII." OSI-1, June 1972. FOIA/Declassified 2008.

Weir, Francis W., Dale W. Bath, and Maurice H. Weeks. "Short Term Inhalation Exposures of Rodents to Pentaborane-9," ASD Technical Report 61-663, Chemical Research and Development Laboratories, Army Chemical Center. Bethesda, MD, December 1961.

Westphal, Otto. *Research on the Production and Immunological Examination of Artificial Antigens Containing Known Sugars (or Oligosaccharides) as the Determinant Groups, in Relation to the Immuno Chemical Analysis of Enterobacterial O-Antigens (Endotoxins).* Frankfurt am Main: Wander Forschungsinstitut, November 30, 1962.

Winfield, R. H. "Preliminary Report of Points of Interest in Aviation Medicine and Physiology in Belgium and France." CIOS Black List Item 24 Medical, File no. XIII-10, January 8, 1945.

Ziemke, Earl F. *The U.S. Army in the Occupation of Germany, 1944–1946.* Center of Military History, United States Army. Washington, D.C.: U.S. Government Printing Office, 1990.

MICROFILM

Records of the United States Nuremberg War Crimes Trials: *United States of America v. Carl Krauch et al.* (Case VI). Washington, D.C.: National Archives Microfilm Publications, 1977.

Records of the United States Nuremberg War Crimes Trials: United States Army Investigation and Trials Records of War Criminals: *United States of America v. Kurt Andrae et al.* (and Related Cases) April 27,

1945–June 11, 1985. Washington, D.C.: National Archives Trust Fund Board National Archives and Record Services, 1981.

ARTICLES

Allen, Charles Jr. "Hubertus Strughold, Nazi in U.S." *Jewish Currents,* September 28, 1974.

"Ambros, Convicted Nazi War Criminal, Abandons Plans to Visit U.S." Jewish Telegraphic Agency, May 3, 1971.

Bach, Julian Jr. "Himmler the Scientist." *Saturday Review,* August 9, 1958.

Beasley, Norman. "The Capture of the German Rocket Secrets Military Intelligence: Its Heroes and Legends." *American Legion Magazine,* October 1963.

Blumenthal, Ralph. "Drive on Nazi Suspects a Year Later: No U.S. Legal Steps Have Been Taken." *New York Times,* November 23, 1974.

Boeri, David. "Looking Out: Nazis on the Harbor." *WBUR,* August 19, 2010.

Campbell, Mark R., et al. "The Controversy of Hubertus Strughold during World War II," http://www.spacemedicineassociation.org/strughold.htm.

————. "Hubertus Strughold: 'The Father of Space Medicine.'" Aviation Space Environmental Medicine, http://www.spacemedicineassociation.org/strughold.htm.

Conason, Joe, with Martin A. Rosenblatt. "The Corporate State of Grace: Reagan's Friends Tries to Bury the Past." *Village Voice,* April 12, 1983.

Central Intelligence Agency. "The Last Days of Ernst Kaltenbrunner: Personal Recollections of the Capture and Show Trial of an Intelligence Chief." September 22, 1993.

Crowther, Bosley. "About von Braun: *I Aim at the Stars* Opens at the Forum." *New York Times,* October 20, 1960.

Day, Matthew. "SS Documents Discovered Near Auschwitz." *Telegraph,* March 23, 2010.

Eliot, George F. "Our Armed Forces: Merger or Coördination?" *Foreign Affairs,* January 1946.

Epstein, Edward J. "The Spy Wars." *New York Times Magazine,* September 28, 1980.

Foggo, Daniel. "Thalidomide 'Was Created by the Nazis.'" *Sunday Times,* February 8, 2009.

"Foreign Relations of the United States, 1950–1955: The Intelligence Community, Document 244." U.S. Department of State Office of the Historian, December 1955.

Gapay, Les. "The Holocaust and the Sins of the Father." *Los Angeles Times,* May 6, 2012.

Hunt, Linda. "U.S. Coverup of Nazi Scientists." *Bulletin of the Atomic Scientists* 41, no. 4 (April 1985).

Lagnado, Lucette. "Space Medicine Group Jousts over Fate of Strughold Prize." *Wall Street Journal,* December 1, 2012.

Lang, Daniel. "A Romantic Urge." *The New Yorker,* April 21, 1951.

Marlow, James. "60 Million Jobs. Late Henry Wallace's Dream Comes True." Associated Press, April 6, 1966.

McBee, Susan. "Interviewing an Accused Spy." *Life,* July 22, 1966.

McCoy, Alfred W. "Science in Dachau's Shadow: HEBB, Beecher, and the Development of CIA Psychological Torture and Modern Medical Ethics." *Journal of the History of the Behavioral Sciences* 43, no. 4 (Fall 2007): 401–17.

Middleton, Drew. "7 Nazis Executed for War Murders." *New York Times,* June 7, 1951.

"Nazi Brains Help U.S." *Life,* December 9, 1946.

"Nazi Doctors to Be Tried Next." *Stars and Stripes,* October 12, 1946.

"Nazi Escape Plan Revealed to Fly Leaders to Greenland." *Scotsman,* December 28, 2003.

New York Times. "Cured in U.S. of Her Ills, Left as Nazi Guinea Pig." March 7, 1952.

O'Donnell, James P. "The Devil's Architect." *New York Times Magazine,* October 26, 1969.

Pearson, Drew. "Air Force Hires Nazi Doctor Linked to Ghastly Experiments." Associated Press, February 14, 1952.

Ravo, Nick. "Dr. Theodor H. Benzinger, 94, Inventor of the Ear Thermometer," Obituary, *New York Times,* October 30, 1999.

Reichhardt, Tony. "First Up? Even Before NASA Was Created, Civilian and Military Labs Were in Search of Spacemen." *Air & Space,* September 2000.

Reuters. "Factbox: Five Facts about Russian Military Intelligence." April 24, 2009.

Roosevelt, Eleanor. My Day. United Feature Syndicate, December 14, 1949.

Rule, Andrew. "Thalidomide, A Wreckage of Innocent Lives." *Daily Telegraph,* June 27, 2011.

Ryan, Cornelius. "G-Gas: A New Weapon of Chilling Terror. We Have It—So Does Russia." *Collier's,* November 1953.

Schmidt, Dana Adams. "Germans on Trial in 'Science Crimes.'" *New York Times,* December 10, 1946.

Shane, Scott. "C.I.A. Knew Where Eichmann Was Hiding, Documents Show," *New York Times,* June 7, 2006.

———. "Md. Experts' Key Lessons on Anthrax Go Untapped." *Baltimore Sun,* November 4, 2001.

Shaw, Herbert. "Wright Field Reveals 'Operation Paperclip.'" *Dayton Daily News,* December 4, 1946.

Shevell, Michael I. "Neurosciences in the Third Reich: From Ivory Tower to Death Camps." *Canadian Journal of Neurological Sciences* 26, no. 2 (May 1999): 75–76.

Siegel, Barry. "Can Evil Beget Good? Nazi Data: A Dilemma for Science." *Los Angeles Times,* October 30, 1988.

Silver, Arnold M. "Questions, Questions, Questions: Memories of Oberursel." *Intelligence and National Security* 8, no. 2 (April 1993): 202–6.

Stoll, Werner A. "Lysergsäure-diethylamid: Ein Phantastikum aus der Mutterkorngruppe." *Schweizer Archiv für Neurologie und Psychiatrie* 60 (1947).

———. "Psychische Wirkung eines Mutterkornstoffes in Ungeewöhnlich Schwacher Dosierung." *Schweizer medizinische Wochenschrift* 79, no. 5 (1949).

Tucker, Jonathan B. "Farewell to Germs: The U.S. Renunciation of Biological and Toxin Warfare, 1969, 1970." *International Security* 27, no. 1 (Summer 2002): 107–48.

U.S. News & World Report. "What's Happening in the Race to the Moon?" June 1, 1964.

Walker, Lester C. "Secrets by the Thousands." *Harper's,* October 1946.

Weeks, Maurice, and Paul Yevich. *American Industrial Hygiene Association Journal* 24, no. 6 (1963): 622–29. Declassified December 27, 2007.

ORAL HISTORIES

Major General Charles E. Loucks, U.S.A., Retired, Oral History Interview by Morris C. Johnson. Senior Office Oral History Program, U.S. Army, 1984.

Major General Harry G. Armstrong, Retired, Oral History Interview by J. Bullard and T. C. Glasgow. U.S. Air Force Oral History Program, 1976.

Hubertus Strughold, Oral History Interview with James C. Hasdorff. Office of the Air Force History. San Antonio, TX. November 25, 1974.

Hubertus Strughold, Oral History Interview by Ingrid Kuehne Kokinda. Institute of Texan Cultures Oral History Collection, University of Texas at San Antonio, 1982.

Paul A. Campbell, Oral History Interview by Frances Kallison. Institute of Texan Cultures Oral History Collection, University of Texas, 1977.

Robert C. Haney, Oral History Interview by Charles Stuart Kennedy. The Association for Diplomatic Studies and Training Foreign Affairs Oral History Project, September 21, 2001.

Kurt H. Debus, Oral History Interview with Walter D. Sohier and Eugene M. Emme. John F. Kennedy Presidential Library and Museum. March 31, 1964.

Wernher von Braun, Interview with Robert Sherrod, NASA, August 25, 1970.

Wernher von Braun, Interview from an unnamed film (discusses the Mittelwerk), Linda Hunt donation to the United States Holocaust Memorial Museum, February 2003.

ONLINE COLLECTIONS

Central Intelligence Agency, Freedom of Information Act Electronic Reading Room.

Nuremberg Trials Project, A Digital Document Collection. Harvard Law School Library, Harvard University.

Library of Congress: Trial of the Major War Criminals before the International Military Tribunal, Nuremberg, November 14, 1945–1946, October 1.

The National Archives, Kew, Richmond, Surrey.

T-Force and Field Information Agency, Technical Archiv und Bibliothek, Fritz Bauer Institut, Frankfurt am Main.

The Avalon Project, Lillian Golden Law Library, Yale Law School, New Haven, CT. Nuremberg Trial Proceedings, Vol. 7, 2008.

Time & Life Pictures Collection, Getty Images.

United States Holocaust Memorial Museum, Washington, D.C.: *Holocaust Encyclopedia; Photo Archives; Encyclopedia of Camps and Ghettos, 1933–1945.*

The Wollheim Memorial, Fritz Bauer Institute, Frankfurt, Germany.

ABOUT THE AUTHOR

Annie Jacobsen was a contributing editor at the *Los Angeles Times Magazine* and is the author of the *New York Times* bestseller *Area 51*. A graduate of Princeton University, she lives in Los Angeles with her husband and two sons.